# NEW RESEARCH ON NANOCOMPOSITES

# NEW RESEARCH ON NANOCOMPOSITES

LUIS M. KRAUSE
AND
JONAS T. WALTER
EDITORS

**Nova Science Publishers, Inc.**
*New York*

For permission to use material from this book please contact us:
Telephone 631-231-7269; Fax 631-231-8175
Web Site: http://www.novapublishers.com



**LIBRARY OF CONGRESS CATALOGING-IN-PUBLICATION DATA**

New research on nanocomposites / Luis M. Krause and Jonas T. Walter (editor).
p. cm.
Includes bibliographical references and index.
ISBN 978-1-60456-799-1 (hardcover)
1. Nanostructured materials. 2. Composite materials. I. Krause, Luis M. II. Walter, Jonas T.
TA418.9.N35N487 2008
620.1'18--dc22

2008024207

*Published by Nova Science Publishers, Inc. ✣ New York*

# CONTENTS

# PREFACE

Nanophase materials and technology is a new application technology. Nanocomposites have attracted tremendous attention due to their potential applications in biomedical, catalytic, separation, chemical sensing, fuel cell, capacitors, microfabrication, tribological, resonant coupling, high flux gas transport etc. Nanocomposite materials are complex nanophase materials which optimize the performance of traditional materials. Various nanocomposites have been synthesized in a wide range of polymerization, sol-gel, deposition, magnetron sputtering, supercritical fluid, sonochemistry, laser, etc.

Among these works, many strategies were addressed to improve nanocomposites mechanical properties by inclusions (fibers, whiskers, platelets, or particles). The embedding of inclusions in a host matrix to make composites, which gives material properties not achieved by either phase alone, has been a common practice for many years. Traditionally, composites were reinforced with micron-sized inclusions. Recently, processing techniques have been developed to allow the size of inclusions to go down to nanoscale.

With the recent developments in the nanoscience and nanotechnology fields, the correlation of composite properties with nanostructure has become a point of great interest. As a result, much of the work is still ongoing and there is yet to be a definite conclusion on the effect of nano-sized structure on nanocomposite systems. This book provides leading edge research from around the globe on this field.

Strained silicon (ε-Si), the fundamental material of integrated circuit, is finding tremendous attention because it boosts speed and reduces power consumptions of electronic devices. However, inhomogeneous distribution of strain in ε-Si layers can degrade performance of electronic devices. Raman spectroscopy is used to study strain fluctuations in silicon because the optical phonons in Raman spectra are strongly influenced by strain. Though silicon are Raman active devices, the Raman efficiency of a nanometer layer of ε-Si is extremely weak and is often eclipsed under the Raman scattering of underlying buffer substrates. Micro Raman measurements show only uniform features in the nano-scale because of averaging effect from diffraction-limited spatial resolution. Here, we utilized surface enhancement in Raman scattering to overcome weak emission problems and to suppress averaging effect. Thin ε-Si layers were covered with thin silver (Ag) layer to invoke surface enhanced Raman spectroscopy (SERS). Results show that SERS effectively enhanced the Raman signal from ε-Si layer and it stands distinctly apart from the Raman signal originating from the buffer layer. This technique is promising but it lacks the spatial resolution in the nano-scale due to diffraction limit from the probing light. In order to achieve nano-scale spectroscopy, point-surface-enhancement was used, rather than a large surface enhancement. We used a silver-coated sharp tip, just like SERS, but only the sample region very close to the tip apex is characterized. This technique, known as the tip-enhanced Raman spectroscopy (TERS), provides nanometric resolution in our measurement. We observed localized strains by employing TERS. The TERS spectra revealed clear nano-scale variation in Raman frequency. Now that we can distinctively separate ε-Si from underlying buffer layer, signal-to-noise ration (SNR) needs further improvement. We improve TERS SNR in two ways: optical field enhancement using different metallic tip and reduction of background signals arising from bulk materials. The tip-enhancement is more important for homogenous nano-materials or for samples with very weak signals whereas the background signal reduction is indispensable for nano-materials that consist of different thin layers with strong signals such as ε-Si or samples with strong signal level. Accordingly, we introduce several approaches mainly for the suppression of background signals arising from other bulk materials. In Chapter 1 we will discuss the utilization of UV light source, specialized tip, sample orientation relative to probing polarization, and depolarization configuration to obtain high contrast Raman signal. The characterization techniques describe above is applicable to other nano-materials.

The delivery system based on stimuli-responsive polymers inside mesoporous silica has been of great interest because it allows a drug to be released in response to the environmental stimulation. This research has demonstrated that several approaches have been developed for the synthesis of thermal-responsive poly(N-isopropyl acrylamide) (PNIPA) inside mesoporous silica with pore size of 7-20 nm. The large pore size of the support and the smart nature of PNIPA can be expected to the potential applications in the controlled release of macromolecular active agents. Chapter 2 reviews the methods of how to introduce multilayer polymers inside mesoporous silica pores, while keeping the pore channels for the adsorption and release of drug molecules. It is interesting to find the resultant delivery system demonstrated a high ibuprofen (IBU) storage capacity. The multilayer polymers inside the pores were considered to form internal cavity for drug molecules in addition to responding to a change in external temperature. Strategies for the preparation of a site-selective and controlled delivery system are presented and the essential properties of the resulting materials with respect to its capability for target and controlled drug release were investigated. Outlook of potential applications and further challenges are also provided.

Bone is one of the most important and fascinating biomineralized nanocomposite structures in nature. On a macroscopic scale, bone provides structural support for our bodies, protects our internal organs, provides a reservoir for the production and storage of mesenchymal and hematopoietic stem cells, and serves as a storage site for calcium, phosphate, and other biologically relevant ionic species. However, the nanoscale features of bone make it truly remarkable from a material's design viewpoint. Biomineralized nanocomposites share the common design motif of organic matrix mediated nucleation and growth of an inorganic phase with controlled chemistry. The process of mineral phase formation is governed by solubility, surface and complexation chemistry, and bioavailability of important elemental and molecular species such as calcium, carbonate, phosphate, magnesium, and silica. Specific aspects of nanocomposite design related to matrix phase chemistry, degree of crystalline order and orientation, phase distribution and microstructure are uniquely controlled during the birth, healing and remodeling of bone. There are many lessons that we can learn from bone structure and biology to improve our ability to synthesize nanocomposites for various applications. The focus of Chapter 3 is describing unique structural characteristics of biomineralized nanocomposites in nature with an emphasis on bone. Key lessons will be extracted from bone that can be used in the design and synthesis of nanocomposites with tailored functionality. An overview of research efforts that have utilized various aspects of bone inspiration will be provided to highlight current advances in the area of bone inspired nanocomposites and indicate areas for further study.

Chapter 4 explores the main characteristics and properties, both structural and mechanical, of nanostructured amorphous carbon; a material with a potential to intermingle properties of carbon nanostructures with those of pure amorphous carbon. Controlling the type and size of the embedded structures opens the possibility to tailor the properties of the amorphous phase. The nanostructures are embedded in an amorphous carbon matrix and are varied both in type and size, ranging from diamond crystallites to nanotubes and concentric-shell graphitic onions bound by dispersion forces, to entirely three-dimensional sp2 covalent conformations, which include porous, open graphene structures with negative curvature. This study is based on classical and semi-empirical simulations, using Monte Carlo schemes within the empirical potential approach and tight-binding molecular dynamics utilizing two different hamiltonians. First, the energetics and stability of the nanocomposites are investigated together with the interaction of the inclusion core with the surrounding matrix and their stress states. Special focus is given to the microscopic character, specifically the behavior of the different carbon hybridization states within these materials, together with the effect of the atomistic features on both the stability and elastic properties. As representative quantities for the latter, the bulk and Young's moduli were calculated for the diamond and nanotube nanocomposites, allowing for comparison to the corresponding values for single phase amorphous carbon. At a final step, the strength under tensile and shear load of dense composites with a diamond inclusion is presented unveiling the fracture mechanism of these highly stable materials.

Ultrananocrystalline diamond/amorphous carbon (UNCD/a-C) films have been deposited by microwave plasma chemical vapour deposition from 17% $CH_4/N_2$ mixtures at temperatures between 450 and 770°C. This process results in a nanocomposite type of films consisting of diamond nanocrystallites of 3-5 nm diameter, which are embedded in an amorphous carbon matrix. The first part of Chapter 5 is devoted to the basic film properties such as crystallinity, composition, and the bonding environment, while the second addresses the nucleation and growth processes and the underlying mechanisms, discussing among others the role of parameters such as the substrate pre-treatment and the substrate temperature.

The final and major part of the contribution addresses properties relevant for future applications of this type of nanocomposite films, e.g. electrical, optical, mechanical, tribological, and bio(techno)logical. Emphasize is laid on a discussion of the influence on the nanocomposite character of these films, i.e. the existence of an amorphous carbon matrix, on these properties. It will be shown that the matrix renders the optical properties for the worse as compared to pure UNCD films but is of advantage e.g. for the mechanical and electrical properties.

Nanostructure and phase configuration are critical to determine mechanical and other properties e.g. thermal stability, oxidation and wear in nanocomposite superhard thin films. Chapter 6 presents our results of the nanostructure evolutions and phase configuration transformations in the formation of nanocomposite thin films prepared by unbalanced magnetron sputtering and their corresponding mechanical and other properties. Two typical ternary nanocomposite thin films Ti-B-N, Ti-C-N and a quaternary thin film Ti-B-C-N were investigated. Their nanostructures were determined by X-ray diffraction and high-resolution transmission electron microscopy, while phase configuration by X-ray photoelectron spectroscopy and Raman spectroscopy, and the hardness was examined by microindentation method. The content includes: (1) introduction; (2) deposition and analyse methods of TiN based nanocomposite thin films; phase configuration and nanostructure in (3) Ti-B-N thin films prepared at room and elevated temperatures, (4) Ti-C-N and (5) Ti-B-C-N thin films deposited at elevated temperature. The section 6 in this chapter refers their mechanical and other properties including hardness, residual compressive stress, thermal stability, oxidation and wear. At the end, a brief summary is outlined.

Carbon - metal (Ni, Ti) nanocomposites were prepared by dc magnetron sputtering. The films were deposited at different deposition temperatures (25 - 800°C) on silicon (001) substrates covered with 300 nm thick siliconoxide. The input power of the magnetrons with Ni target was between 5 W and 25 W, and the magnetrons with Ti target between 10 W and 40 W. The input power of carbon target was 150 W.

As presented in Chapter 7, we found correlation between structural, electrical and mechanical properties of films, and developed the optimal structure of films for protective applications. Transmission electron microscopy and high resolution electron microscopy were used to determine the structure of films. The standard four probe technique was used for electrical measurements. The mechanical properties of films, e.g. hardness, modulus of elasticity were measured by nanoindentor.

In both cases (C-Ni, C-Ti), the highest hardness (C-Ni ~ 11 GPa, C-Ti ~ 18 GPa) and highest modulus of elasticity (C-Ni ~ 114 GPa, C-Ti ~ 205 GPa) were obtained for the coating preparet at 200°C consisting of columnar ($Ni_3C$ and TiC) crystallites and 2 nm thin amorphous and graphite-like carbon matrix. The hardness of the coating is independent on the metal content but is influenced by the structure of thin carbon matrix.

The area of periodically arranged organic-inorganic nanocomposites (OINCs) prepared by self-assembly in the presence of organic molecules underwent dramatic growth. Compared with large amounts of heavy research on the oxide-based materials, such as silicate-base materials, non-oxide-based OINCs, metal-chalcogenide-based OINCs, are particularly much more attractive. Although the considerable challenge of the construction of non-oxide analogues which is caused by the unstable precursor anions and poor understanding of their condensation behavior in solution, metal-chalcogenide-based OINCs offer new potential opportunities in materials sciences, photo-electric conversion, photocatalysis, separation, and environmental remediation. All the reported metal-chalcogenide-based OINCs can be regarded as periodically arranged organic-inorganic nanocomposites, and divided into two categories according to the size of the periodically repeated units: (1) mesostructured (2-50 nm) surfactant-chalcogenide OINCs, and (2) microstructured (< 2 nm) open-framework organic molecule-chalcogenide OINCs. Chapter 8 reviews the current state of the art in chemical synthesis, characterization, and applications of the mesostructured (2-50 nm) surfactant-chalcogenide OINCs.

As explained in Chapter 9, development of organic-inorganic nanocomposites have in the recent years have generated tremendous potential for the use of commodity polymers in more engineering applications. This enhancement is attributed to the significant improvements in the properties of the polymers at minimal inorganic filler volume fractions without affecting the optical density or transparency of the parent polymer matrices. However, the property enhancements have been generally limited to the polar polymers like polyurethanes, epoxies, polyamides etc. owing to the unfavourable thermodynamic and kinetic forces in the case of non-polar polymers like polypropylene and polyethylene leading to poor interfacial compatibility and mixing of the two phases. Conventional surface treatments of the inorganic filler like long alkyl chain ammonium ions are not sufficient to obtain the nanoscale mixing. The use of compatibilizers (low molecular weight polyethylene or polypropylene grafted with maleic anhydride) is common approach to achieve interfacial compatibility, however, the low molecular weight of the compatibilizer has negative affect on the mechanical properties. Recently, more advanced techniques have been developed to circumvent these concerns. Grafting of polymeric oligomers 'to' and 'from' the surface of filler particles helps to achieve longer chain lengths ionically bound to the filler surface. These long chains lead to higher basal spacing and hence increase the susceptibility of the filler exfoliation in the polymer matrix during shearing. The architecture and chemical composition of the chains grafted on the surface can be accurately controlled for the system specific requirement. Controlled living polymerization techniques like nitroxide mediated polymerization or atom transfer radical polymerization can also be employed to achieve very high molecular weights on the surface. Another simpler approach includes filler treatment with reactive modifications (eg. terminal hydroxyl or carboxyl groups) and subsequent reactions (eg. esterification) on the surface. Polymerization of ethylene or propylene itself with initiator immobilized on the filler surface have also been achieved to more or less success. All these approaches require no compatibilizer and can easily be characterized with the conventional techniques. Apart from that, the developed techniques can also be employed to a wide spectrum of fillers like layered silicates or aluminosilicates, multi-walled nanotubes and even polymer particles.

Transparent conductive oxides (TCO) are widely used transparent conducting coating for optoelectronic devices such as flat panel displays, solar cells, and touch screens. Also, it is an essential material for the automotive, train and aircraft industries. Therefore, the numerous research programs are going on to develop different types of TCO films. However, those ceramic TCO films are brittle, which is a mechanical limitation for their use in the flexible optoelectronic devices. Again, we could find the demanding use of plastic optoelectronic devices because of their ease of processing, mechanical flexibility, and promise for low cost fabrication of large areas. Therefore, we need to find out the flexible alternate transparent conducting film of brittle TCO films. For substituting the delicate TCO films, polymer-indium tin oxide (PI) nanocomposite films with various indium tin oxide (ITO) concentrations have been fabricated by electro-spinning in the Chapter 10. These flexible films look like a mat with tangled nanofibers, and are optically transparent (85 - 90%) in the visible region of the solar spectrum (400-700 nm). The electronic properties of those transparent films were studied. The highest conductivity achieved is $1.04 \times 10^3$ $Scm^{-1}$. The promising optoelectronic properties present the great promise of PI nanocomposite to be a top layer of the next generation flexible photo devices, like solar cells and light emitting diodes (LED).

In: New Research on Nanocomposites
Editors: Luis M. Krause et al, pp. 1-35
ISBN: 978-1-60456-799-1


*Chapter 1*

# NANO-SCALE CHARACTERIZATION AND SPECTROSCOPY OF STRAINED SILICON

***Norihiko Hayazawa** and Alvarado Tarun***
Nanophotonics Laboratory, RIKEN, 2-1 Hirosawa, Wako, Saitama 351-0198, Japan

## Abstract

Strained silicon (ε-Si), the fundamental material of integrated circuit, is finding tremendous attention because it boosts speed and reduces power consumptions of electronic devices. However, inhomogeneous distribution of strain in ε-Si layers can degrade performance of electronic devices. Raman spectroscopy is used to study strain fluctuations in silicon because the optical phonons in Raman spectra are strongly influenced by strain. Though silicon are Raman active devices, the Raman efficiency of a nanometer layer of ε-Si is extremely weak and is often eclipsed under the Raman scattering of underlying buffer substrates. Micro Raman measurements show only uniform features in the nano-scale because of averaging effect from diffraction-limited spatial resolution. Here, we utilized surface enhancement in Raman scattering to overcome weak emission problems and to suppress averaging effect. Thin ε-Si layers were covered with thin silver (Ag) layer to invoke surface enhanced Raman spectroscopy (SERS). Results show that SERS effectively enhanced the Raman signal from ε-Si layer and it stands distinctly apart from the Raman signal originating from the buffer layer. This technique is promising but it lacks the spatial resolution in the nano-scale due to diffraction limit from the probing light. In order to achieve nano-scale spectroscopy, point-surface-enhancement was used, rather than a large surface enhancement. We used a silver-coated sharp tip, just like SERS, but only the sample region very close to the tip apex is characterized. This technique, known as the tip-enhanced Raman spectroscopy (TERS), provides nanometric resolution in our measurement. We observed localized strains by employing TERS. The TERS spectra revealed clear nano-scale variation in Raman frequency. Now that we can distinctively separate ε-Si from underlying buffer layer, signal-to-noise ration (SNR) needs further improvement. We improve TERS SNR in two ways: optical field enhancement using different metallic tip and reduction of background signals arising from bulk materials. The tip-enhancement is more important for homogenous nano-materials or for samples with very weak signals whereas the background signal reduction is indispensable for nano-materials that consist of different thin layers with strong signals such

* E-mail address: hayazawa@riken.jp. Phone: +81-48-467-9339, Fax: +81-48-467-9170
(Author for Correspondence)

as ε-Si or samples with strong signal level. Accordingly, we introduce several approaches mainly for the suppression of background signals arising from other bulk materials. We will discuss the utilization of UV light source, specialized tip, sample orientation relative to probing polarization, and depolarization configuration to obtain high contrast Raman signal. The characterization techniques describe above is applicable to other nano-materials.

# 1. Introduction

Strained Si (ε-Si) on silicon germanium ($Si_{1-x}Ge_x$) substrate has now been widely used in state-of-the-art electronic devices such as Intel dual core products primarily due to the mobility enhancement of electrons and holes in the ε-Si layer [1,2]. As ε-Si technologies become more complicated and transistors become smaller, in the order of nanometer (~45nm), issues involving strain defects and the characterization of key properties in silicon substrates are expected to increase. Thus, development of advanced techniques for nanoscale characterization and investigation of localized strain in the ε-Si substrate are important in the current semiconductor technology. Raman spectroscopy has been widely used for the investigations of strain in semiconductors [3-5] because of its nondestructive capability to observe lattice vibrations sensitive to strains. However, due to the low Raman scattering efficiency, it becomes difficult to observe the localized strain in nanometer scale. Moreover, in case of thin ε-Si layer fabricated on $Si_{1-x}Ge_x$ substrate, the Raman scattering of Si-Si vibrational mode from $Si_{1-x}Ge_x$ buffer layer can easily eclipse the vibrational signals originating from the ε-Si layer. Because of these two difficulties, conventional Raman scattering is not suitable for strain measurements in these thin samples. Nanometer scale and surface selective sensitivities are required, and hence surface enhanced Raman scattering (SERS) [6] and tip-enhanced Raman spectroscopy (TERS) [7-13] turn out to be exceptionally promising spectroscopic techniques. In this chapter, we demonstrate how effectively and selectively enhance the Raman signal from the thin ε-Si layer on a thick underlying $Si_{1-x}Ge_x$ layer using SERS. We show that the proposed technique can be straightforwardly used for TERS microscopy to investigate localized nano-scaled strain in ε-Si. In section 2, we discuss SERS from the thin ε-Si covered with thin Ag layer. Results show that SERS effectively enhanced the Raman signal originating from ε-Si layer and it stands distinctly apart from the Raman signal originating from the buffer layer. This technique is promising but it lacks the spatial resolution in the nano-scale due to diffraction limit from the probing light. In section 3, in order to achieve nano-scale spectroscopy, point-surface-enhancement was used, rather than a large surface enhancement. We used a sharp Ag-coated tip placed over the ε-Si and only the sample region very close to the tip apex is characterized. This technique, known as the tip-enhanced Raman spectroscopy (TERS), provides nanometric resolution in our measurement. We observed localized strains by employing TERS. The TERS spectra revealed clear nano-scale variation in the Raman frequency. Now that we can distinctively separate ε-Si from underlying buffer layer, signal-to-noise ratio (SNR) needs further improvement. We improve TERS SNR in two ways: optical field enhancement using different metallic tip and background signals reduction arising from bulk materials. The tip-enhancement is more important for homogenous nano-materials or for samples with very weak signals whereas the background signal reduction is indispensable for nano-materials that consist of different thin layers with strong signals such as ε-Si or samples with strong

Raman signal level. Accordingly, wc introduce several approaches mainly for the suppression of background signals arising from other bulk materials. We will discuss the utilization of UV light source and specialized tip in section 4 and then introduce depolarization configuration in TERS for the efficient suppression of background signals in section 5.

## 2. Surface Sensitive Detection by Surface Enhanced Raman Scattering (SERS)

Recent developments in surface enhanced Raman scattering (SERS) has gained attention not only due to enormous ($10^{14}$) level of Raman enhancement capable of single-molecule detection but also for surface sensitive characterization of nanometer structured surfaces. It was observed that the huge increase in the Raman cross section was the result of excitation of surface plasmons in metals. The most popular metal used in SERS studies is Ag due to intense plasmon resonance excitation produced by nanometer sized Ag in the visible wavelength. In this section, we utilize SERS to obtain strain information in a thin layer, ~30nm, of ε-Si over silicon germanium substrate.

### 2.1. Experimental Configuration for SERS

Figure 1 illustrates the concept of the SERS experimental system for highly sensitive detection of strain in the ε-Si surface. The substrate used for growing ε-Si layer was prepared by doping Ge in a pure ⟨100⟩ oriented silicon substrate. The concentration of Ge in the host Si lattice was gradually increased up to 25 % to expand the lattice constant of the host Si lattice. In order to minimize inhomogeneous strain, a layer with 1.0 μm thick $Si_{1-x}Ge_x$ buffer layer was grown on the prepared substrate at constant 25% Ge concentration. Finally, a 30-nm-thick Si layer was epitaxially grown on the buffer layer to fabricate ε-Si. The crystal face for both the buffer and silicon layer was maintained at ⟨100⟩. Due to lattice parameter mismatch at the interface, the thin Si layer was under a constant strain. The thickness of this ε-Si layer has to be low otherwise the induced strain could be gradually relaxed with increasing the thickness. To detect Raman from a very thin layer, we invoked SERS effects by evaporating Ag with 8-10 nm thickness on the ε-Si layer under vacuum (~$10^{-6}$ Torr). The topographic image is shown in the inset of Fig. 1. This image was obtained by contact-mode atomic force microscope (AFM) using a silicon cantilever (Nanosensor ATEC-CONT). The Ag film consists of plenty of Ag grains, with typical grain size of 20~40 nm in diameter and 10 nm in height. These fractal-like "pancake" structures of Ag grains provide almost uniform surface enhancement factor averaged within the diffraction limited focused spot of micro-Raman experiments. A CW blue laser (wavelength: 488 nm, power: 9 mW) was focused by an oil-immersion high numerical aperture (NA) objective lens (NA=1.4; 100x). The gap of Ag film layer and the sample was filled with oil that has a matching refractive index (1.515) as the objective lens. The Raman and SERS signals from the sample were collected by the same objective lens and detected by confocal Raman microscope (pinhole: 30 μm), in which the spectrometer (focal length: 560 mm, 1800 grooves/mm, blaze wavelength: 500 nm) was equipped with both, a cooled charge coupled device (CCD) camera for Raman spectra measurements (such as the results in Fig. 2), and an avalanche photodiode for Raman image acquisitions (such as in Fig. 3).

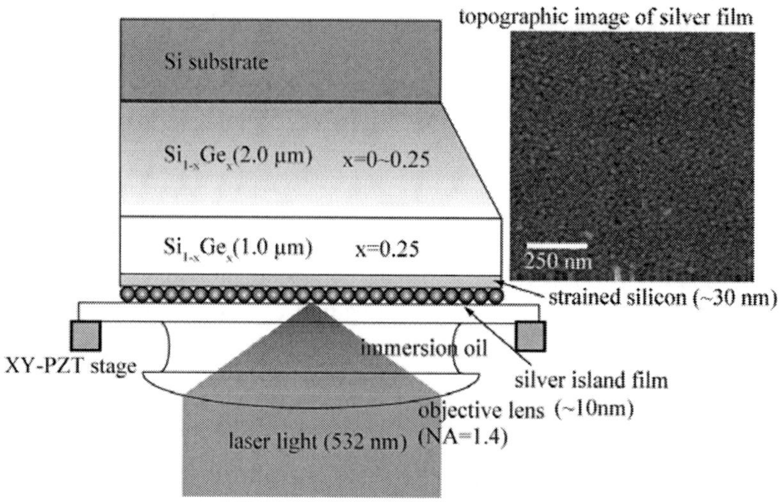


Figure 1. Schematic of SERS microscopy for ε-Si and AFM topographic image of the Ag island film coated on the ε-Si.

## 2.2. Comparison of Raman and SERS Spectra

Vibrational modes located at the center of the Brillouin zone can only be observed in the first-order Raman scattering due to the momentum conservation. Further, the selection rules for the ⟨100⟩ oriented Si crystal in backscattering geometry allow only LO phonon to be Raman active. Therefore, we expect to observe only LO phonon in first-order Raman scattering experiments, which should appear at 520 $cm^{-1}$ for an unstrained or pure Si crystal. Since the sample consists of a ε-Si layer above Ge-doped Si buffer layer, we expect shift from the original frequency of LO phonon in both layers. The LO phonons in doped crystals can be easily modified depending on the atomic mass as well as the amount of the dopant. In Ge-doped Si crystal, the LO phonon can have three different vibrations, namely the Si-Si mode arising from the vibrations of neighboring Si atoms, the Ge-Ge mode arising from the vibrations of neighboring Ge atoms, and the Si-Ge mode arising from the vibrations involving the bonds between a Si and a Ge atom. Since the amount of Ge is much smaller than the amount of Si in our samples, the LO mode corresponding to the Si-Si vibration is dominant. We need to concentrate only on Si-Si mode because it interferes with the Si-Si vibration coming from the thin ε-Si layer in the experiments and we will be discussed throughout the chapter. The vibrational frequency of this mode decreases almost linearly with increasing Ge concentration [14]. The bonds between two neighboring Si atoms are stretched when Ge is doped introducing strain in the lattice. This strain is responsible for the shift of LO mode frequency originating from the Si-Si vibration in the buffer layer. Similarly, when a thin Si layer is grown on $Si_{1-x}Ge_x$ buffer layer, the Si-Si bonds in Si layer are stretched due to the lattice mismatch at the interface. If the thickness of the layer is not too large, the strain produced in the Si layer is transferred vertically throughout the surface. The amount of developed strain at surface can be different from that of the strain in the buffer layer even if the strain in both layers is caused by the presence of Ge atoms in the buffer layer.

The penetration depth of blue-green probing light in Si is much larger than the thickness of the ε-Si layer in our sample [15]. Therefore, in Raman scattering experiments, the probing laser penetrates deep through the buffer layer. As a result, Raman scattering in the buffer layer is strongly excited. The intensity of LO phonon mode coming from the buffer layer is much more strong than the one coming from the ε-Si layer. This is a problem if the vibrational frequencies of the two phonons are close to each other. It becomes difficult to clearly observe the LO phonon corresponding to the ε-Si layer. To overcome this problem, we used the SERS technique to selectively enhance Raman at the surface and compared the results from those obtained by normal Raman spectroscopy. Figure 2 shows SERS (with a Ag film deposited on ε-Si layer) and Raman spectra (without a Ag film) of the sample. Both spectra were fitted by double Lorentzian functions as shown by dashed lines. The lower and higher Raman modes in both spectra are the LO phonon modes arising from the Si-Si vibration in $Si_{1-x}Ge_x$ (x=0.25) and in ε-Si, respectively. The Raman spectrum is dominated by a strong peak at 504.9 $cm^{-1}$, corresponding to the background LO phonon signal from $Si_{1-x}Ge_x$ substrate, together with a very weak shoulder at 513.8 $cm^{-1}$ that arise from the LO phonon in ε-Si [16-20]. Since the Raman signal from the ε-Si is very close to the $Si_{1-x}Ge_x$ buffer layer and considering 30 nm thickness of the ε-Si layer, the Raman signal from ε-Si is almost overwhelmed by the signal from the $Si_{1-x}Ge_x$ layer. On the other hand, in SERS spectrum, the Raman signal from ε-Si is remarkably enhanced. The signal becomes even stronger than the $Si_{1-x}Ge_x$ buffer layer due to the SERS effect. Even though the ε-Si Raman signal in SERS is enhanced relative to the signal from the buffer layer, we noticed reduction in the scattering efficiency in SERS with a signal bias of ~20000 counts/60s compared to the Raman spectrum. The reduction in overall scattering efficiency is due to the lower transmission of the probing light caused by Ag island film whereas the bias signal is due to the white continuum emission from the Ag island film, which is often accompanied in SERS measurements [21]. As seen from Fig. 2, we resolved Raman signal from the ε-Si in SERS. This suggests that SERS effect is very powerful to observe the surface condition of the thin ε-Si layer. Both SERS and Raman measurements were done in the same condition except the existence of Ag

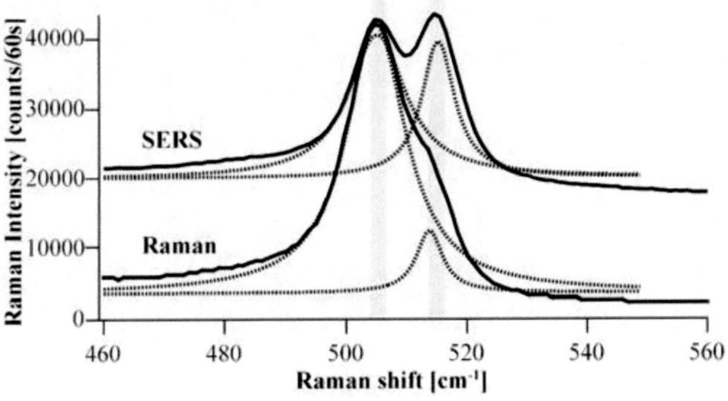


Figure 2. SERS (with Ag film) and Raman (without Ag film) spectra of ε-Si. Both spectra are fitted by double Lorentzian functions representing Raman modes of $Si_{1-x}Ge_x$ (lower frequency) and ε-Si (higher frequency). The meshed areas are used for Raman imaging in Fig. 3.

island film to invoke SERS. It should be noted here that we did not use any analyzer for polarization dependent detection in either SERS or Raman detection. The ratio of the intensities between $Si_{1-x}Ge_x$ and ε-Si would change when measured through an analyzer, which will be discussed later in Section 5.

## 2.3. Comparison of Raman and SERS Images

Next, we carried out confocal micro-Raman imaging of the same sample at specific vibrational modes from $Si_{1-x}Ge_x$ and ε-Si. Figure 3(a) shows the reflection confocal image of the sample from a region close to the edge of the evaporated Ag film. As expected, the Ag covered layer appears to be brighter in the reflection image. Figure 3(b) was obtained by setting the spectrometer at the lower frequency in Fig. 2, so that the avalanche photo diode detects only the Raman signal from $Si_{1-x}Ge_x$ buffer layer. Since the intensity levels of SERS and Raman from $Si_{1-x}Ge_x$ buffer layer is almost similar from Fig. 2, Raman image in Fig. 3(b) also reflects low contrast between Ag coated and uncoated area. However, when the spectrometer is set to detect only the higher frequency shift arising from ε-Si (see Fig. 2), the enhanced Raman signal from Ag coated ε-Si has high contrast as depicted in Fig. 3(c). Fig.

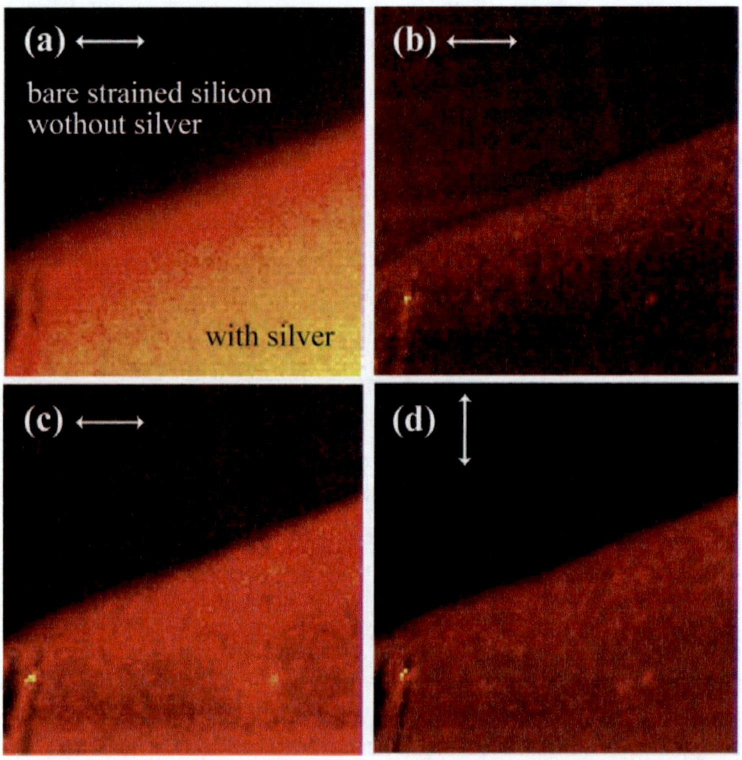


Figure 3. (a) A reflection confocal and Raman images obtained at Raman shift of (b) $Si_{1-x}Ge_x$ and (c) ε-Si as indicated by meshed area in Fig. 2. (d) is obtained at Raman shift of ε-Si, the same as (c) but with different polarization of excitation light. An arrow in the figure indicates polarization of an excitation light in each image. The dimension of each image is 45 μm by 45 μm consisting of 100 pixels by 100 pixels.

3(c) does not reflect any particular information about the possible nanometric strain distribution in ε-Si, which could be varying within areas smaller than the focal spot. This is because the enhanced scattering signal in this experiment is averaged within the diffraction limited focused spot. Accordingly, when polarization of the excitation light is rotated 90 from the previous polarization as indicated in the Fig 3(d), no significant change in the contrast could be observed. The entire signal level in Fig. 3 (d) is decreased compared to that in Fig 3(c) due to the polarization dependence of crystalline silicon ⟨100⟩ facet [22,23] and also due to the different detection efficiency of the experimental system. Here, we note that the temperature dependence of Raman scattering in silicon [24] is not observed in our experiments under room temperature conditions. However, the temperature dependence is also an important factor that can influence the properties of ε-Si substrates and could be observed under the experimental conditions of SERS.

## 3. Surface Sensitive and High Spatially Resolved Detection by Tip-Enhanced Raman Scattering (TERS)

In SERS measurement, high surface sensitivity and signal enhancement was successfully achieved, however, the spatial resolution is still diffraction limited. Moreover, SERS is an irreversible or destructive detection process because it requires Ag coating on the sample. For non-destructive and higher spatial resolution measurement capable of detecting localized nanometric strain, a sharpened metallic tip [25-27] can be utilized instead of Ag island film. In this section we introduce the use of "tip-enhancement" instead of "surface enhancement" to go beyond the diffraction limit of probing light. The concept of TERS is rather straightforward that can be understood by considering the sharp apex of a metallic tip as a point source surface enhancer [7-10].

### 3.1. Experimental Configuration of Reflection-Mode TERS

Figure 4 shows the concept of reflection mode TERS microscopy [28-31] for opaque sample such as ε-Si layer assembled on thick substrate. There are generally two configurations for TERS, transmission mode and reflection mode. In transmission mode [7-13,32], the metallic probe tip is illuminated through the sample using objective lens. This configuration has been widely used in TERS because of its high efficiency in both illumination and collection, especially when using a high numerical aperture (NA) objective lens. However, transmission mode TERS cannot be used for opaque or thick samples. In reflection mode TERS [28-31], the tip is illuminated from the same side in which it is approached towards the sample (Fig. 4). This is advantageous and promising for observing opaque samples such as silicon substrates.

A schematic of the experimental configuration is shown in Fig. 5 [33,34]. An incident CW YAG laser (532 nm, JDS Uniphase) is directed to the optical set-up through a single mode optical fiber. The light coming out of fiber is set to p-polarization, parallel to the tip axis, for optimum tip-enhancement [28] from the combination of a half-waveplate and a polarizer. The beam diameter of the laser is then expanded to ~10 mm. This expanded and

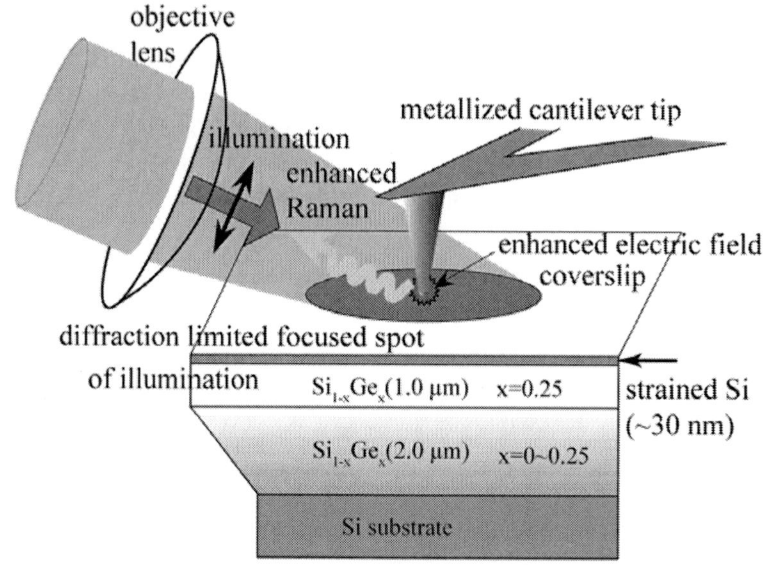


Figure 4. Schematic of reflection-mode TERS.

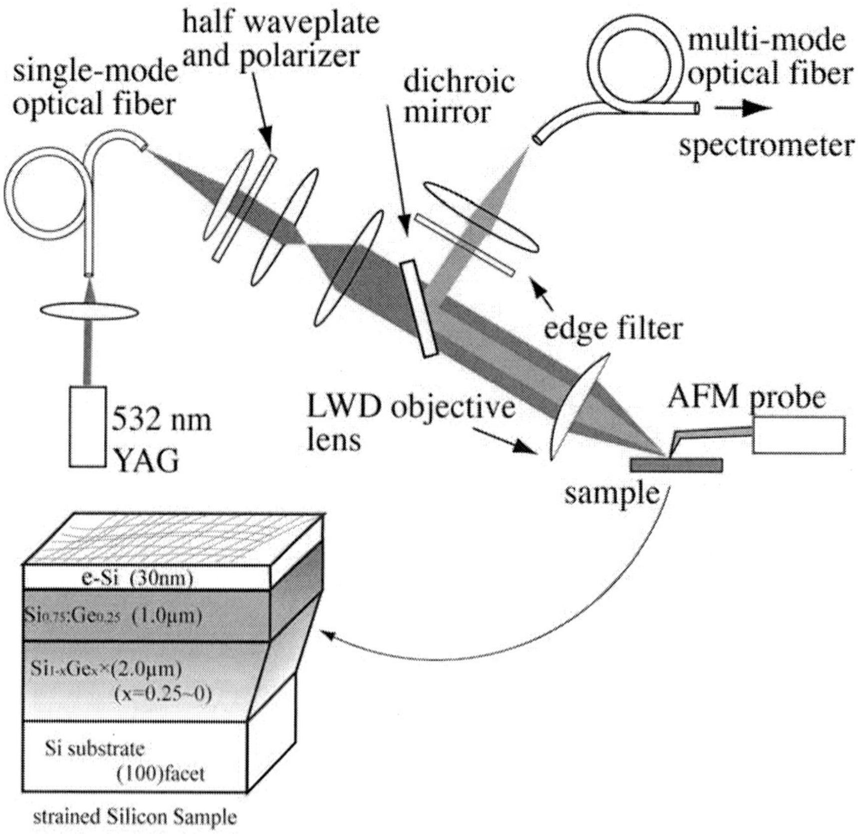


Figure 5. Experimental setup of reflection-mode TERS. The inset shows the structure of ε-Si on $Si_{1-x}Ge_x$ (x = 0.25).

polarized light is focused on the tip and sample using a long-working-distance (LWD) objective lens (Mitsutoyo, NA=0.28, 20x, WD=30.5 mm). The illumination optics mentioned above, including the LWD objective lens, is precisely positioned by three-axis actuators (New Focus Inc. Picomotor, accuracy <30nm). The apex of the cantilever (Nanosensor ATEC-CONT) is adjusted onto the focused spot using the same principle as a contact-mode operation of AFM. The power of the incident laser light is set to deliver 1 mW on the sample. The same LWD objective lens collects the TERS signal. A dichroic mirror (or a non-polarized beam splitter cube) directs the output Raman signal via a multimode optical fiber to a spectrometer (Photon Design PPDT3-640). The spectrum is detected using a liquid nitrogen-cooled CCD camera (Princeton Instruments). Raman signal was obtained with an integration time of 120 seconds. To attain an efficient tip-enhancement effect in the reflection-mode TERS, the probe tip is prepared by depositing Ag on commercially available silicon cantilever used for AFM. This is done by two-step thermal evaporation process. First, a 5-nm-thick gold palladium alloy (AuPd) is deposited on a tip to modify the surface of silicon, and then Ag is evaporated with the thickness of 30~50 nm at a rate of 0.5 Å/sec.

## 3.2. Comparison of Raman and TERS Spectra

The reflection mode TERS described above has been successfully applied to nanoscale localization of ε-Si using dichroic mirror and Ag coated Si-tips with an excitation laser wavelength of 532nm [33]. Figure 6(a) shows the Raman signal from 532nm probing light under Ag coated Si-tip (Nanosensor, ATEC-CONT-20). Figure 6(b) shows the background corrected signal, which is the difference between the TERS and far-field (tip withdrawn) Raman signal. In order to determine the Raman peaks, we decomposed the corrected spectrum in Fig. 6(b), into three Lorentzian functions. The three peaks are assigned to Si-Si vibrations in the SiGe (503 $cm^{-1}$) bulk layer, Si-Si phonon mode in ε-Si (514 $cm^{-1}$) layer and Si-Si in Si-tip (520 $cm^{-1}$). The corrected spectrum does not contain the peaks around 480 $cm^{-1}$, which was later found to be coming from PC monitor. This background corrected TERS spectrum shown by the spectrum in Fig. 6(b), contains nanometer scale information of localized strains. In the far-field spectrum, the ε-Si peak is recognized only as a shoulder of the strong background signal generated from the underlying $Si_{1-x}Ge_x$ (x=0.25) layer. The TERS spectrum, however, shows a clear and distinct peak owing to tip-enhancement. Since the ε-Si layer is in the order of several tens of nanometers, a sufficient amount of incident light penetrates into the $Si_{1-x}Ge_x$ buffer layer, and hence the background signal from the underlying buffer layer is still significant. The shift of LO phonon mode originating from the ε-Si with respect to the original position of the LO phonon for an unstrained Si (520 $cm^{-1}$) [3] provides us with plenty of information, such as the local strain in the layer [35-38]. The estimated enhancement factor for ε-Si is $2.2 \times 10^4$[33], which is calculated by considering the size of the diffraction limited focused spot (ϕ: 3 μm) and the enhanced electric field corresponding to the tip diameter (ϕ: 40 nm).

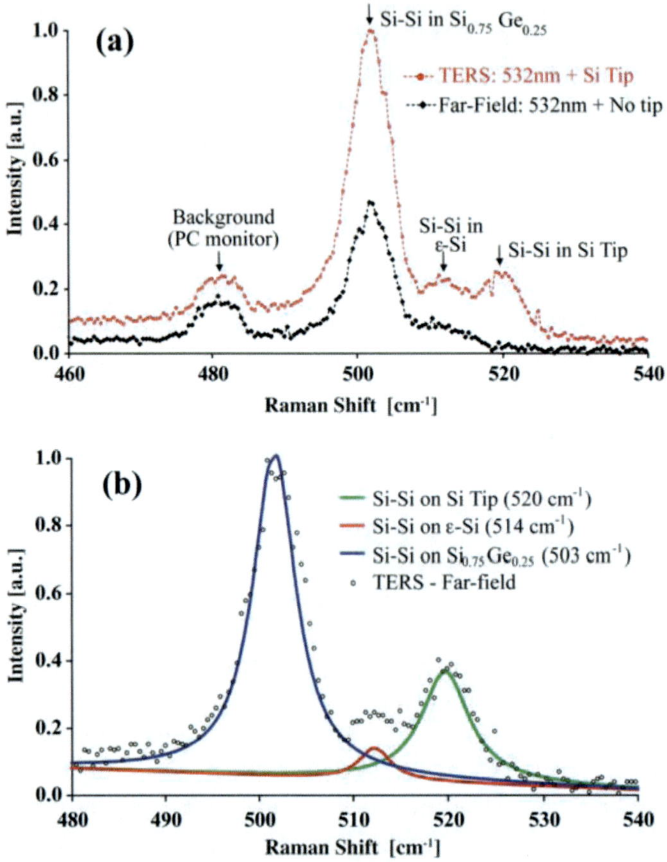


Figure 6. (a) TERS (red) and far-field (black) spectra of ε-Si obtained using 532nm probing light with Ag coated Si-tip. (b) Background corrected near-field spectrum that is broken down into three Lorentzian curves for band peak identification. The peaks at 503, 514, and 520 $cm^{-1}$ are assigned to phonon modes from SiGe layer, ε-Si layer and Si-tip, respectively.

## 3.3. TERS Spectra Mapping

Nanoscale lateral heterogeneity of strain in ε-Si / $Si_{1-x}Ge_x$ structures is a serious problem in fabrication of high-mobility electronic devices. Therefore, it is necessary to investigate the heterogeneity of strain at nanoscale. A position dependent TERS measurement can be used for a detailed characterization. Figure 7(a) shows the bright-field microscopic image of the ε-Si surface. The crosshatch patterns induced by lattice-mismatch are seen on the epitaxial surface of the ε-Si. Fig. 7(b) illustrates the contact AFM image (3 μm x 3 μm) of the crosshatch pattern. The six red crosses indicate positions where the tip-enhanced measurements were carried out. Fig. 7(c) shows near-field (TERS) spectra of ε-Si extracted by Lorentzian curve fitting. From Fig. 7(c), we can observe that the peak position and the intensity of the ε-Si change. The changes are due to variations in thickness of the ε-Si layer based on an underlying surface topography in Fig. 7(b). Fig. 7(d) corresponds to the $Si_{1-x}Ge_x$ Raman spectra. The spectral pattern in the $Si_{1-x}Ge_x$ under layer depends both on the Ge concentration and presence of any residual defects in the $Si_{1-x}Ge_x$ due to lattice mismatch during heteroepitaxial growth[36,37]. The dark areas, *c* and *b* have lower ε-Si intensity compared to bright areas, *a, d, e,* and *f*, in the AFM image. If the intensity of ε-Si (Fig. 7(c)) is weak, the intensity of $Si_{1-x}Ge_x$ (Fig. 7(d)) is strong. From the cross sectional view of Fig. 7(b), the step height between the bright to the dark area is ~16 nm. Figure 7(c) also illustrates the slight shift (~ 2 $cm^{-1}$) of the ε-Si peaks. While the average deviation of the peak of the silicon tip stays within a range 520 ± 0.04 $cm^{-1}$ (data not shown), its value in the ε-Si spectrum is 514 ± 0.8 $cm^{-1}$. Raman shift of Si-Si vibration basically provides a measure of the inter-atomic Si-Si spacing in the ε-Si layer. This spacing causes fluctuation of channel strains, and subsequently affects the performance of the device. It has been reported that lateral strain variations is mostly associated with the crosshatched pattern and its variation is in the order of ~1 $cm^{-1}$ [36]. Calculated from the shift in Raman peaks, the strain imposed on the capped ε-Si layer is about 1.5 ± 0.2 GPa [37].

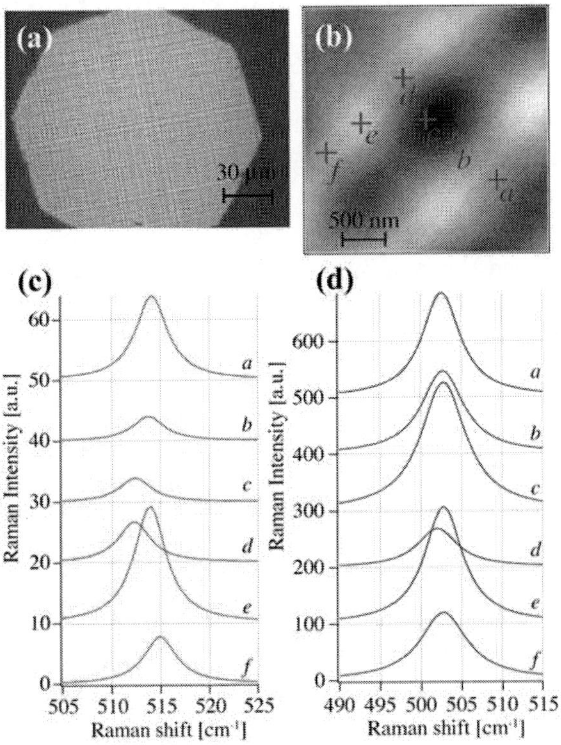


Figure 7. (a) Reflection microscopic image of the ε-Si surface. (b) AFM image of the ε-Si surface. Red crosses indicate the tip positions where TERS mapping was carried out. (c) and (d) are Lorentzian fitting results of position dependent TERS spectra of ε-Si and $Si_{1-x}Ge_x$, respectively. *a-f* correspond to the positions shown in (b).

Figure 7(d) shows the subtracted near-field spectrum for the LO phonon mode arising from $Si_{1-x}Ge_x$ buffer layer. The LO phonon mode in the underlying $Si_{1-x}Ge_x$ layer did not show much spectral shift during the spectra mapping except a slight shift observed in the spectrum *d*. This can be attributed to the local density of doped Ge, which could be higher around the position *d*. This is also evident from Fig 7(c) point *d*, which exhibits the highest spectral shift for the LO phonon mode in ε-Si layer. The nearby spectra obtained at *c* and *e* also have slight spectral shifts. This means that, at position *d*, we have high concentration of defects associated to the heterogeneity in the Ge-concentration in the underlying layer. This is the direct evidence of the fact that the Si-Si bond length in ε-Si is affected by the crystalline structures and concentration of Ge in the underlying $Si_{1-x}Ge_x$ layer on nanoscale. However, the observed spectral shifts in the ε-Si (~2 $cm^{-1}$) are higher than those in the $Si_{1-x}Ge_x$ layer (~1 $cm^{-1}$). This is understandable, because even though the strains in both layers are caused by the presence of Ge atoms in the buffer layer, the amount of strain may not be the same.

All these observed nanometer scale information on localized strain cannot be detectable by micro SERS experiments due to averaging effect within a diffraction limited focused spot [34]. Using TERS, however, we could successfully reveal the localized strain hidden in nanometer scale.

# 4. Managing the Excitation Wavelength and the Tip Material for TERS

So far, we have demonstrated selective enhancement of LO phonon signals arising from the thin ε-Si layer by SERS and TERS in the presence of strong background scattering from $Si_{1-x}Ge_x$ buffer layer to distinctly observe Raman signal and successfully revealed localized strain in nanometer scale using TERS. However, as can be seen from Fig. 6, the LO phonon from the ε-Si layer is sandwiched between two comparatively strong unwanted peaks arising from the $Si_{1-x}Ge_x$ buffer layer and Si-tip material. In order to have higher sensitivity and to obtain high contrast image of the strain distribution in silicon layer by TERS, it is important to get rid of these two unwanted background scattering modes. We address this by using shorter wavelength and using different tip material.

## 4.1. Ultraviolet Light Excitation for Background Suppression from Underlayers

Since the strong scattering from the buffer layer is excited due to the longer penetration depth of the probing light, this problem can be overcome only if the penetration depth of the probing light in silicon can be reduced. It has been reported [15] that by varying the excitation wavelength from 532 to 351 nm, the penetration depth of the probing light in silicon can be changed from 400 to 5 nm. For example, the penetration depth is 5 nm, 47 nm, 108 nm, 241 nm, and 406 nm for the excitation wavelength at 351 nm, 405 nm, 442 nm, 488 nm, and 532 nm, respectively [39]. Moreover, resonant Raman scattering in Si has been reported [40] for excitation around 365 nm (3.4 eV). Thus, an increase in the phonon intensity due to the resonance effect can also be observed for lower excitation wavelengths, which is an additional advantage in using the lower excitation wavelengths. However, the use of Ag-coated tip in TERS limits the wavelength that can be effectively used for enhancement. This is because the field enhancement near the tip apex comes from the plasmon resonance [41] within the Ag layer coated on the tip. By reducing the excitation wavelength, we move away from the plasmon resonance of the tip, and hence lose the enhancement of the field at the tip apex. Therefore, for an efficient TERS detection from the thin layer of ε-Si, we need to find a compromise between the two opposite effects of reducing the excitation wavelength. One, a positive effect by reducing the penetration depth together with increasing Raman efficiency due to resonance Raman effect, and the other, a negative effect by reducing the field enhancement at the tip apex. To compromise for the negative effect of using shorter wavelength, we performed far-field Raman measurements in different excitation wavelength. Figure 8(a) shows the far-field Raman spectra (without tip) obtained from sample. Each spectrum is normalized and then decomposed into two Lorentzian function (dashed lines) for peak intensity analysis. The relative intensity of the Si-Si in ε-Si relative the Si-Si in $Si_{1-x}Ge_x$ buffer layer increases with decreasing wavelength. The total intensity for each phonon mode in $Si_{1-x}Ge_x$ and ε-Si were obtained by calculating the area under the two Lorenztian curves.

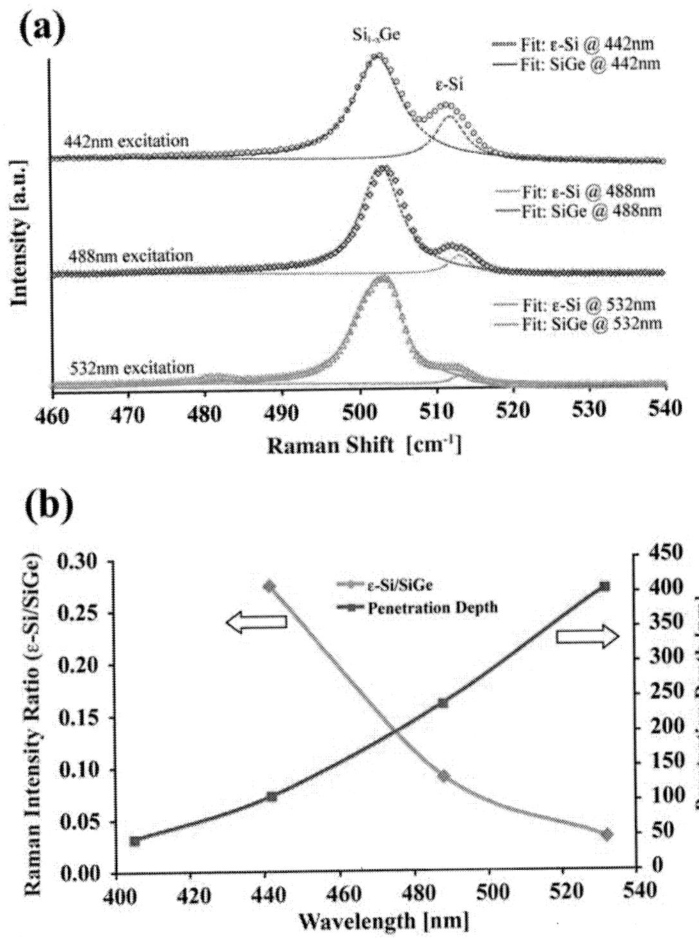


Figure 8. (a) Far-field Raman intensity of ε-Si layer grown in the SiGe layer at different wavelength. The Lorentzian functions under Si-Si in $Si_{1-x}Ge_x$ (x=0.25) peaks at 502.6, 503.4 and 502.6 $cm^{-1}$, which corresponds to phonon modes from 442nm, 488nm and 532nm, respectively. (b) Total Raman intensity ratio (blue) between ε-Si and SiGe and penetration depth (red) as a function of excitation wavelength. The Raman intensity ratios were obtained by calculating the area under the Lorentzian curves.

The relative intensity of the LO phonon mode arising from ε-Si layer is increased by shortening the excitation wavelength. Figure 8(b) shows the total Raman intensity ratio of ε-Si over the $Si_{1-x}Ge_x$ at varying wavelengths. The secondary y-axis in Fig. 8(b) is the calculated penetration depth for each excitation wavelength used. As the penetration depth decreases with decreasing wavelength, the total Raman intensity ratio between ε-Si and $Si_{1-x}Ge_x$ increases. At 442nm excitation wavelength, the penetration is ~108nm and still bigger than the 30nm thickness of ε-Si sample. A laser with 405nm excitation wavelength may be a good candidate because the penetration is ~47nm and near the 30nm ε-Si thickness. At 442nm, the Raman intensity from ε-Si is increased to half the intensity of $Si_{1-x}Ge_x$. This is a significant improvement compared to 532nm light where only a small ε-Si Raman signal is observed (see Fig. 6). The incurred loss in enhancement due to shorter wavelength was offset by the gain in the ratio between ε-Si and SiGe signal. Reducing the excitation wavelength allows us to obtain effectively background-free scattering signal from the ε-Si layer, leading to a feasible imaging of strain in ε-Si substrates. The field enhancement at the tip apex is still reasonable for the He-Cd laser with excitation wavelength of 441.6 nm (data not shown here), so we selected this wavelength in further TERS aimed towards higher sensitivity detections.

### 4.2. Selection of Tips for Background Suppression from Silicon Tips

In order to further improve the sensitivity in TERS, we still need to address the other problem arising from the presence of the LO phonon mode from the tip itself. Since the tip is made of silicon and coated with Ag, it works as a perfect enhancer for the Si-Si scattering in the tip itself. This is because all the silicon atoms are adsorbed in the enhanced field created at the tip apex. Hence, it is impossible to suppress the scattering from the tip. The best way to address the unwanted peak arising from Si-Si vibration mode in the Si-tip that interferes with the Raman signal from ε-Si layer, is to change the tip material. We need to select a material that will not interfere with sample vibrations. Unfortunately, most of the commercially available AFM cantilever tips are made of silicon, which is not a good tip-material for detecting TERS from silicon samples. However, cantilever tips made of silicon nitride ($Si_3N_4$) are now commercially available. The Raman vibrational modes of $Si_3N_4$ are distinctly separated from those of Si, which is a good tip-material for observing TERS from Si samples. Aside from $Si_3N_4$ tips, oxidized silicon tips [42,43] or silicon dioxide ($SiO_2$) tips such as a tapered optical fiber probes can be also useful for the base material for metallized tips. If all Si atoms of the tip are oxidized, Raman vibration of Si from the tip can be completely suppressed, and hence this tip can be effectively use. It has been recently reported [44] that tips made of materials with lower refractive index, such as $Si_3N_4$ ($n$=2.05) or $SiO_2$ ($n$=1.5) could show higher enhancement factors especially in shorter wavelength region.

Figure 9 shows the transmission electron microscope (TEM) image of the thermally oxidized silicon cantilever probes at 1100°C temperature under steam atmosphere for 3 minutes. As seen in the figure, $SiO_2$ layer with a thickness of ~100 nm is formed on a silicon-based tip without degrading the original sharpness. This oxidized silicon tips is also promising for tuning the plasmon resonance of metals coated on the tip by controlling the thickness of the $SiO_2$ layer. Instead, for our experiments, we utilized the commercially available $Si_3N_4$ cantilever tips, after coating them with Ag. Figure 10 shows the SEM image

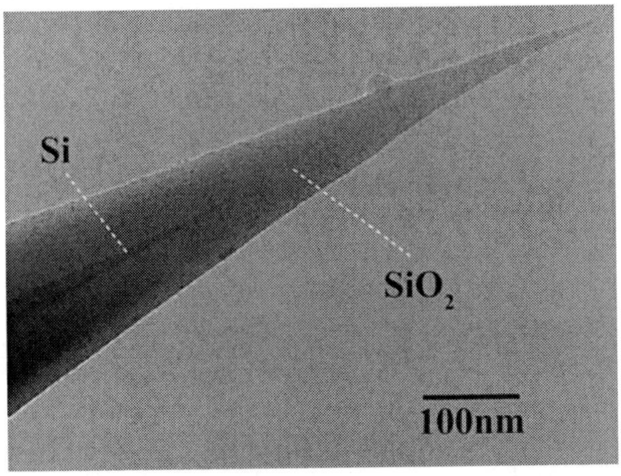


Figure 9. A TEM image of an oxidized silicon cantilever tip.

of the $Si_3N_4$ tip (Budget Sensors, BS-SiNi) coated with 50-nm-thick Ag layer by thermal evaporation process. The diameter of the tip apex after Ag coating was about 30 nm, which is relatively smaller than the 50nm thickness of coated Ag because of the very sharp tip shape. Fig. 10 (b) is the SEM image (bottom view) of the cantilever. In order to illuminate the tip by a LWD objective lens with the incident angle of 60 degrees, part of the cantilever is removed by focused ion beam (FIB), otherwise it creates an unwanted shadow on the sample [28]. This was not necessary for the previous silicon tip, because the shapes of the bottom structures for the Si and $Si_3N_4$ tips are different. Even after chopping off a part of the tip, some shadow of the tip still existed for $Si_3N_4$ tips. It was impossible to reduce this shadow any further, so the overall intensity of Raman scattering was slightly reduced for TERS compared to the far-field measurements in the TERS experiments involving $Si_3N_4$. For reflection mode TERS, the geometrical accessibility of the tip for illumination should also be taken into consideration to minimize excitation power loss and fake scattering signals from the protruding parts of most commercially available tips.

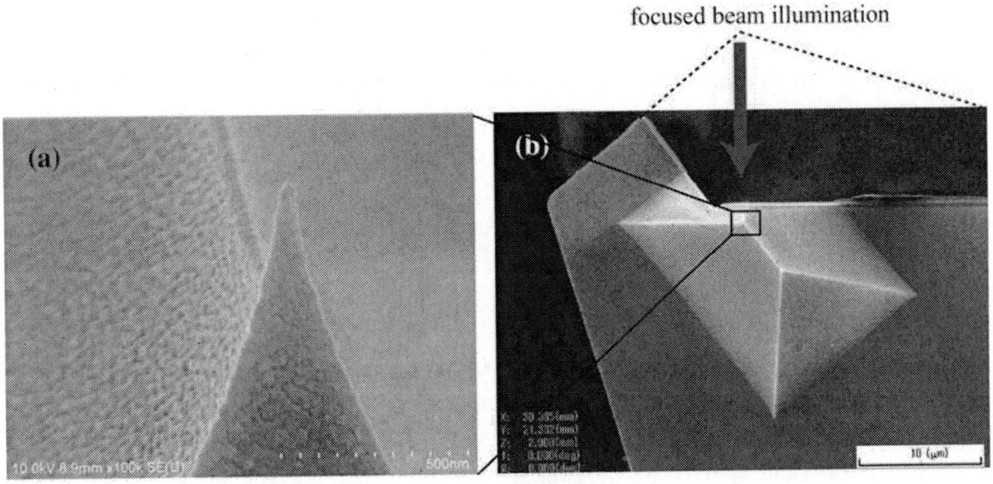


Figure 10. SEM images of Ag coated $Si_3N_4$ tip. The area surrounded by dashed lines is removed by FIB for an efficient light delivery onto the tip by a LWD objective lens.

### 4.3. TERS Imaging

In this section, we demonstrate TERS imaging on ε-Si layer using He-Cd laser with 441.6 nm wavelength and Ag coated $Si_3N_4$ cantilever tip. Figure 11 (a) shows the comparison between TERS (with a tip) and far-field Raman spectrum (without a tip). As expected, the Si-Si background signal seen from the tip at 520 $cm^{-1}$ in Fig. 6 is completely eradicated. Furthermore, the penetration depth of the excitation light at 441.6 nm in the underlying $Si_{1-x}Ge_x$ layer is drastically reduced compared with that at 532 nm. The far field Raman signal from the thin ε-Si layer is relatively increased to the half of the intensity of the background signal from the $Si_{1-x}Ge_x$ layer. This is a significant change compared to the Fig. 6(a), which uses 532nm, where only a small shoulder corresponding to the LO phonon arising from the ε-Si layer was observed. In the TERS spectrum in Fig. 11 (a), the background signal from $Si_{1-x}Ge_x$ layer is slightly reduced because of the shadowing effect of the tip. Even with shadowing effect by the tip, the signal from the ε-Si layer is increased because of the efficient tip-enhancement effect. In addition, Fig. 11 (b) shows the comparison between SERS (with Ag film; thickness ~ 8nm) and far-field Raman spectrum (without Ag film). The Raman spectrum without Ag film is the same without Ag tip in Fig. 11 (a). The presence of thin Ag layer reduces the intensity of excitation light reaching the sample due to absorption. This explains the reduction of the overall intensity of the Raman scattering in SERS spectrum compared to the far-field Raman spectrum. However, in the SERS spectrum, the relative intensity of the LO phonon arising from the ε-Si layer compared to the LO phonon intensity of the buffer layer is increased. This is once again a good demonstration of selective enhancement of the LO phonon vibration arising from the ε-Si layer. Comparing the experimental results of TERS and SERS in Fig. 11, the enhancement in the relative intensity of ε-Si with respect to the scattering from $Si_{1-x}Ge_x$ buffer layer achieves almost same value. But, the scattering volume involved in TERS (about the same size as tip apex) is much smaller than that in SERS (same as the diffraction limited focused spot), which suggests that tip-enhancement with a Ag coated $Si_3N_4$ tip at 442 nm excitation works much more efficiently compared with surface enhancement at the same wavelength.

Since the combination of a $Si_3N_4$ tip and 442 nm excitation shows much higher sensitivity in TERS measurements compared to the previous experiments using 532nm and Si tip, we employed this combination to image the nano-scale strain distribution on the surface of the ε-Si layer through near-field Raman imaging. In order to realize the nano distribution of the strain, the near-field Raman image was also compared with a far-field image from the same area of the sample. Figure 12(a) shows a topographic image (128 pixels by 128 pixels) and Figs. 12(b), and 12(c) show the strain images extracted from near-field and far-field Raman spectra, respectively, obtained from the same position (32 pixels by 32 pixels). In order to obtain these Raman images, the LO phonon modes from each position where fitted with Lorentzian functions, and then the peak positions were estimated from the fittings, which were plotted with respect to the position. The near-field image was constructed by the above-mentioned process applied to the subtracted near-field spectra obtained by moving the tip at different positions. The far-field image on the other hand was constructed by applying the process to the far-field spectra by moving the center of the focused laser spot at different positions. Since the shift in the frequency position of the LO phonon mode is a direct

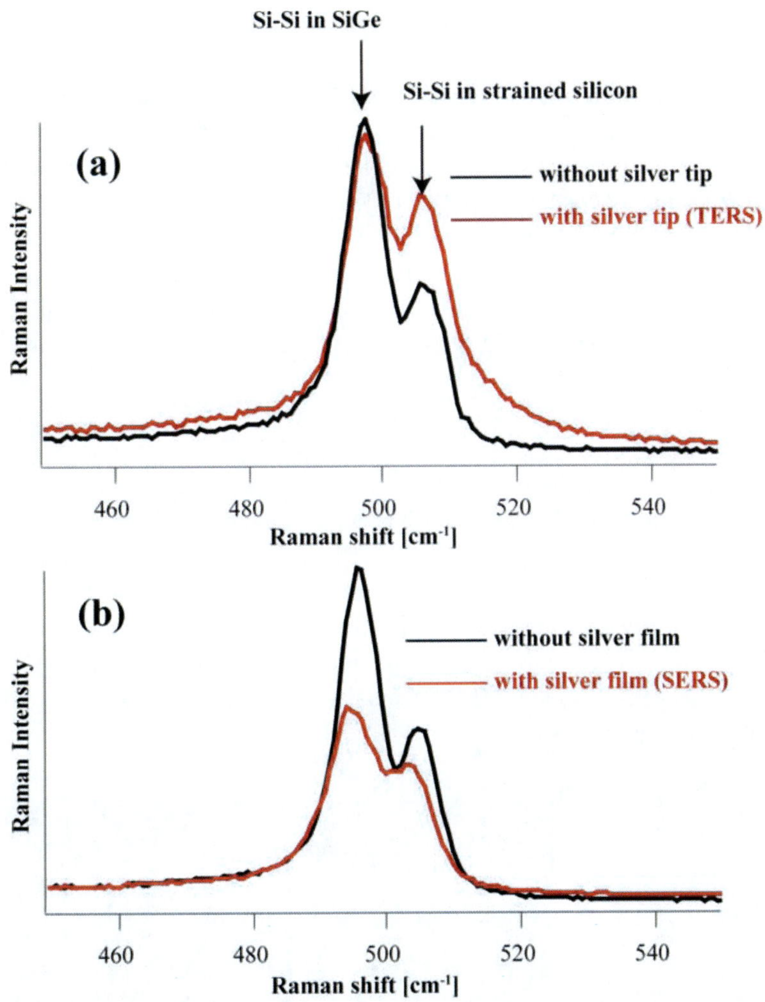


Figure 11. (a) Comparison of TERS and far-field Raman spectra, (b) Comparison of SERS and far-field Raman spectra.

reflection of the local strain, the images shown in Fig. 12(b) and 12(c) correspond to the distribution of the local strain in the sample. The near-field image clearly reveals strain embedded in nanoscale, which cannot be seen in the far-field image that shows almost uniform contrast. This confirms that there is nanometric distribution of the local strain in the sample, which could not be seen by the far-field imaging. Figures 12(d) and 12(e) show the plots of the number of pixels as a function of Raman peak position for the near-field image in Fig. 12(b) and for the far-field image in Fig. 12(c), respectively. Narrow Gaussian distribution produced in the far-field image is due to averaging effect within the diffraction limit while the widely broadened distribution in the near-field image is due to the observation of nano-scale strain distribution. This meets our expectation simply because near-field image in Fig. 12(b) shows large fluctuation of contrast compared to Fig. 12(c). It would be important to point out that the LO phonon peak positions in Fig. 12(d) is distributed in lower Raman shifts compared to Fig. 12(e), as can be seen from the shift of the Gaussian distributions in the two figures. There are three possible mechanisms for this effect; one is related to the chemical effects taking place between the Ag atoms of the coated tip and silicon atoms of the strained layer, and the others, due to the local pressure and heating [45, 46] applied on the sample by the tip. Further investigation is required for confirmation. A careful observation of SERS spectrum in Fig. 11 reveals that there was a small downshift of Raman mode in SERS compared to far-field spectrum. It is also well know that Raman modes in SERS measurements are often downshifted [47-50]. The primary reason for such a shift is the chemical bonding between the Ag atoms and the sample atoms, which effectively modifies Raman vibrations. This can also happen for TERS, when Ag molecules of a coated tip come in close contact with the sample molecules. Since we used the contact mode operation of AFM to control the tip position in TERS experiment, the tip can apply a unidirectional force on the sample at the contact point. Even though the value of this force is not large, the resultant pressure per atom could be large enough due to the nano-sized of the tip apex. This effect was recently observed in various nanomaterials [47,51,52] under controlled application of pressure using tip. In our case, the tip applied pressure works in a direction perpendicular to sample surface. In the equilibrium condition, we can expect that the pressure to increase the bond lengths in the direction parallel to the sample surface. As a result, it adds to the strain present in the silicon layer due to the lattice mismatch between the silicon and the buffer layers. Hence, Raman modes are further shifted to the lower frequencies. For detailed characterization of strain distribution, some analytical investigations of the tip-pressure and heating such as finite element method (FEM) are also very important factors together with the development of experimental system.

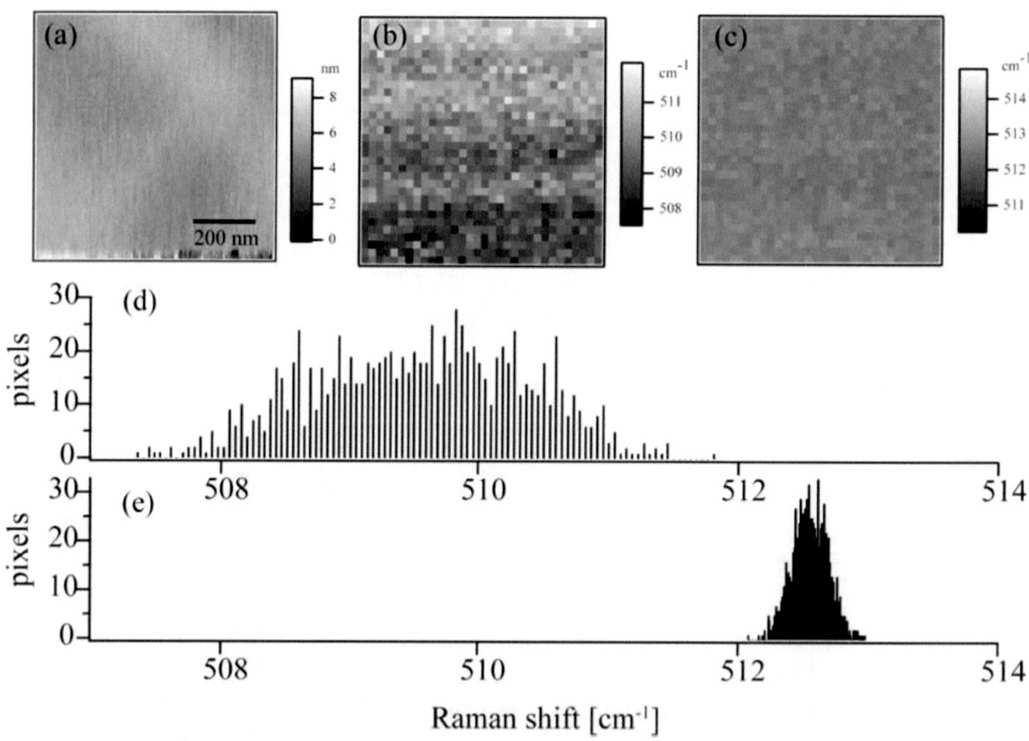


Figure 12. (a) topographic image and strain images of (b) near-field and (c) far-field, respectively. (d) and (e) are Raman peak distributions in (b) and (c), respectively.

# 5. Controlling the Polarization in Detection

At this point, we tried to improve sensitivity by modifying the tip material, using shorter wavelength to reduce penetration depth and focus spot, and manipulating the well-defined polarization of crystalline sample along with the polarization conditions of probing light. The key issue is how to improve the sensitivity of the weak near-field signal at the nanoscale level to be detectable from the strong background coming from the total illuminated area. Most of the research works done have focused on electric field enhancement, either by Ag coated Si and $Si_3N_4$ tip or by pure metallic tip [11,29,32,53,54]. Various arguments were made regarding tip form and material [55-57] and the control of polarized excitation light [58,59]. Recently, researchers started working on reducing the far-field background signal without comprising the field enhancement to obtain high contrast near-field Raman image. One way is manipulating the well-defined polarization properties of crystalline sample relative to the polarization and propagation of the incident light. Ossikovski et al, proposed a simple model that describe polarized TERS to characterize tip-enhancement process [60]. The model is based on tip-amplification tensor describing the interaction of tip on both the incident and scattered optical field. From their results, by appropriately setting the polarizer and analyzer relative to the material crystalline axis, far-field signal can be reduced. Poborchii et al, proposed far-field suppression by depolarization configuration [61]. In this configuration, the tip depolarizes the incident beam from p-polarized light to s-polarized light or vise versa. The resulting depolarized light will incur depolarized Raman signal that can be detected through an analyzer while blocking the Raman signal from the originally polarized light to minimize the far-field Raman signals [62]. Sokolov's group optimizes this depolarization configuration to obtain high contrasts Raman image on silicon material with lateral resolution of 20nm [63]. However, the authors noted that the optimum polarizer and analyzer condition, which gives highest contrast, changes with the morphology and geometry at the tip end.

In this section, we used the silicon Raman tensors to estimate at which combination of incident polarization and sample orientation will offer minimum far-field Raman intensity. We also introduce different analyzer setting to further reduce the far-field background by taking into account the depolarization effect mentioned above. We performed SERS experiment to validate the numerical results and we found good agreement with the Raman tensor calculation. The configuration that gives minimum far-field background signals with high contrast in SERS was utilized in TERS experiment to obtain high contrast near-field Raman signal. In this approach, we fixed the polarizer and analyzer along with sample orientation based on SERS to ease the difficulty of optimizing the polarizer and analyzer setting for different tip geometry. Both the field enhancement effect and depolarized detection were considered to obtain a high signal-to-noise TERS signal. We found out that for Raman active thin samples assembled in bulk crystalline materials, depolarization effect outweighs the field enhancement effect in getting high-contrast Raman signal.

## 5.1. Theoretical Calculation of Raman Intensity by Side Illumination Optics

The geometrical model used in the Raman intensity calculation is shown in Figure 13. This model was formulated for reflection-mode TERS system where a metallic tip is placed in contact with the ε-Si sample. There are two Cartesian coordinate systems, the sample (*xyz*) and the light *(x'y'z')*. The laser beam is introduced at $\phi$=60 degrees relative to the sample normal (*z*-axis). The sample rotation angle (azimuth) is denoted as $\theta$. The sample azimuth is varied to obtain the maximum and minimum Raman intensity level for both *p*- and *s*-polarization of incident light.

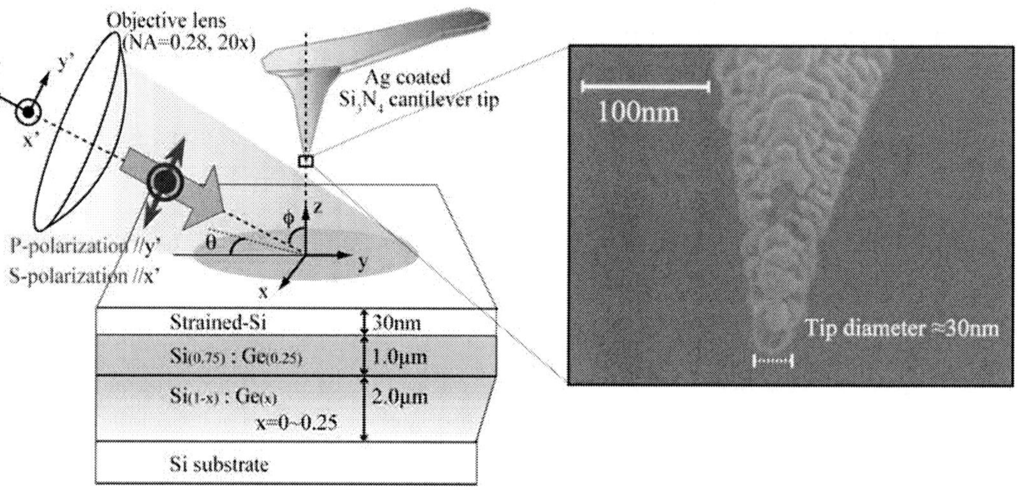


Figure 13. Schematic of reflection-mode TERS. TERS were defined with two Cartesian coordinate systems, the sample (*xyz*) and the light *(x'y'z')* reference frames. Laser beam is introduced at $\phi$ relative to the sample's normal (*z*-axis) and $\theta$ denotes the rotation angle (azimuth). The thickness of the strained Si is ~30nm. Electron micrograph of the Ag coated tip shows a tip-diameter of ~30nm.

The far-field Raman scattering intensity ***I*** is generally expressed as [22,64]:

$$\boldsymbol{I} \propto \sum_j \left| \boldsymbol{e}_s^T \boldsymbol{R}_j \boldsymbol{e}_i \right|^2 \qquad \text{(eq.1)}$$

where ***R*** is Raman tensor for any *j* active phonon, and $\boldsymbol{e}_i$, $\boldsymbol{e}_s$ are the unit vector of incident and scattered electric field, respectively. The superscript *T* denotes transpose of the scattered electric field vector. The derivation of Raman tensor ***R*** for *c*-silicon in the (100) plane by back-scattering geometry (normal incidence) is discussed in detail by R. Loudon [22]. The *c*-silicon Raman tensor for the phonon modes in the sample reference frame (*xyz*) is:

$$\mathrm{TO}_1 \Rightarrow \boldsymbol{R}_1 = \begin{pmatrix} 0 & 0 & 0 \\ 0 & 0 & d \\ 0 & d & 0 \end{pmatrix}, \ \mathrm{TO}_2 \Rightarrow \boldsymbol{R}_2 = \begin{pmatrix} 0 & 0 & d \\ 0 & 0 & 0 \\ d & 0 & 0 \end{pmatrix}, \ \mathrm{LO} \Rightarrow \boldsymbol{R}_3 = \begin{pmatrix} 0 & d & 0 \\ d & 0 & 0 \\ 0 & 0 & 0 \end{pmatrix} \qquad \text{(eq.2)}$$

in which $\boldsymbol{R}_1$, $\boldsymbol{R}_2$, and $\boldsymbol{R}_3$ are respectively, the $TO_1$, $TO_2$, and LO, phonon modes of *c*-silicon and *d* is the constant that is dependent on the Raman polarizability of the sample. Note that in Raman spectra of Si, there is only one very strong one-phonon peak at 519 $cm^{-1}$ (at 305 K), corresponding to the zone-center optical phonons in Si, in which the TO and LO phonons are degenerate at zone center in diamond-type crystals [65]. The *d* value for the *c*-silicon can be different from ε-Si but for qualitative Raman intensity analysis, using *c*-silicon Raman tensor can provide a good approximation. Nevertheless, the Raman tensor of unstrained silicon expressed in eq. 2 is different from ε-Si because of the inhomogeneous tensile stress induced in the ε-Si layer. The amount of strain is maximized at the boundary of SiGe layer and ε-Si layer, and then gradually decreases and minimized at the surface layer of ε-Si. In order to obtain the qualitative comparison of Raman intensity in ε-Si at different incident and scattered polarizations with sample azimuth, we simply employ the Raman tensor of unstrained Si. To utilize the Raman tensors and scattered Raman intensity in eq.1 in reflection-mode TERS studies, the Raman tensors must be transformed to the laser illumination reference frame (*x'y'z'*) at angle $\phi$. This is because the Raman tensors in eq. 2 were obtained in normal incidence. First, we need to rotate the sample azimuth about its normal (along the *z*-axis) in the xyz coordinate, and then formalize the z-component of the transformation matrix, $T_z(\theta)$. The Raman polarizability tensor of the sample rotated by an angle $\theta$ would become,

$$R_j'(\theta) = T_z^T(\theta) R_j T_z(\theta) \qquad \text{(eq.3)}$$

where, $T_z(\theta)$ is the rotational matrix for a sample and given by,

$$T_z(\theta) = \begin{pmatrix} \cos(\theta) & -\sin(\theta) & 0 \\ \sin(\theta) & \cos(\theta) & 0 \\ 0 & 0 & 1 \end{pmatrix} \qquad \text{(eq.4)}$$

Next step is to transform the Raman tensor in the *x'y'z'* reference frame. In this case, rotation is done along the *z'*-axis. The transformed Raman tensor $R''(\phi, \theta)$ of the sample in the (*x'y'z'*) reference frame at an angle $\phi$ is given by

$$R_j''(\phi,\theta) = T_x^T(\phi) R_j'(\theta) T_x(\phi) \qquad \text{(eq.5)}$$

where, $T_x(\phi)$ is the *x*-component of the rotational matrix converting (*x'y'z'*) to (xyz) and expressed as

$$T_x(\phi) = \begin{pmatrix} 1 & 0 & 0 \\ 0 & \cos(\phi) & -\sin(\phi) \\ 0 & \sin(\phi) & \cos(\phi) \end{pmatrix} \qquad \text{(eq.6)}$$

Consequently, the transformed scattering intensity $I'$ for the transformed Raman tensor in far-field takes the form similar to (eq.1) as

$$I' \propto \sum_j \left| e_s^T R_j{}''(\phi, \theta) e_i \right|^2 \qquad \text{(eq.7)}$$

Another factor, that we need to consider, is the change in amplitude of the electric field at the surface, such as air/ε-Si and strained Si/SiGe interfaces. The transmission and reflection coefficients for the p- and s-polarized light at different media of known refractive index can be easily be derived from Snell's law and Fresnel equations given as [66]:

$$n_1 \sin\phi_1 = n_2 \sin\phi_2 \qquad \text{(eq.8)}$$

$n_1$ and $n_2$ are refractive indexes. $\phi_1$ and $\phi_2$ are the angle of incidence and refraction of light, respectively.

$$T_p = \frac{n_2\cos\phi_2}{n_1\cos\phi_1}\left|t_p\right|^2 \qquad t_p = \frac{2n_1\cos\phi_1}{n_2\cos\phi_1 + n_1\cos\phi_2} \qquad \text{(eq.9)}$$

$$R_p = \left|r_p\right|^2 \qquad r_p = \frac{n_1\cos\phi_1 - n_2\cos\phi_1}{n_1\cos\phi_2 + n_2\cos\phi_1} \qquad \text{(eq.10)}$$

$$T_s = \frac{n_2\cos\phi_2}{n_1\cos\phi_1}\left|t_s\right|^2 \qquad t_s = \frac{2n_1\cos\phi_1}{n_1\cos\phi_1 + n_2\cos\phi_2} \qquad \text{(eq.11)}$$

$$R_s = \left|r_s\right|^2 \qquad r_s = \frac{n_1\cos\phi_2 - n_2\cos\phi_2}{n_1\cos\phi_1 + n_2\cos\phi_2} \qquad \text{(eq.12)}$$

The summary of the transmittance ($T_p$, $T_s$) and reflectance ($R_p$, $R_s$) obtained from the calculation are illustrated in Figure 14. Refractive indexes of $n_2$ (Si) and $n_3$ (SiGe) were quoted from the reference [67,68]. However, the strain is gradually relaxed from the boundary of strained Si/SiGe to strained Si/air. Thus, the strain at the surface is minimum and is very close to ordinary silicon, which means that the refractive index of ε-Si approximately equal to unstrained Si. Detailed characterization of the refractive index can be analyzed by ellipsometry. It should be noted that when the incident angle $\phi_1$ is 60 degrees, the angle of incidence inside underneath the ε-Si layer, $\phi_2$, is 10.5 degrees. This fact should be taken into account when calculating the far-field Raman contributions inside the material. Taking into account all the factors above, the dependence of the far-field Raman intensities with sample azimuth was calculated at four combinations of polarizer and analyzer. The four combinations are P-P-θ°, P-S-θ°, S-P-θ°, and S-S-θ°, which represent the polarizer-analyzer-sample azimuth in degrees, respectively. In the next section, numerical results will be compared with SERS experimental data.

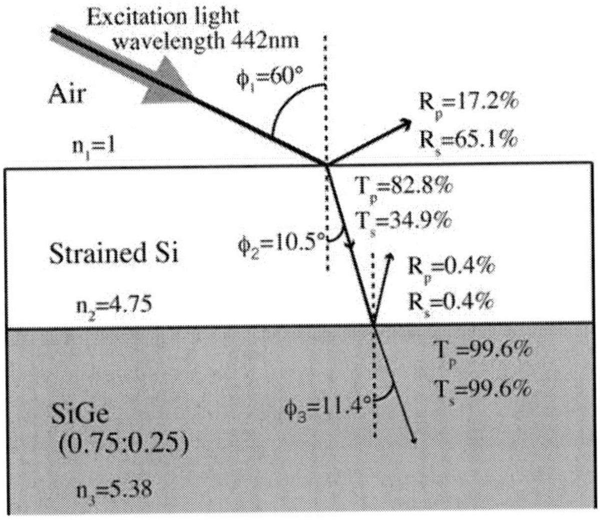


Figure 14. Schematic of transmittance and reflectance ratio in ε-Si. $T_p$, $T_s$ are transmission coefficient for p and s polarization at $\phi$=60°. $R_p$, $R_s$ are reflection coefficient for p and s polarization at $\phi$=60°. The index of refraction $n_1$, $n_2$ and $n_3$ are for air, ε-Si and SiGe, respectively. Index of refraction values were obtained in Ref.67,68.

## 5.2. Sample Azimuth Dependence of Polarization Raman Measurement : Depolarization Effect in SERS

The left most part in Figure 15(a)P-P-θ° (b)P-S-θ°, (c)S-P-θ° and (d)S-S-θ° shows the calculated normalized far-field Raman intensities as a function of sample azimuth relative to y-axis using *c*-silicon Raman tensor at different polarization states of polarizer and analyzer. Fig. 15(e)P-P-0°, (i)P-P-45°, (f)P-S-0° (j)P-S-45°, (g)S-P-0°, (k)S-P-45°, (h)S-S-0°, and (l)S-S-45° are the experimental results of the far-field Raman and SERS spectra at 0 and 45 degrees sample azimuth, which indicate the minimum or maximum far-field Raman intensity for (a)P-P-θ°, (b)P-S-θ°, (c)S-P-θ° and (d)S-S-θ°. All the spectra in Fig. 15 were obtained with the same exposure time of 10 minutes and laser power of 9.8 mW before the objective lens (NA:0.28) in a side illumination optics oriented at an angle $\phi_1$=60 degrees from the sample normal. The CCD dark counts are subtracted from all the spectra. For SERS spectroscopy, a 10 nm thick Ag film is evaporated at the rate of ~0.7Åm/s on the ε-Si surface under the ~$10^{-6}$ Torr vacuum pressure. The SiGe Raman peak intensities at ~503 $cm^{-1}$ in SERS (with Ag) are reduced to almost half of the far-field (without Ag) Raman intensity. This is because the illumination light has to transmit through the Ag films, in turn losing some probing power. Besides, no surface enhancement effect is expected to reach the underlying SiGe layers because the enhanced field decays strongly and is localized within ~20 nm from the surface [30,69]. From the calculation results, the amplitude of sinusoidal Raman intensities in (a)P-P-θ° is to found to be greater than (b)P-S-θ°, (c)S-P-θ° and (d)S-S-θ°. Fig. 15 (b)P-S-θ° and (c)S-P-θ° have the same amplitude. This amplitude of the Raman intensity is dependent on the excitation transmittance at the sample surface, e.g., the detection efficiency of P-P-θ° is 67 % (0.82 x 0.82: incident x scattered), P-S-θ°and S-P-θ° is 28% (0.82 x 0.34=0.34 x 0.82), and

S-S-θ° is 12% (0.34 x 0.34). The detection efficiency was summarized in Fig. 14. In the P-P polarizer-analyzer setting in Fig. 15, the far-field Raman intensity level of (e)P-P-0° can be suppressed down to 1/4 when compared to the (i)P-P-45° where sample is rotated 45 degrees. This observation is consistent with the calculation result wherein the ratio of the maximum amplitude intensity over the minimum amplitude intensity is eight because the offset intensity marked in the red circle in Fig. 15(a) is ~0.12. Note that this discrepancy between experiments and calculations is because experimental results are obtained as an average of wide variety of incident angles using an objective lens. The offset of ~0.12, marked in the red circle in Fig. 15(a), is the sample azimuth where Raman intensity was obtained in (e)P-P-0° setting. However, in (j)P-S-45°, (k)S-P-45° and (h)S-S-0° cases, the far-field Raman intensity offset available for detection is suppressed down to almost zero. The reason why there exists a strong Raman background signals in (e)P-P-0° measured in side illumination, can be justified from the contributions of $\boldsymbol{R}_1$ ($TO_1$), $\boldsymbol{R}_2$ ($TO_2$), and $\boldsymbol{R}_3$ (LO) to the total detected Raman intensity. However, when measured by backscattering illumination optics with normal

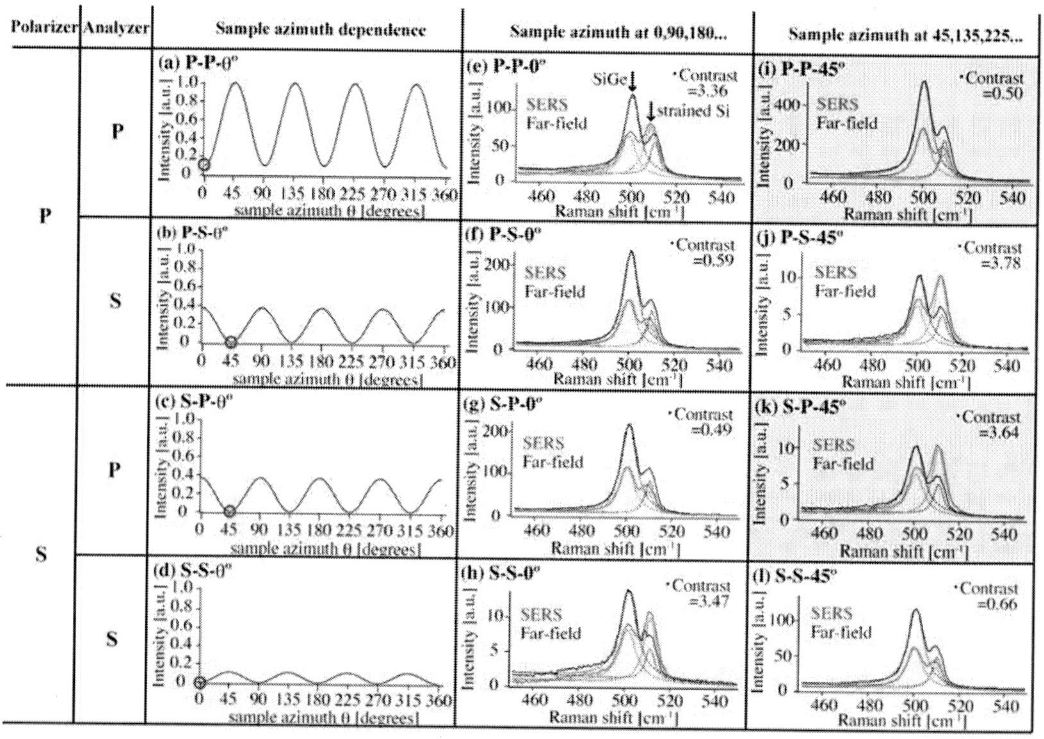


Figure 15. (a-d) Calculated Raman intensity as a function of sample azimuth at a fixed polarizer and analyzer setting using the transformed silicon Raman tensor. (e-l) Comparison of far field and SERS Raman spectra at various combinations of polarizer, analyzer and sample azimuth. The black line spectra are far field Raman spectra, and the red line spectra are SERS spectra. Dashed lines are fitted Lorentz function spectra for Raman intensity analysis. SERS contrast value at each combination was obtained and indicated. (i) and (k) shaded with gray are the two conditions used in the TERS experiment. Red circles in the (a-d) indicate the sample angle at which minimum Raman intensity is observed.

incidence ($\phi$=0; normal incidence optics), only $\boldsymbol{R}_3$ (LO) becomes Raman active and can be totally suppressed by an analyzer [23]. On the other hand, at incident light angle of $\phi \neq 0$ (side illumination optics), both $\boldsymbol{R}_1$ ($TO_1$) and $\boldsymbol{R}_2$ ($TO_2$) also becomes Raman active and contributes to the total Raman intensity because it's not suppressed by the analyzer. In other words, Raman signal of (e)P-P-0° are mainly coming from TO phonon modes but not from the LO phonon modes.

Strictly, the TO modes may not be overlapped with the stretched LO phonon mode. For example, in Fig. 15(e), Raman peak of ε-Si is observed at 511.1 $cm^{-1}$, which is mainly due to the TO modes, while SERS peak of ε-Si is slightly shifted and observed at 509.4 $cm^{-1}$, which is due to stretched LO mode. These facts suggest that the TO modes are considered as background signal for LO mode in our experiment and should be suppressed for higher sensitivity. Total intensities from TO and LO phonon modes are plotted in Fig. 15, (a)P-P-θ°, (b)P-S-θ°, (c)S-P-θ° and (d)S-S-θ°. To clearly understand the origin of background Raman signal, Figure 16 shows the calculated Raman intensities at each LO, $TO_1$, and $TO_2$ phonon mode in both normal and side illumination optics. Fig. 16(a) shows the normal incidence and Fig. 16(b), (c), and (d) correspond to P-P-θ°, P-S-θ° (or S-P-θ°), and S-S-θ° in the side illumination optics tilted at $\phi_1$=60 degrees incidence, respectively. When the incident angle is 0 degree (normal incidence), the minimum intensity of LO phonon mode is zero at the sample azimuth of 0 and multiples of 90 degrees (see inset in Fig. 16(a)), while the maximum intensity occurs at 45, 135, 225 and 315 degrees because the TO phonon mode is not active at all. However, in cases of (b) and (c), the TO phonon modes at the minimum LO phonon mode is not zero. This is because the two TO phonon modes are 90 degrees out of phase, making the sum a non-zero constant value at minimum LO phonon mode sample azimuth. Only in the case of (d), the TO phonon modes are not active. This suggests that selective LO phonon mode measurement is feasible. The issue is that (d)S-S-θ° in Fig. 15 is not optimal situation for spectroscopy. In particular, for microscopic image acquisition the Raman signal level is extremely weak. According to the individual Raman intensities assigned to LO, $TO_1$ and $TO_2$ phonon modes shown in Fig. 16, the residual (offset) far-field Raman intensities of (e)P-P-0°, (j)P-S-45° and (k)S-P-45°, red circles in Fig. 15, are all due to TO phonon modes because LO phonon modes are completely rejected by the analyzer. On the contrary, far-field Raman intensity of S-S-0° should not have a contribution from the TO phonon modes, however, the far-field Raman spectrum in Fig. 15 (h)S-S-0° has the signal intensity similar to the (j)P-S-45° and (k)S-P-45°. This residual far-field Raman signal observed in (h)S-S-0° is experimentally due to the wide variation of the incident angle and depolarization effect of the objective lens itself; the background signal level becomes almost the same as (k)S-P-45°. In this sense, S-S-0° is not advantageous over S-P-45°. Using this polarization dependence, we suggest a method to collect the near-field Raman signals with high efficiency in P-P-0°, P-S-45°, and S-P-45° by tip-depolarization effect [61-63]. The far-field Raman signal that comes from the large focused spot is linearly polarized while the near-field Raman signal that comes from the proximity of the tip can be depolarized by the tip geometry. The depolarized near-field Raman signal can pass through the analyzer by choosing proper conditions of incident angle, polarizer, analyzer, and sample azimuth. Therefore, SERS spectra were studied first by coating 8nm thickness Ag film on surface of ε-Si before TERS. The depolarization effect in SERS occurs at the entire focused spot with the aid of roughened Ag thin films. In TERS, the depolarization is localized and takes place at the tip apex while in SERS, the depolarization occurs at the entire focus. Hence, the main difference in the depolarization effect is the scattering volume. The SERS and far field Raman spectra are shown in Fig. 15(e)P-P-0°, (i)P-P-45°, (f)P-S-0°, (j)P-S-45°, (g)S-P-0°, (k)S-P-45°, (h)S-S-0°, and (l)S-S-45° which correspond to either minimum or maximum conditions for each polarizer-analyzer set.

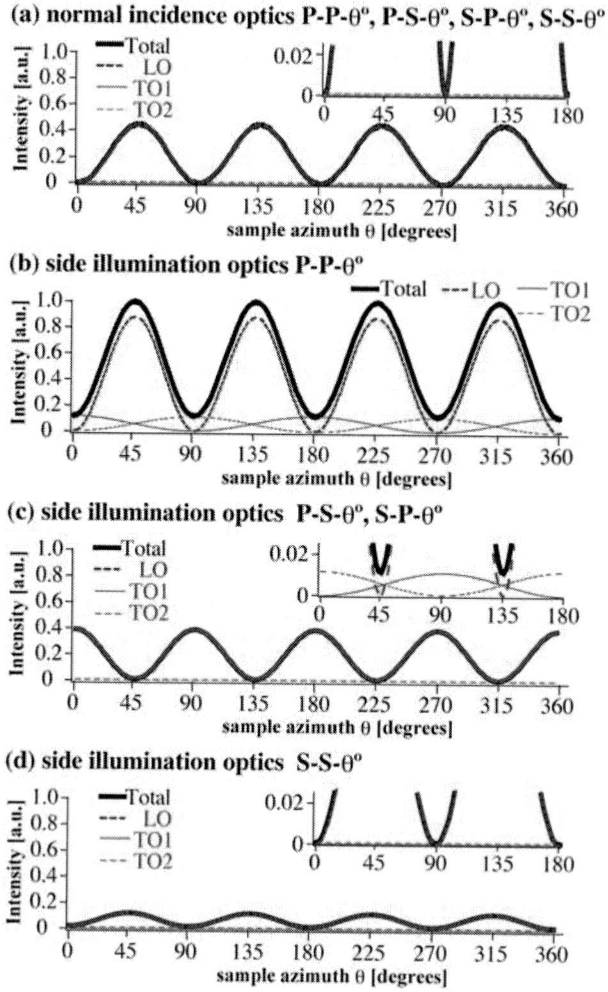


Figure 16. Raman intensity dependence of different Raman modes, LO (red) and TO (blue), as a function of sample azimuth at different polarizer and analyzer setting. (a) Normal incidence at P-P, P-S, S-P and S-S polarizer-analyzer polarization setting and side illumination (oblique incidence) (b)P-P, (c)P-S and S-P, (d)S-S polarizer-analyzer polarization configuration. Solid and dashed blue lines are Raman signal intensity of $TO_1$ and $TO_2$ phonon modes, respectively. Red solid line is Raman signal intensity from LO phonon mode. Black line is total Raman signal intensity. (inset: zoomed plot showing the zero Raman intensity).

Based on the SERS spectra, the Raman intensities of ε-Si is higher compared to their far-field Raman intensities at (e)P-P-0°, (j)P-S-45°, (k)S-P-45°, and (h)S-S-0° conditions. The reason is that, at these conditions, the far-field Raman is at minimum intensity as indicated by red circles in the calculation results of Fig. 15 and both depolarization and surface enhancement effect works efficiently with Ag film. To quantify the improvement, we calculated the parameter, *contrast* that is defined as the ratio of near-field signal intensity $I_{near}$ to the far-field signal intensity $I_{far}$ expressed as [63]:

$$Contrast = \frac{I_{near}}{I_{far}} = \left( \frac{I_{total}}{I_{far}} - 1 \right)$$

$I_{total}$ is the measured intensity that is the sum of far-field and near-field Raman intensities. Since the Raman signal from the sample consists of both the target ε-Si layer and the underlying SiGe layer, which generates strong Raman signal as background component, we focused on the relative intensity (*I(ε-Si /SiGe)*) between ε-Si and SiGe layer. In this case, the contrast can be expressed as:

$$Contrast = \left( \frac{I_{total}}{I_{far}} - 1 \right) = \left( \frac{I_{SERS}\left(\frac{\varepsilon\text{-}Si}{SiGe}\right)}{I_{far}\left(\frac{\varepsilon\text{-}Si}{SiGe}\right)} - 1 \right)$$

Since the Raman signal from the sample consists of both the target ε-Si layer and the underlying SiGe layer, which generates strong Raman signal as background component, we focused on the relative intensity (*I(ε-Si /SiGe)*) between ε-Si and SiGe layer. The Raman spectra were break down into Lorentz function spectra for peak intensity analysis. The total intensity for each peak, ε-Si and SiGe, were obtained from the area under the fitted Lorentz function spectrum. The calculated *contrast*s for each combination of polarizer, analyzer and sample azimuth are labeled in the experimental results of Fig. 15. The highest *contrast* (~3.7) is obtained in the (j)P-S-45° and (k)S-P-45° conditions. We can expect the same effect in TERS experiment. Following the SERS experiments, in order to show that the background suppression is more important than electromagnetic field enhancement for contrast improvement in Raman active and bulk crystalline materials such as silicon, we performed TERS measurement under the conditions in P-P-45° and S-P-45° (shaded with gray in Fig. 15). In the former case, P-P-45°, Raman signal level is strongest because tip-enhancement effect is expected in both for the incident and scattered light. In the latter case, S-P-45°, background signal is well suppressed due to the s-polarized incident light. Moreover, tip-enhancement effect is not expected to happen in S-P-45° configuration because the incident light is s-polarized. However, tip-enhancement is expected for the scattered light. It is important to point out that the only difference between P-P-45° and S-P-45° is the incident polarization and the other parameters, analyzer and sample azimuth, are the same. These facts suggest that S-P-45° benefits on the tip-enhancement effect from the scattered light whereas P-P-45° utilizes tip-enhancement effect also from the incident light. This is because it is generally accepted that p-polarized illumination is better for the efficient incident tip-enhancement effect [28,57]. Besides, the S-S-0° condition actually showed the best background suppression in the calculation, however, we exclude the use of S-S-0° in TERS because as discussed above the observed residual far-field Raman signal in S-S-0° is mainly due to the wide variation of the incident angle and depolarization effect caused by the objective lens itself. In this context, S-S-0° is not advantageous over S-P-45°.

### 5.3. Experimental Observation of Depolarization Effect in TERS

In order to compare contributions of field enhancement and depolarization effect included in P-P-45° and S-P-45° spectra, TERS measurements were carried out with and without Ag coated $Si_3N_4$ tip. In pure $Si_3N_4$ tip (without Ag), field enhancement is lower when compared to Ag coated $Si_3N_4$ tip that is expected to have higher field enhancement. The TERS result obtained from the Ag coated $Si_3N_4$ tip is shown in Figure 17. In the P-P-45° condition, the far-field Raman signals from LO phonon mode can efficiently pass through the analyzer. The decrease in the peak Raman intensity of SiGe when Ag coated tip is in contact can be attributed to the depolarization effect and the incident intensity loss caused by shadowing effect [28,63] in the tip. The influence of the shadowing due to tip insertion is almost negligible because tip diameter (~30nm) is much smaller than the area of an ellipsoidal focused spot (~2.6 x 1.5$\mu m^2$) generated by LWD objective lens. The Raman intensity loss in SiGe associated with depolarization effect is the dominant factor because the Raman signal from LO phonon mode arising from the depolarized incident light, s-polarized in the P-P-45° case, is blocked by the analyzer. In order to accurately analyze the field enhancement and the depolarization effect, the discussion below are based on the following assumptions: (1) the diameter of the near-field scattering volume resulting from field enhancement is comparable to the tip diameter (~30nm) and the same as the thickness of the ε-Si layer (~30 nm), (2) the tip-enhancement factor for SiGe layer is negligible [70] because the thickness of ε-Si is 30 nm and set to 1x. Next, the 0.89x TERS depolarization effect was derived from simple intensity ratio between with Ag and without Ag coated tip in Fig. 17(a). This is because the reductions in SiGe Raman signal level is mainly attributed to the depolarization effect. These values are summarized in Fig. 17(c). On the other hand, the ε-Si Raman intensity exhibits a signal increment. Since depolarization effect was observed for SiGe layer, it should be also applicable to ε-Si layer in the same manner because both layers have the same facet (100). The depolarization effect factor on ε-Si is set to 0.89x in Fig. 17(c). The ratio of the total TERS intensity from ε-Si with Ag tip is 1.64 times higher than without Ag coated tip. The tip-enhancement factor is set to 1.84x in Fig. 17(c), and was calculated from 1.64/0.89. This fact supports that the tip actually enhanced the incident electric field intensity. Due to the existence of high background Raman signals from the entire focused spot, the estimated tip-enhancement contribution is small, ~1.84x. It should be noted that only the surface (ε-Si) is selectively enhanced and SiGe signal is reduced. This signifies that the result is not based on the signal enhancement from the multiple reflections, which is necessary in evaluating field enhancement on multilayered sample as discussed in C. Georgi et al, studies [71]. In the case of S-P-45°, the obtained depolarization effect value is 1.34x. The reason why the value turned-out to be more than one (signal is increased) is based on the fact that the polarized LO phonon (background) signals were rejected by the analyzer and the depolarized LO phonon signals by the tip passed through the analyzer efficiently. Using the value of the depolarization effect, tip-enhancement value can be also calculated in the same way we calculate for P-P-45°. Results are tabulated in Fig. 17(d). In S-P-45°, the tip-enhancement is very large, ~11.7x, compared to P-P-45°, ~1.84x, despite the s-polarized incident polarization and p-polarized scattered light detection. In this case, tip-enhancement is expected to occur only on the Raman signal at the detection. These results strongly suggest that for bulk sample and Raman active crystals with well-defined crystalline polarization such as silicon, far-field (background) Raman signal suppression, which is likely to happen in s-polarized illumination, is more important and should be carefully considered in improving the signal-to-noise ratio (SNR) in TERS experiment. For microscopic application and for faster data acquisitions, P-P-45° is recommended because of higher Raman signal level. The Raman signal from thin ε-Si layer was detected with much higher SNR in S-P-45° than the case in P-P-45° in the presence of strong background signals from the thick underlying SiGe layer.

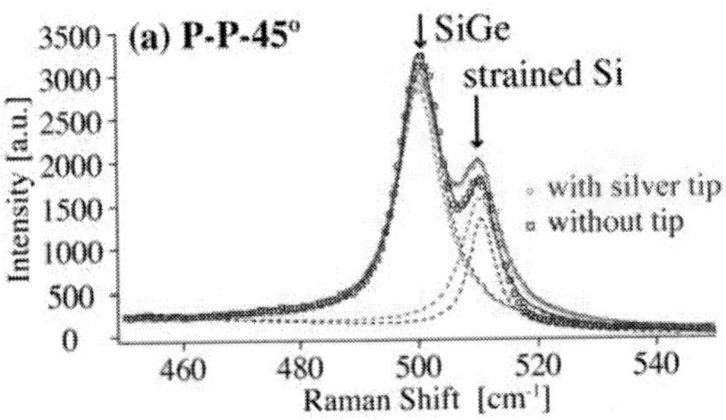


| (c) P-P-45° | SiGe peak | Strained Si peak |
|---|---|---|
| Tip-Enhancement | 1x | 1.84x |
| Depolarization effect | 0.89x | 0.89x |

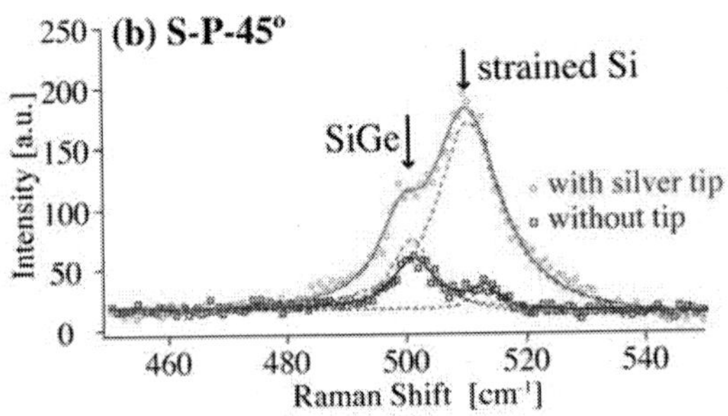


| (d) S-P-45° | SiGe peak | Strained Si peak |
|---|---|---|
| Tip-Enhancement | 1x | 11.7x |
| Depolarization effect | 1.34x | 1.34x |

Figure 17. TERS (red) and far field (blue) Raman spectra with Ag coated silicon nitride ($Si_3N_4$) tip at (a)P-P-45° and (b)S-P-45° polarizer-analyzer-sample azimuth setting. (c) and (d) tabulates the calculated values for tip enhancement and depolarized effect contribution coming from P-P-45° and S-P-45° setting, respectively.

To further demonstrate the importance of depolarization effect, TERS using pure $Si_3N_4$ tips (without Ag) was carried out and the results are shown in Figure 18. The far-field Raman spectra in Fig. 18(a) and (b) and the ones of Fig. 17 (a) and (b) are the same. The summary of enhancement factors induced by depolarization effect and tip-enhancement effect are listed in Fig. 18(c) and (d). Depolarization effect by the $Si_3N_4$ tip in P-P-45° case also caused reduction in the SiGe Raman intensity. This time, the depolarization effect is a bit higher, ~0.97x in Fig. 18(c) compared to 0.89x from the previous TERS experiment using Ag coated tip in Fig. 17(c). This could be due to the slight difference in the tip geometry between Ag coated $Si_3N_4$ tips and uncoated $Si_3N_4$ tips. On the other hand, the field enhancement value at the ε-Si is 1.24x, smaller than TERS with Ag coated tip. This means that tip-enhancement is still present even without the Ag coating. This can be due to the large contribution of an image dipole (the polarizability of image dipole: $\beta=(\varepsilon_s-1)/(\varepsilon_s+1)$, $\varepsilon_s$: dielectric constant of material [72]) and efficient gap mode formed by a high refractive-index sample, in this study, ε-Si. Furthermore, in S-P-45°, the depolarization effect value (2.60x) is larger than Ag coated tip. This can be again due to difference in the tip geometry discussed above. The tip-enhancement value is 2.61x and smaller than the case with Ag coated tip, but it is more than twice as high as Fig. 18 (a)P-P-45° with an uncoated tip. Even with uncoated $Si_3N_4$ tip and s-polarization incidence, the ε-Si signal level got almost the same signal level as SiGe by the depolarization effect, and the SNR is much better than Fig. 17 (a) P-P-45° with a Ag coated tip.

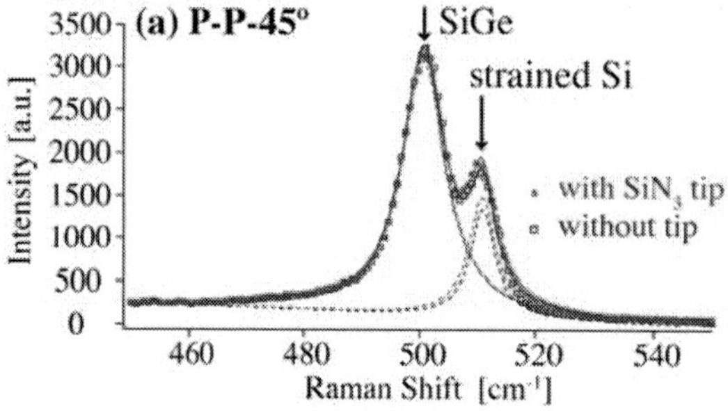


| (c) P-P-45° | SiGe peak | Strained Si peak |
|---|---|---|
| Tip-Enhancement | 1x | 1.24x |
| Depolarization effect | 0.97x | 0.97x |

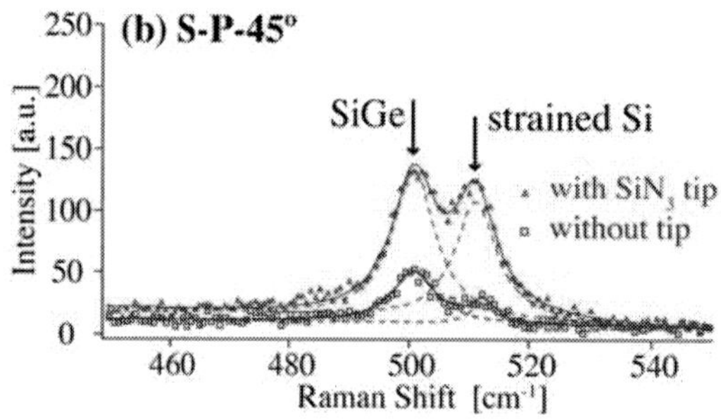


| (d) S-P-45° | SiGe peak | Strained Si peak |
|---|---|---|
| Tip-Enhancement | 1x | 2.61x |
| Depolarization effect | 2.60x | 2.60x |

Figure 18. TERS (red) and far field (blue) Raman spectra with pure silicon nitride ($Si_3N_4$) tip at (a)P-P-45° and (b)S-P-45° polarizer-analyzer-sample azimuth setting. The tip enhancement and depolarized effect contributions in the P-P-45° and S-P-45° setting are listed in (c) and (d), respectively.

# 6. Conclusion

We introduced the utilization SERS to selectively enhance the LO phonon vibrations originating from a thin silicon layer, which is under strain due to lattice mismatch with underlying Ge-doped silicon buffer layer. The selective enhancement makes it possible to efficiently detect the weak Raman signal originating from a thin layer, in the presence of a strong background signal originating from comparatively thick buffer layer. This selective enhancement could be localized to nanometric scale by utilizing TERS, where a metal-coated sharp tip is used for point-surface enhancement, which turned out to be an effective tool to investigate the nanometric variation of strain within the ε-Si layer. Further, we suggested some optimization and improvements in TERS for a higher sensitivity of the measurements. This included the selection of shorter wavelength for excitation, proper selection of tip material, and consideration of tip-originated effects, such as the tip-applied pressure that can add up in the local strain of the sample. In order to further improve the signal-to-noise ratio, we calculated the far-field Raman intensity of ε-Si (100) based on Raman tensor, Snell's law, and Fresnel's equations with the incident angle of 0 (normal incidence) and 60 degrees. Numerical results suggest that for normal incidence, only LO phonon mode is observed while in side illumination, incidence angle of 60 degrees, both TO and LO phonon modes are also observed. Because of TO phonon modes contributions, background Raman signal cannot be completely suppressed in cross-polarized configuration. However, LO phonon mode can be suppressed. SERS experiments suggest that P-P-45° gives the highest signal intensity and good for faster data acquisition in microscopic Raman studies. The P-S-45° and S-P-45° show the highest contrast between ε-Si and SiGe layers. In order to demonstrate the importance of background signal reduction for TERS, we compared experimentally the TERS spectra between P-P-45° and S-P-45° whose only difference is the incident polarization. Results show that S-P-45° condition strongly reduces the background signal but also reduces the tip-enhancement effect. However, much higher signal contrast is observed between near field and far field compared to P-P-45° condition where high tip-enhancement effect is higher but also with strong far field background. Hence, for imaging Raman active and bulk crystalline materials such as silicon, background signal suppression (s-illumination) needs to be carefully considered as well as the field enhancement with strong far field Raman signal level (p-polarization). Accordingly, depending on the specific geometry of each tip, which affects both depolarization effect and field enhancement effect, the proper combination of s- and p-illuminations show the highest contrast in the experiments while minimizing the background contributions from the sum of $TO_1$, $TO_2$, and LO modes through optimize analyzer and sample azimuth setting.

In addition to the development of the experimental system described above, the method to analyze the obtained data is strongly required by considering the tip-induced effects such as chemical, pressure, and heating effects.

## References


[1] Welser, J., Hoyt, J. L., and Gibbons, J. F. (1994). Electron-mobility enhancement in strained-Si N-type metal-oxide-semiconductor field-effect transistors. *IEEE, Electron Lett.* **15**, 100-102.

[2] Tezuka, T., Sugiyama, N., and Takagi, S. (2001). Fabrication of strained Si on an ultrathin SiGe-on-insulator virtual substrate with a high-Ge fraction. *Appl. Phys. Lett.* **79**, 1798-1800.

[3] Englert, Th, Abstreiter, G, and Pontcharra, J. (1980). Determination of existing stress in silicon films on sapphire substrate using Raman-spectroscopy. *Solid-State Electronics.* **23**, 31-33.

[4] Webster, S., Batchelder, D. N., and Smith D. A. (1998). Submicron resolution measurement of stress in silicon by near-field Raman spectroscopy. *Appl. Phys. Lett.* **72**, 1478-1480.

[5] Bonera, E., Fanciulli, M., and Batchelder, D. (2002). Raman spectroscopy for a micrometric and tensorial analysis of stress in silicon. *Appl. Phys. Lett.* **81**, 3377-3379.

[6] Chang, R. K. and Furtak, T. E. (1981). *Surface Enhanced Raman Scattering*. Plenum Press.

[7] Inouye, Y., Hayazawa, N., Hayashi, K., Sekkat, Z., and Kawata, S. (1999). Near-field scanning optical microscope using a metallized cantilever tip for nanospectroscopy. *Proc. SPIE Int. Soc. Opt. Eng.* **3791**, 40-48.

[8] Stöckle, R. M., Suh, Y. D., Deckert, V., and Zenobi, R. (2000). Nanoscale chemical analysis by tip-enhanced Raman spectroscopy. *Chem. Phys. Lett.* **318**, 131-136.

[9] Anderson, M. S. (2000). Locally enhanced Raman spectroscopy with an atomic force microscope. *Appl. Phys. Lett.* **76**, 3130-3132.
[10] Hayazawa, N., Inouye, Y., Sekkat, Z., and Kawata, S. (2000). Metallized tip amplification of near-field Raman scattering. *Opt. Commun.* **183**, 333-336.
[11] Hayazawa, N., Inouye, Y., Sekkat, Z., and Kawata, S. (2001). Near-field Raman Scattering Enhanced by a Metallized Tip. *Chem. Phys. Lett.* **335**, 369-374.
[12] Hayazawa, N., Inouye, Y., Sekkat, Z., and Kawata, S. (2002). Near-field Raman Imaging of Organic Molecules by an Apertureless Metallic Probe Scanning Optical Microscope. *J. Chem. Phys.* **117**, 1296-1301.
[13] Hartschuh, A., Sánchez, E. J., Xie, X. S., and Novotny, L. (2003). High-Resolution near-field Raman microscopy of single-walled carbon nanotubes. *Phys. Rev. Lett.* **90**, 095503.
[14] Brya W. J. (1973). Raman-scattering in Ge-Si alloys. *Solid Stat. Commun.* **12**, 253-257.
[15] Holtz, M., Duncan, W. M., Zollner, S., Liu, R. (2000). Visible and ultraviolet Raman scattering studies of Si1-xGex alloys. *J. Appl. Phys.* **88**, 2523-2528.
[16] Sugii N, (2001). Thermal stability of the strained-Si/Si0.7Ge0.3 heterostructure. *J. Appl. Phys.* **89**, 6459-6463.
[17] Cedeira, F., Buchenauer, C. J., Pollak, F. H., and Cardona, M. (1972). Stress-Induced Shifts of First-Order Raman Frequencies of Diamond- and Zinc-Blende-Type Semiconductors. *Phys. Rev. B* **5**, 580-593.
[18] Lockwood D. J., and Baribeau, J. M. (1992). Strain-shift coefficients for phonons in Si1-xGex epilayers on silicon. *Phys. Rev. B* **45**, 8565-8571.
[19] Sui, Z., and Herman I. P. (1993). Effect of strain on phonon in Si, Ge, and Si/Ge heterostructures. *Phys. Rev. B* **48**, 17938-17953.
[20] Anastassakis, E., Cantarero, A., and Cardona, M. (1990). Piezo-Raman measurements and anharmonic parameters in silicon and diamond. *Phys. Rev. B* **41**, 7529-7535.
[21] Nie, S. and Emory, S. R. (1997). Surface-enhanced resonance Raman spectroscopy of Rhodamine 6G adsorbed on colloidal silver. *Science* **275**, 1102-1106.
[22] Loudon, R. (1964). The Raman Effect in Crystals. *Adv. Phys.* **13**, 423-482.
[23] Nakashima, S., and Mitsuishi, A. (1984). *OYOBUTSURI.* **53**, 558-578 (in Japanese).
[24] Hart T. R., Aggarwal, R. L., and Lax, B. (1970). Temperature Dependence of Raman Scattering in Silicon. *Phys. Rev. B* **1**, 638-642.
[25] Fischer, U. Ch. and Pohl, D. W. (1989). Observation of single-particle plasmons by near-field optical microscopy. *Phys. Rev. Lett.* **62**, 458-461.
[26] Inouye, Y., and Kawata, S. (1994). Near-field Scanning Optical Microscope using a Metallic Probe Tip. *Opt. Lett.* **19**, 159-161.
[27] Bachelot, R., Gleyzes, P., and Boccara, A. C. (1995). Near-field optical microscope based on local perturbation of a diffraction spot. *Opt. Lett.* **20**, 1924-1926.
[28] Hayazawa, N., Tarun, A., Inouye, Y., and Kawata, S. (2002). Near-field enhanced Raman spectroscopy using side illumination optics. *J. Appl. Phys.* **92**, 6983-6986.
[29] Pettinger, B., Ren, B., Picardi, G., Schuster, R., and Ertl, G. (2004). Nanoscale Probing of Adsorbed Species by Tip-Enhanced Raman Spectroscopy. *Phys. Rev. Lett.* **92**, 096101.
[30] Mehtani, D., Lee, N., Hartschuh, R. D., Kisliuk, A., Foster, M. D., Sokolov, A. P., and Maguire, J. F. (2005). Nano-Raman spectroscopy with side-illumination optics. *J. Raman Spectrosc.* **36**, 1068-1075.

[31] Roy, D., Leong, S.H., and Welland, M. E. (2005). Dielectric contrast imaging using apertureless scanning near-field optical microscopy in the reflection mode. *J. Korean Phys. Soc.* **47**, S140-S146.
[32] Bulgarevich, D. S., and Futamata, M. (2004). Apertureless tip-enhanced Raman microscopy with confocal epi-illumination/collection optics. *Appl. Spectrosc* **58**, 757-761.
[33] Saito, Y., Motohashi, M., Hayazawa, N., Iyoki, M., and Kawata, S. (2006). Nanoscale characterization of strained silicon by tip-enhanced Raman spectroscope in reflection mode. *Appl. Phys. Lett.* **88**, 143109.
[34] Hayazawa, N., Motohashi, M., Saito, Y., and Kawata, S. (2005). Highly sensitive strain detection in strained silicon by surface enhanced Raman spectroscopy. *Appl. Phys. Lett.* **86**, 263114.
[35] Tsang, J. C., Mooney, P. M., Dacol, F., and Chu, J. O. (1994). Measurements of alloy composition and strain in thin GexSi1-x layers. *J. Appl. Phys.* **75**, 8098-8108.
[36] Goodman, G. G., Pajcini, V., Smith, S. P., Merrill, P. B. 2005. Characterization of strained Si structures using SIMS and visible Raman. *Mat. Sci. Semiconductor Processing* **8**, 255-260.
[37] Miura H, (2005). *Saishin Silicon Device to Kesshou Gijyutsu* (in Japanese), (pp.325-330) Tokyo, Realize Science & Engineering.
[38] Atkinson, A., and Jain, S. C. (1999). Spatially resolved stress analysis using Raman spectroscopy. *J. Raman Spectrosc.* **30**, 885-891.
[39] Aspnes, D. E., and Studna, A. A. (1983). Dielectric functions and optical-parameters of Si, Ge, GaAs, GaSb, InP, InAs, and InSb from 1.5 to 6.0eV. *Phys. Rev. B* **27**, 985-1009.
[40] Renucci, J. B., Tyte, R. N., and Cardona, M. (1975). Resonant Raman-scattering in silicon. *Phys. Rev. B* **11**, 3885-3895.
[41] Johnson, P. B., and Christy, R. W. (1972). Optical-constants of noble-metals. *Phys. Rev. B* **6**, 4370-4379.
[42] Akamine, S., and Quate, C. F. (1992). Low-temperature thermal-oxidization sharpening of microcast tips. *J. Vac. Sci. Technol. B* **10**, 2307-2310.
[43] Ono, A., Masui, K., Saito, Y., Sakata, T., Taguchi, A., Motohashi, M., Ichimura, T., Ishitobi, H., Tarun, A., Hayazawa, N., Verma, P., Inouye, Y., and Kawata, S. (2008) Active Contorl of the Oxdization of a Silicon Cantilever for the Characterization of Silicon-based Semiconductors. *Chem. Lett.* **37**, 122-123.
[44] Yeo, B. S., Zhang, W., Vannier, C., and Zenobi, R. (2006). Enhancement of Raman signals with silver-coated tips. *Appl. Spectrosc.* **60**, 1142-1147.
[45] Balkanski, M., Wallis, R. F., and Haro, E. (1983) Anharmonic effects in light scattering due to optical phonons in silicon, *Phys. Rev. B*, **28**, 1928-1934.
[46] McCarthy, B., Zhao, Y., Grover, R., and Sarid, D. (2005) Enhanced Raman scattering for temperature measurement of a laser-heated atomic force microscope tip, *Appl. Phys. Lett.* **86**, 111914.
[47] Watanabe, H., Ishida, Y., Hayazawa, N., Inouye, Y., and Kawata, S. (2004). Tip-enhanced near-field Raman analysis of tip-pressurized adenine molecule, *Phys. Rev. B* **69**, 1-11.
[48] Watanabe, H., Hayazawa, N., Inouye, Y., and Kawata, S. (2005). DFT vibrational calculations of Rhodamine 6G adsorbed on silver: Analysis of tip-enhanced Raman spectroscopy. *J. Phys. Chem. B* **109**, 5012-5020.

[49] Hayazawa, N., Watanabe, H., Saito, Y., and Kawata, S. (2006). Towards atomic site-selective sensitivity in tip-enhanced Raman spectroscopy. *J. Chem. Phys.* **125**, 244706.

[50] Giese, B., and McNaughton, D. J. (2002). Surface-enhanced Raman spectroscopic and density functional theory study of adenine adsorption to silver surfaces. *J. Phys. Chem. B* **106**, 101-112.

[51] Verma, P., Yamada, K., Watanabe, H., Inouye, Y., and Kawata, S. (2006). Near-field Raman scattering investigation of tip effects on C60 molecules. *Phys. Rev. B* **73**, 045416.

[52] Yano, T., Inouye, Y., and Kawata, S. (2006). Nanoscale uniaxial pressure effect of a carbon nanotube bundle on tip-enhanced near-field Raman spectra. *Nano Lett.* **6**, 1269-1273.

[53] Novotny, L., Bian, R. X., and Xie, X. S. (1997). Theory of nanometric optical tweezers. *Phys. Rev. Lett.* **79**, 645-648.

[54] Ren, B., Picardi, G., and Pettinger, B. (2004). Preparation of gold tips suitable for tip-enhanced Raman spectroscopy and light emission by electrochemical etching. *Rev. Sci. Instrum.* **75**, 837-841.

[55] Cui, X., Zhang, W., Yeo, B., Zenobi, R., Hafner, C., and Erni, D. (2007). Tuning the resonance frequency of Ag-coated dielectric tips. *Opt. Exp.* **15**, 8309-8316.

[56] Martin, Y. C., and Wickramasinghe, H. K. (2002). Strength of the electric field in apertureless near-field optical microscopy. *J. Appl. Phys.* **89**, 5774-5778.

[57] Sánchez, E. J., Novotny, L., and Xie, X. S. (1999). Near-field fluorescence microscopy based on two-photon excitation with metal tips. *Phys. Rev. Lett.* **82**, 4014-4017.

[58] Novotny, L., Sánchez, E. J., and Xie, X. S. (1998). Near-field optical imaging using metal tips illuminated by higher-order Hermite-Gaussian beams. *Ultramicroscopy.* **71**, 21-29.

[59] Hayazawa, N., Saito, Y., and Kawata, S. (2004). Detection and characterization of longitudinal field for tip-enhanced Raman spectroscopy. *Appl. Phys. Lett.* **85**, 6239-6241.

[60] Ossikovski, R., Nguyen, Q., and Picardi, G. (2007). Simple model for the polarization effects in tip-enhanced Raman spectroscopy. *Phys. Rev. B.* **75**, 045412.

[61] Poborchii, V., Tada, T., and Kanayama, T. (2005). Subwavelength-resolution Raman microscopy of Si structures using metal-particle-topped AFM probe. *Jpn. J. Appl. Phys.* **44**, L202-L204.

[62] Hayazawa, N., Motohashi, M., Saito, Y., Ishitobi, H., Ono, A., Ichimura, T., Verma, P., and Kawata, S. (2007). Visualization of localized strain of a crystalline thin layer at the nanoscale by tip-enhanced Raman spectroscopy and microscopy. *J. Raman Spectrosc.* **38**, 684-696.

[63] Lee, N., Hartschuh, R. D., Mehtani, D., Kisliuk, A., Maguire, J. F., Green, M., Foster, M. D., and Sokolov, A. P. (2007). High contrast scanning nano-Raman spectroscopy of silicon. *J. Raman Spectrosc.* **38**, 789-796.

[64] Turrell, G. (1989) in D. J. Gardiner and P. R. Graves (Eds.), *Practical Raman Spectroscopy*, Berlin: Springer-Verlag.

[65] Yu, P. Y. and Cardona, M. (2005). *Fundamentals of Semiconductors*. Berlin: Springer.

[66] Born, M., and Wolf, E. (1999). *Principles of optics.* Cambridge: Cambridge University Press.

[67] Palik, E. D. (1985). *Handbook of optical Constants of Solids*. Boston: Academic Press.

[68] Palik, E. D. (1998). *Handbook of optical Constants of Solids,***3** Boston: Academic Press.
[69] Yano, T., Ichimura, T., Taguchi, A., Hayazawa, N., Verma, P., Inouye, Y., and Kawata, S. (2007). Confinement of enhanced field investigated by tip-sample gap regulation in tapping-mode tip-enhanced Raman microscopy. *Appl. Phys. Lett.* **91**, 121101.
[70] Zhang, W., Yeo, B. S., Schmid, T., and Zenobi, R. (2007). Single molecule tip-enhanced Raman spectroscopy with silver tips. *J. Phys. Chem. C.* **111**, 1733-1738.
[71] Georgi, C., Hecker, M., and Zschech, E. (2007). Raman intensity enhancement in silicon-on-insulator substrates by laser deflection at atomic force microscopy tips and particles. *Appl. Phys. Lett.* **90**, 171102.
[72] Knoll, B., and Keilmann, F. (1999). Near-field probing of vibrational absorption for chemical microscopy. *Nature* **399**,134-137.

In: New Research on Nanocomposites
Editors: Luis M. Krause et al, pp. 37-56
ISBN: 978-1-60456-799-1


*Chapter 2*

# STIMULI-RESPONSIVE MESOPOROUS SILICA MATERIALS FOR CONTROLLED DRUG RELEASE

***Shenmin Zhu*[1] *and Di Zhang*[2]**
State Key Laboratory of Metal Matrix Composites, Shanghai Jiao Tong University, 800 Dongchuan Road, Shanghai 200240, P. R. China

## Abstract

The delivery system based on stimuli-responsive polymers inside mesoporous silica has been of great interest because it allows a drug to be released in response to the environmental stimulation. This research has demonstrated that several approaches have been developed for the synthesis of thermal-responsive poly(N-isopropyl acrylamide) (PNIPA) inside mesoporous silica with pore size of 7-20 nm. The large pore size of the support and the smart nature of PNIPA can be expected to the potential applications in the controlled release of macromolecular active agents. We review the methods of how to introduce multilayer polymers inside mesoporous silica pores, while keeping the pore channels for the adsorption and release of drug molecules. It is interesting to find the resultant delivery system demonstrated a high ibuprofen (IBU) storage capacity. The multilayer polymers inside the pores were considered to form internal cavity for drug molecules in addition to responding to a change in external temperature. Strategies for the preparation of a site-selective and controlled delivery system are presented and the essential properties of the resulting materials with respect to its capability for target and controlled drug release were investigated. Outlook of potential applications and further challenges are also provided.

## Introduction

Controlled drug-delivery systems (DDSs) have been considered to be one of the most promising applications for human health and care, for controlling the rate of drug delivery at the specific issues (1). In this case, DDSs exhibiting the properties of biocompatible bioceramics would be highly desirable. Among the biomaterials that display biocompatibility

[1] E-mail address: smzhu@sjtu.edu.cn; Tel.: +86-21-34202584; Fax: +86-21-34202749 (Corresponding author.)
[2] E-mail address: zhangdi@sjtu.edu.cn

are the silica-based systems. The biocompatibility of silica nanospheres that interact with cellular systems has been depicted by researchers (2-3). In 1992, the discovery of mesoporous silica opened up new possibilities in many areas of chemistry and material science (4). These materials possess high specific surface areas (up to ca. 1500 $m^2g^{-1}$), high specific pore volumes (up to ca. 1.5 $cm^3g^{-1}$), and well-ordered pore structures with uniform mesopores adjustable in diameter from about 2 to 30 nm (5). Several applications have been proposed in the fields of catalysis, sensors and solar cells, indicating their superior adsorption capacities (6). In the biomedical field, the application of mesoporous silica as drug vehicles in drug delivery systems has also been explored (7). Obviously, mesoporous materials have a clear advantage over microporous zeolites and zeotype molecular sieves for the adsorption and transformation of large organic molecules, which renders them ideal candidates as hosts for biomolecules. Moreover, variation of the synthesis conditions enables the researchers to tailor the inorganic host so that the encapsulation of a variety of biological molecules is feasible.

Unfortunately, the use of bulk mesoporous silica materials in many applications, especially in controlled and target drug delivery systems suffer from some inherent limitations. Their drug-storage capacity is relatively low and the drug delivery is erratic. The development of drug delivery systems exhibiting controlled and target release is still urgent.

Stimuli-responsive polymers (SRPs) that are sensitive to environment or to externally applied impetus, such as temperature, pH, or salt concentration (8), could play a fundamental role in enabling controlled task. However, one of the main problems of DDSs based on pure polymers is the loss of activity of several drugs before reaching the target tissue as a result of premature degradation of the active agent (9). The rational design of stimuli responsive systems often needs well-structured porosity and high surface areas to support the chemical species that are to control the drug release. Mesoporous silica is no doubt a suitable support because it fulfills many of the requirements for drug delivery systems, such as hydrophilic character, water insolubility, chemical and thermal stability, mechanical strength, suitable particle form, regenerability, and toxicological safety (10). In addition, the silica shell of these materials provides an appropriate surface to achieve stability at physiological pH. Surface-functionalized mesoporous silica materials with stimuli-responsive polymers have been demonstrated to be excellent hosts of molecules of various sizes, shapes, and functionalities for controlled release (11).

As for drug targeting, reports on surface-functionalized, superparamagnetic iron oxide nanoparticles have demonstrated their feasibility for use in a variety of biological applications including drug delivery (12). Most of these systems comprise magnetic nanoparticle "cores" coated with organic or inorganic "shell" constituents (13). In the presence of an external magnetic field, these core–shell materials are attracted toward the magnet. This magnetic motor effect is indeed attractive for the development of site-specific drug delivery systems. Recent reports on the design of capped and gated mesoporous silica have shown a great promise in the generation of controlled-release systems (14). The combination of ordered mesoporous silica with magnetic nanoparticles is expected to make a great contribution to achieve site-selective and controlled drug release (15). A delivery system based on MCM-41 nanorods capped with superparamagnetic iron oxide nanoparticles was reported by Li et al. Guest molecules that are smaller than 3 nm, such as fluorescein, could be encapsulated and released from the delivery system by using cell-produced antioxidants (e.g. dihydrolipoic acid) as triggers in the presence of an external magnetic field (16). Unfortunately, the systems based on mesoporous MCM-41 supports with pore size of 2-4 nm, are not suitable for encapsulation drugs with large molecular weights involved. Moreover, the magnetic particles are generally formed independent of the self-assembly of mesoporous silica, resulting in poor control of the resultant morphology. Mesoporous silica with pore sizes of 7-20 nm makes it more attractive for the easy modification inside and for the storage and release of large molecules. PH-responsive carrier system was designed and synthesized by Xiao and co-workers based on polycations grafted to anionic, carboxylic acid modified SBA-15 by ionic interactions (17). The polycations could act as closed gates to store the drug within the mesopores. When the ionized carboxylic acid groups are protonated in response to a change in pH value, the polycations are detached from the surface and the drug can be released from the mesopores. However, the mono-stimuli-responsive mesoporous silica systems could be hardly applicable for site-specific drug delivery for the lack of magnetic property.

In this context, in the first section, we present our recent developed approaches for thermo-responsive polymers functionalized macrosopores for controlled drug delivery (18). Secondly, strategies for the preparation of a site-selective and controlled delivery system are demonstrated and the essential properties of the resulting materials with respect to its capability for target and controlled drug release are discussed (19). Finally, outlook of potential applications and further challenges are also provided.

## Delivery System Based on Thermo-responsive Polymer Inside Mesoporous Silica Using ATRP

The synthesis of systems that combine the silica inorganic phase with thermally active polymers has attracted much attention in DDSs. In 2005, Chang developed a kind of hybrid systems by combining disordered silica inorganic phase with thermoresponsive poly(Nisopropylacryl amide) (PNIPA), to produce spongelike phases (20). The sponge phases are formed by self-assembly of amphiphilic templates during the formation of mesoporous inorganic materials. The systems consist of a three-dimensional random packing of a multiple connected bilayer of surfactant and cosurfactant, which divides the space into two subspaces filled with solvent. The distance between adjacent silicate layers in the porous structure were controlled by changing the hydrophilic domain. Drugs could be loaded into the spongelike mesoporous domains, and drug release is controlled by cyclic thermally induced polymer shrinkage, which regulates the aperture of the pores. As for ordered mesoporous materials, Lopez reported the control of molecular transport based on PNIPA polymerized inside MCM-41 (21), which responded to environmental changes and, thus, modified the adsorption and release characteristics. The adsorption of molecules in the mesoporous matrix is reported to be governed by size selectivity. It is well known the pore size of mesoporous materials determines the size of the molecule that can be adsorbed into mesopores. Generally, pore diameters should be slightly larger than the drug molecules. Therefore, the systems based on mesoporous MCM-41 supported with pore size of 2-4 nm, are not suitable for encapsulation drugs with large molecular weights involved. Mesostructure cellular foam materials (MCF) with large pore diameters of 17-42 nm and large surface areas are attractive to host drugs with relative large molecules (22). Infiltration, then polymerization, is considered to be a facile method to insert polymers inside mesoporous pores. In comparison to the traditional radical polymerizations, Atom transfer radical polymerization (ATRP) is very useful for the efficient modification of pore surfaces (23). Polymerization is restricted to the surface that no polymer forms in solution. As a result, predictable molar masses and controlled composites will be obtained. The most important is that the multilayer polymers inside the pores would be expected to form internal cavity for drug molecules in addition to respond to a change in external temperature.

The reaction process is described as shown in Figure 1. The mesoporous siliceous foam (MCFs) with large pore size was prepared using 1, 3, 5-trimethylbenzene (TMB) as a swelling agent as described. Then the pore surface was modified with 3-aminopropyltriethoxysilane (APTES), enabling the initiator to be grafted onto the surface. ATRP of NIPA was conducted in the presence of CuBr and PMDETA. Nanocomposite delivery system of MCF-PNIPA was thus obtained.

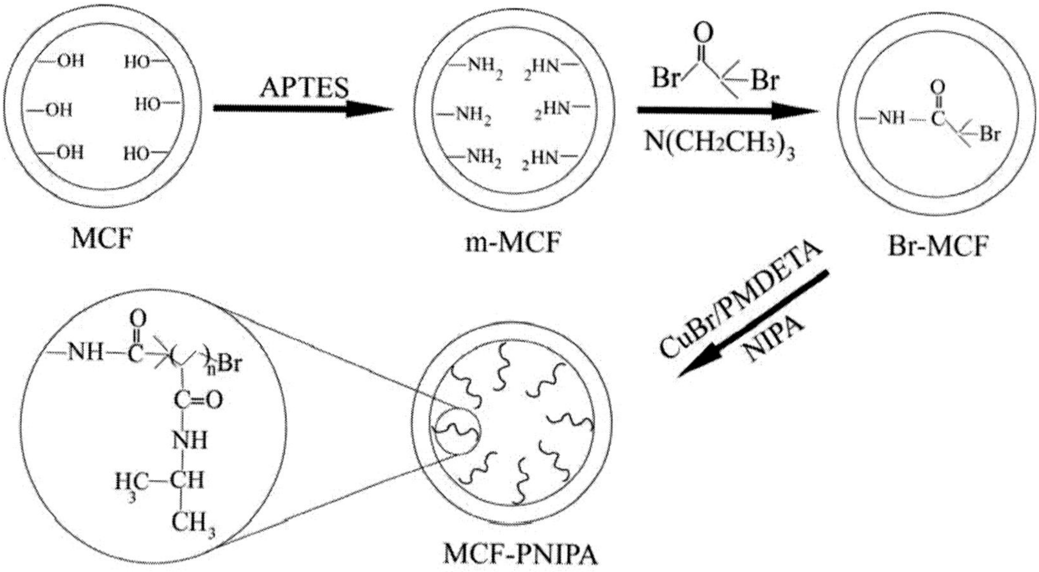


Figure 1. Design of drug delivery system based on modified mesoporous siliceous foam (MCF) and N-isopropyl acrylamide (NIPA) polymerized via Atom Transfer Radical Polymerization (ATRP) inside the pore channels as the thermo-responsive polymer. Reproduced from ref 18.

The structures of the obtained materials were characterized by using Fournier transform infrared (FT-IR) spectroscopy, which showed a shift of carboxyl band at 1707 $cm^{-1}$ (IBU) to 1720 $cm^{-1}$ corresponding to the sample of MCF-PNIPA-IBU. This slight blue shift is owing to the hydrogen bonding of PNIPA with the carboxyl groups of IBU, which was further characterized by $^1H$ NMR measurement in $D_2O$. When analyzed, only one peak at small angles was detected for the sample MCF in small X-ray diffraction (XRD) pattern (not shown), attributing to the less long-range ordered structures of MCF. After ATRP of NIPA inside MCF, no diffraction peaks were detected at small angles, possibly owing to the less contrast between PNIPA and the pores. Nitrogen adsorption/desorption isotherms show large hysteresis, which is in conjunction with BdB pore size analyses possessing ink-bottle-type pores in which large cells are connected by narrower windows. The sizes of the cells and windows of m-MCF were determined to be 22.1 nm and 7.0 nm, respectively. The representative isotherm and the corresponding pore size distributions are summarized in table

1. Before grafting PNIPA inside MCF, a Brunauer-Emmett-Teller (BET) surface area of 443 $m^2g^{-1}$ and a pore volume of 1.23 $cm^3g^{-1}$ (m-MCF) were presented. After polymerization of NIPA inside the pores (MCF-PNIPA), the resultant surface area and pore volume were found to be 99.4 $m^2g^{-1}$ and 0.4767 $cm^3g^{-1}$, respectively. Correspondingly, the cavity size was observed decrease from 22.1 nm to 19.2 nm determined by nitrogen adsorption, suggesting the increase of the wall thickness after PNIPA formed inside MCF. Collectively, these data confirm that the pores are grafted with PNIPA, resulting in the decrease of surface area and pore volume.

**Table 1. Textural parameters of the various samples[18]**

| sample | BET Surface area ($m^2g^{-1}$) | Pore volume ($cm^3g^{-1}$) | Cavity size (nm) | Entrance size (nm) |
|---|---|---|---|---|
| m-MCF | 443.5 | 1.232 | 22.1, | 7.0 |
| MCF-PNIPA | 99.4 | 0.4767 | 19.2 | 6.5 |
| MCF-PNIPA-IBU | 15.697 | 0.0582 | 14.8 | |

[a]Calculated from the adsorption branches of the N2 sorption isotherms based on the BdB sphere model.
[b]Calculated from the desorption branches of the N2 sorption isotherms based on the BdB sphere model.

TEM images of MCF before and after polymerization of NIPA inside their porous networks via ATRP are displayed in Figure 2. Although the diameter of the white disks is not simply correlated to the pore size, the edge of the particle might be very thin, and the image contrast is dominated by mass contrast. In this case, the pore size can be estimated to be larger than 20 nm, coincidence with the $N_2$ sorption measurements. The m-MCF with a cell diameter of about 25 nm displays local and less long-rang order. As for the MCF-PNIPA, uniformly sized and large spherical cells that are interconnected by uniform windows to create a continuous 3-D pore system are demonstrated. The pore wall size of m-MCF before grafting is found to be about 3.2 nm, but after grafting, the pore wall size is detected to be 4.1 nm, confirming the graft polymerization of NIPA on the wall. Thermal gravimetric analysis (TGA) revealed about 18 wt% of PNIPA was polymerized inside the pores. In contrast to homogeneous PNIPA, a broad temperature transition was detected from 30-40 °C measured by DSC for the sample MCF-NIPA, owing to steric hindrance of PNIPA chains by silica host. This result is in coincidence with previous theoretical and experimental studies that have concluded the solubility transition for polymer brushes in constrict geometrics is broad compared to that observed for free polymerization (24).

Control of drug release through the porous network was investigated by measuring uptake and release of ibuprofen (IBU, an analgesia and anti-inflammatory drug) as monitored by UV spectroscopy and thermogravimetry (TG). It is noted 58 wt% (IBU/silica) of IBU can be adsorbed inside MCF-PNIPA, which is almost 57 % higher than that reported by Xiao based on carboxylic acid modified SBA-15 (17). Kawi reported a higher loading of functional groups on the pore surface would lead to a higher loading of drugs in his study of functionalized SBA-15. However, the ratio of NIPA/silica around 0.9 $mmolg^{-1}$ herein is not

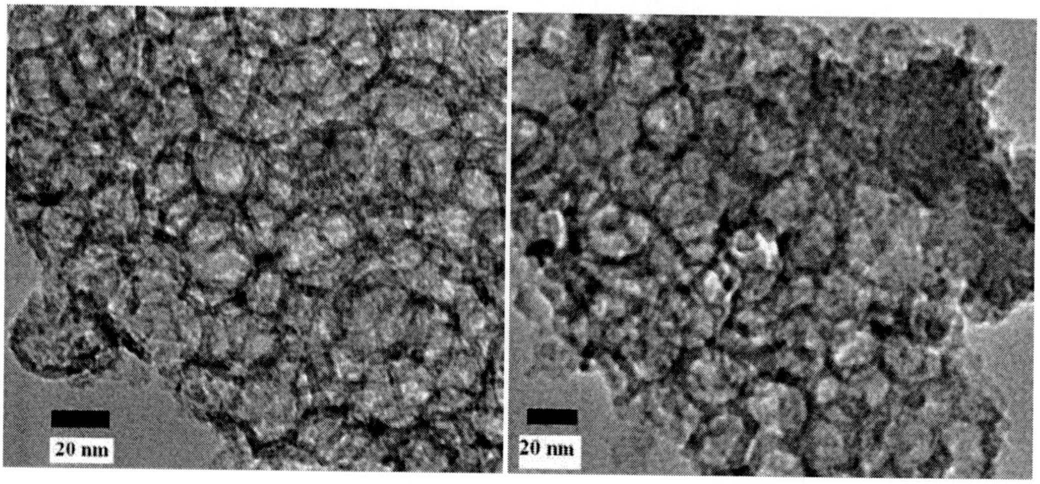


Figure 2. TEM images of the resultant m-MCF and MCF-PNIPA. Reproduced from ref 18.

enough to result in such an improved adsorption capacity. The structure nature of MCF-PNIPA seems to play a crucial role for the high adsorption capacity. Apart from increasing the interaction with IBU, the extended polymer chains of PNIPA would act as holder to retain the drugs inside the pores as verified by the nitrogen isotherm showing the complete filling of the pores after adsorption. The in vitro delivery of the drug was performed by soaking the samples in a solution simulating body fluid composition (SBF), maintaining pH at 7.4. Even though the range of temperature over which the transition takes place is quite broad (30-40 °C), a transition is detected at 32 °C for the MCF-PNIPA-IBU. That is to say, PNIPA inside MCF can respond to the thermo-stimuli from environment. Drug release can be controlled in the system of MCF-PNIPA.

Therefore, large macromolecules can be entrapped into cavities suitable for their adsorption and release using modified MCF as reservoirs. The development of functional cagelike or "plumberKs nightmare" pore structures will play an important role in controlled release of drugs with large molecular weights. New mesoporous carriers able to host, protect, and target these kind of products at the appropriate site could play an essential role in these new therapies.

## Site-Specific and Controlled Drug Delivery System: Magnetic, Thermal-Responsive, and Ordered Mesoporous

Ordered mesoporous drug-delivery systems that respond to external magnetic fields can be obtained by direct encapsulation of magnetic nanoparticles into the mesoporous silica. Such an approach yields materials are usually based on magnetite or even iron, covered with mesoporous silica (25). Another interesting approach involves capping mesopores with magnetic nanoparticles that can be alternately placed and removed (16). The mesoporous materials were synthesized by cocondensation with mercaptopropyl silanes and subsequent linkage through the SH groups with 2-carboxyethyl-2-pyridyl disulfide to yield acid-functionalized mesoporous silica. The mesopore entrances of the MCM-41 materials were then closed with magnetic $Fe_3O_4$ nanoparticles by placing the acid-functionalized silica in a suspension with magnetite nanoparticles. The controlled release was realized through the reduction of the disulfide linkage between magnetic $Fe_3O_4$ nanoparticles and the thiol-functionalized silica mesoporous material by reducing agents such as dihydrolipoic acid or dithiothreitol. Mesopore entrances can also be modified by large organic functionalities to develop supramolecular mesoporous materials that respond to several chemical signals. Lin and co-workers have also reported pore capping with polyamidoamine (PAMAM) dendrimers (26). The PAMAM caps served as a nonviral gene transfection reagent to enable plasmid DNA of an enhanced green fluorescence protein (Aequorea victoria) to cross a cell membrane through endocytosis and then be released. As the best to our knowledge, the site-specific and controlled drug delivery system based on Magnetic, thermal-responsive, and ordered mesoporous are reported seldom. One of the biggest obstacles lies in the low drug storage capacity, because the polymers inside the mesopores reduces the effective pore size of the matrix and subsequently hinders the adsorption of molecules into the pore voids as well as their release from them. Fujiwara and co-workers developed a photocontrolled release system for direct-drug release applications based on pore-entrance modification with coumarin groups. These groups undergo reversible dimerization upon irradiation with UV light at wavelengths longer than 310 nm and return to the monomer form by subsequent irradiation at shorter wavelengths (ca. 250 nm). The dimer form of the coumarin, when grafted on the surface of mesoporous silica systems, such as MCM-41, reduces the effective pore size of the matrix and subsequently hinders the adsorption of molecules into the pore voids as well as their release from them. How to arrange magnetic nanoparticles, thermo-responsive polymers and highly ordered mesoporous silica efficiently while keeping a high drug storage capacity is the key point to prepare target and controlled delivery systems. Several approaches have been developed as described below:

## Nanocasting Then Sonication for the Multifunctional Nanocomposite

Thermoresponsive PNIPA and magnetic nanoparticles iron oxide were inserted into the pore channels of SBA-15 combining nanocasting method with ultrasonic chemistry. The concept and the fabrication procedure of the proposed multifunctional nanocomposites are schematically illustrated in Figure 3. The first step in the procedure was to functionalize SBA-15 with the amino group of 3-aminopropyltrithioxysilane (APTES) to ensure the absorption of NIPA monomer in the internal surface of SBA-15, the sample is thus obtained as m-SBA-15. Then, NIPA monomer, azoisobutyronitrile and N,N'-methylene bisacrylamide were absorbed in the pore channels of m-SBA-15, and PNIPA nanocasting SBA-15 can be obtained via in situ free-radical polymerization under nitrogen, the powder thus prepared was designated as SBA-PNIPA. Next, pre-prepared SBA- PNIPA, iron pentacarbonyl, and decalin were mixed together, and then the mixture was irradiated with a high-intensity ultrasonic horn under argon atmospheres for 3 h, resulting in the materials of multifunctional nanocomposites SBA-PNIPA-$Fe_2O_3$.

Nanocasting of polymer inside the pores was characterized through $N_2$ adsorption/desorption. The incorporation of $Fe_2O_3$ inside the channels of SBA-15 was detected directly from TEM images. The novelty of the method lies in combination of ultrasonication with nanocasting method in realizing multifunctional nanocomposites with

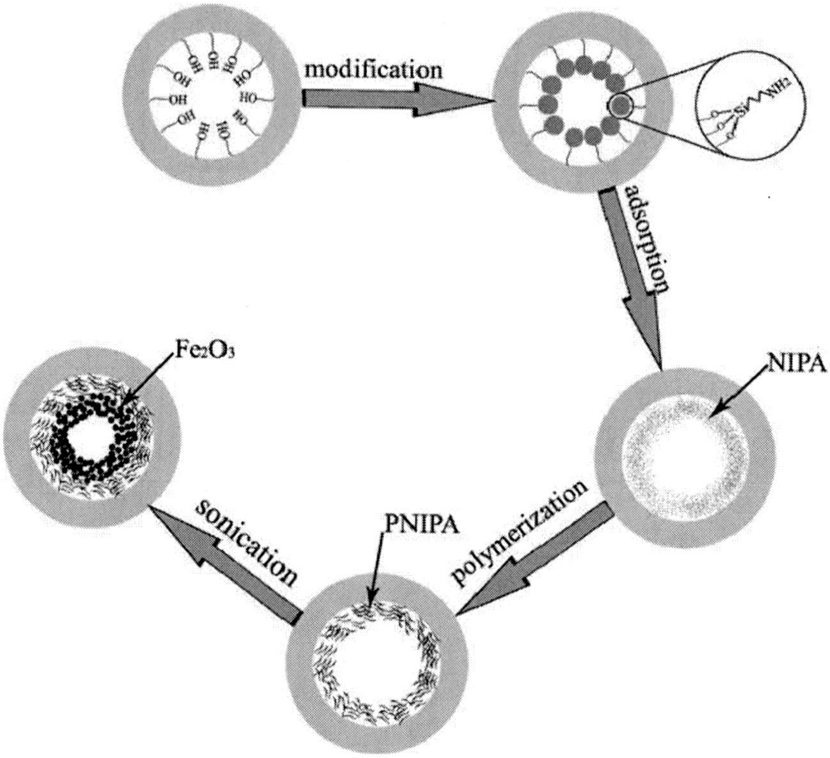


Figure 3. Schematic illustrations for fabrication of the multifunctional nanocomposites with well-defined mesoporosity by nanocasting then sonication approach. Reproduced with permission from ref 19a.

well-defined mesoporosity, which has a potential application in drug delivery system. Quantitative determination of organic content by thermogravimetric analysis (TGA) shows that around 57 wt% of PNIPA was contained in SBA-NIPA. The relatively low surface area (146 $m^2$ $g^{-1}$) and small pore size (3.7 nm) for SBA-NIPA with respect to m-SBA-15 (229 $m^2$ $g^{-1}$, 4.5 nm) is attributed to the large amount of polymers inside the pores. From $N_2$ sorption isotherm, no distinct hysteresis was demonstrated for the product SBA-PNIPA-$Fe_2O_3$, and this together with the great decrease of pore volume from 0.14 to 0.07 $cm^3g^{-1}$ suggested the formation of nanoparticles $Fe_2O_3$ inside the mesopores. That's to say, even though this approach is interesting to insert polymers and nanoparticles inside mesopores, the low volume of the resultant nanocomposite is a great obstacle for this system to be used as drug reservoir.

## Assembly of Controlled Release System with $FeCl_2$ as Magnetic Precursor and PNIPA Formed Inside the Pores in Situ

Irradiated by the one-step synthetic route to prepare ordered mesoporous silica monoliths with metal oxide nanoparticles formed in their channels (27), we employed hydrophilic $FeCl_2$ as an additional precursor to fabricate magnetic particles inside SBA-15 via surfactant template sol-gel process as demonstrated in Figure 4, the sample thus prepared is denoted to be m-SBA. In comparison with others, the novelty of the system is the control of drug release through stimuli-responsive polymer inside mesopores instead of outside the supports. Therefore, the polymer layers formed inside could act as storage gate and release switch in response to the stimuli from environment. Moreover, the magnetic particles formed in situ not only result in system with high surface area (653 $m^2$ $g^{-1}$) and large pore diameter (5.7 nm) but also provide the possibility of magnetic label.

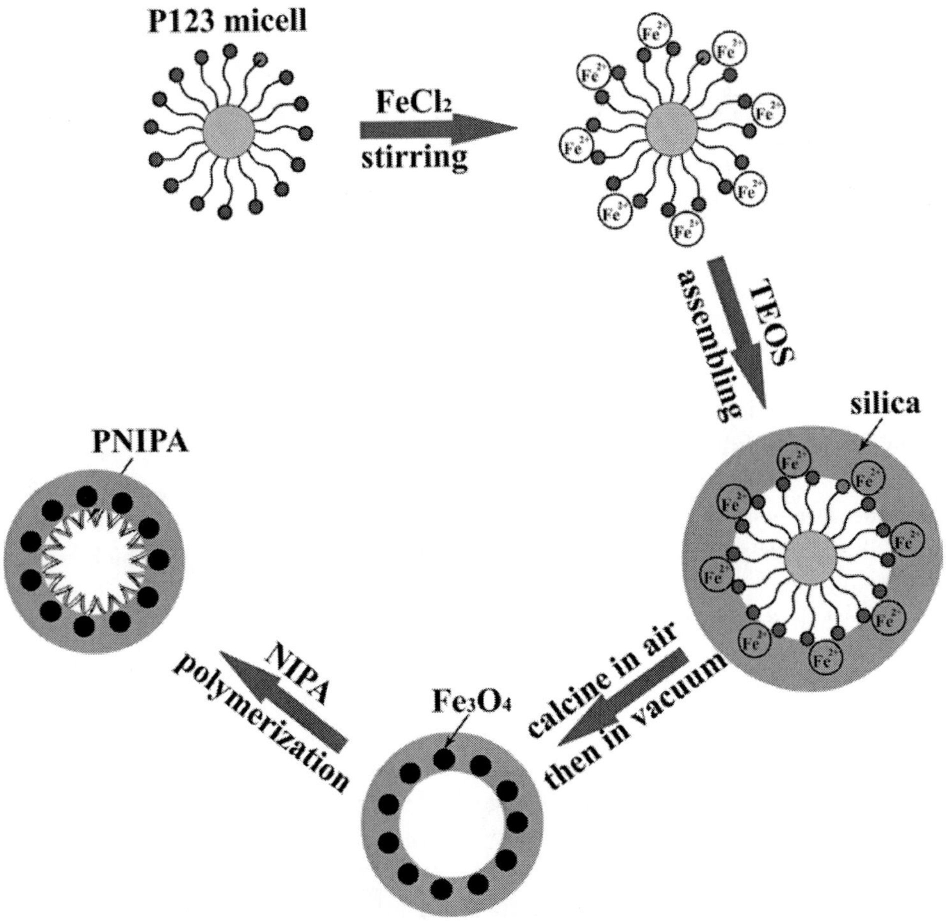


Figure 4. Design of delivery system based on SBA-15 with magnetic particles formed in situ and thermo-responsive PNIPA polymerized inside the pore channels by using hydrophilic $FeCl_2$ as an additional precursor. Reproduced with permission from Elsevier from ref 19b.

Small-angle XRD patterns showed a highly ordered mesoporous structures for the sample of m-SBA. For the sample m-SBA-PNIPA-IBU, the shift of the diffraction peaks to lower angle would attribute to introducing polymers or loading IBU drugs inside the pores. The formation of iron oxide was investigated by the energy-dispersive X-ray (EDX) measurement, which demonstrated the existence of the elements Fe, O, Si, C, and N in the sample m-SBA-PNIPA. The clear dark spots detected by the high-resolution TEM images represented the existence of $Fe_3O_4$ nanoparticles formed inside the pores of SBA-15 (Figure 5).

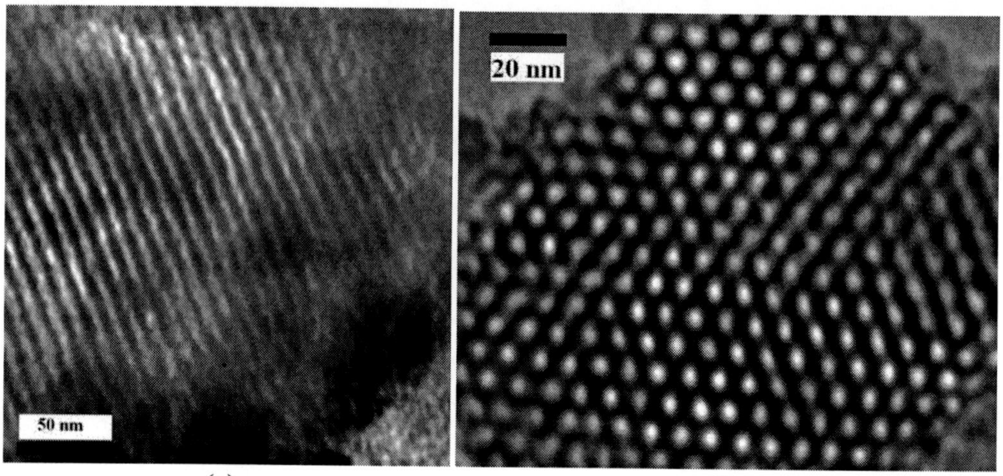


(a) (b)

Figure 5. TEM images of the samples m-SBA along (a) and perpendicular (b) to the direction of the hexagonal pore arrangement.

However, no distinct wide-angle diffraction peaks could be detected for m-SBA, owing to the low content of iron oxide assembled into the mesostructured materials during the sol-gel process. The formation of iron oxide nanoparticles would be further confirmed by the magnetic measurement. $N_2$ adsorption/desorption isotherms of m-SBA and m-SBA-PNIPA showed both samples yielded type IV isotherms with a $H_1$ hysteresis loop, which are typical for mesoporous materials with two-dimensional structures. The delivery system held an average pore diameter of 3.8 nm with surface area of 344 $cm^2g^{-1}$ even though 29.5 wt% PNIPA formed inside the pores. The pore volume was found to be 0.47 $cm^3g^{-1}$, which is more than three times as that reported by Lopez in polymerization of NIPA inside mesoporous MCM-41 (0.133 $cm^3g^{-1}$) (21).

The LCST of PNIPA captured in m-SBA channels was measured only to be 17 °C, which could be possibly explained by quantum confined effect as described by Lopez (28). The amount of drug adsorbed was monitored by thermogravimetry (TG), revealing 27 wt% (IBU/silica) drugs had been loaded in m-SBA-PNIPA system. The cumulative percentage release of IBU from m-SBA-PNIPA at different temperatures was investigated. When the temperature was during 7-11 °C, a drug release of 40.5 % from the m-SBA-PNIPA was observed, probably due to the small size of IBU. Then, as the temperature rising to 15 °C, the polymer coating shrank gradually, leading to the progressive increase of the release rate. A distinct transition at around 15 °C to 20 °C was detected, during which the release amount increased quickly. The release percentage between 15 °C to 20 °C was about 35 %, which was about three times as that of 11-15 °C. Finally, the release percentages kept almost unchangeable above 20 °C. Thus, the delivery system thus prepared showed a typical thermo-sensitive release property.

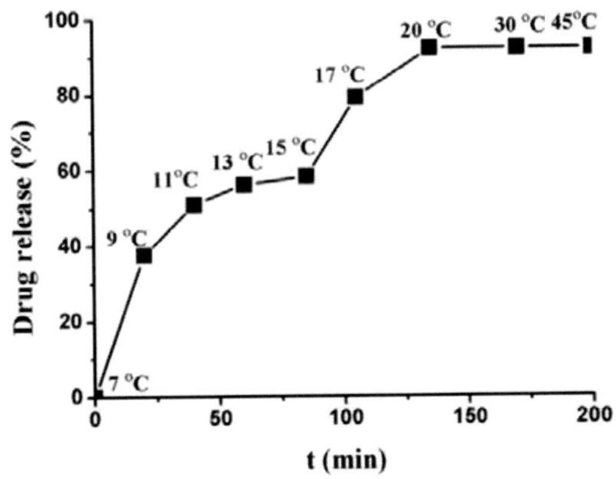


Figure 6. The IBU-release curve of m-SBA-PNIPA multi-responsive delivery system in phosphate buffer (pH=7.4) at different temperatures.

According to the magnetic characterization by a vibrating sample magnetometer at room temperature, m-SBA displayed a hysteresis loop with a saturation magnetization (Ms) of 0.40 $emug^{-1}$. The magnetism would be too weak to be detected in practical use. The measured weak magnetism was possibly induced by the small amount of Fe contained (1.7 wt %) and nano-sized length of the obtained magnetic particles. During the sol-gel process, small amount of $Fe^{2+}$ could be assembled into P123 micelles for the hydrophilic effect of $FeCl_2$, resulting in a small amount of magnetic particles in the silica channels.

## Assembly of Controlled and Target Drug Release System with Hydrophobic $Fe(CO)_5$ as Magnetic Precursor and PNIPA Polymerized Inside the Pores

For magnetic drug targeting, the drug is designed to be immobilized on a microparticulate carrier possessing magnetic properties. How to introduce efficient magnetic particles inside the mesoporous silica without decrease the pore size and volume is the biggest challenge for synthesis of controlled and target drug release system. Inspired by the functionality of 1,3,5-trimethylbenzene (TMB) which is a swelling agent in preparing mesoporous materials with large pore size, we choose hydrophobic $Fe(CO)_5$ instead of hydrophilic $FeCl_2$ as the magnetic precursor source. As described in Figure 7, this method is based on the formation of ordered mesoporous silica with magnetic particles formed via surfactant template sol-gel, and control of transport through polymerization of N-isopropyl acrylamide (NIPA) inside the pores. Thus, in addition to act as the source of the resultant magnetic particles, $Fe(CO)_5$ can play as a swelling agent, which can guarantee a high surface area and large pore size of the result. Comparing with others, another specialty of this system is the control of drug release through stimuli-responsive polymers (Poly(N-isopropylacrylamide), PNIPA) inside mesopores rather than outside the magnetic particles. As a result, the multilayer polymers inside the pores act as not only a switch role in response to the environmental temperature but also a hold for drug molecules.

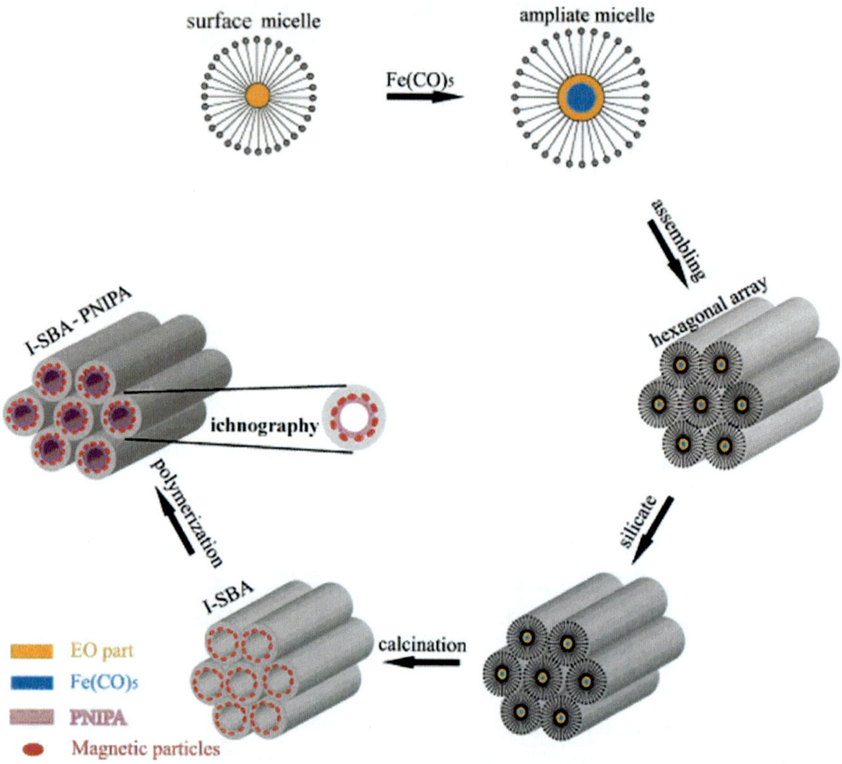


Figure 7. Design of drug delivery system based on SBA-15 with magnetic particles formed in situ using $Fe(CO)_5$ as swelling agent and NIPA polymerized inside the pore channels as the thermal-responsive polymer. Reproduced with permission from ref 19c.

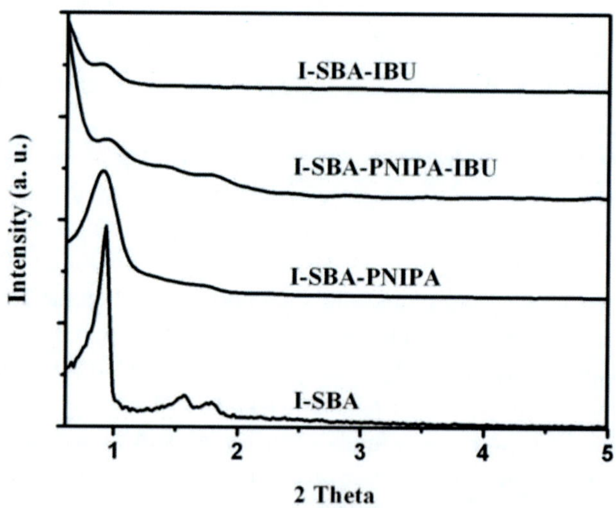


Figure 8. Low-angle powder XRD patterns for mesoporous I-SBA, I-SBA-IBU, I-SBA-PNIPA, I-SBA-PNIPA-IBU and I-SBA-PNIPA after delivery of IBU. Reproduced with permission from ref 19b.

When analyzed, the sample thus prepared (I-SBA) exhibits an XRD pattern of well-resolved diffraction peaks, confirming the ordered hexagonal structures of the porous networks (Figure 8). No distinct wide-angle diffraction peaks are observed for the sample of I-SBA, suggesting the amorphous nature of the magnetic particles formed. The clear dark spots detected from TEM images perfectly demonstrate the magnetic nanoparticles formed, which will be further confirmed by Mössbauer and magnetic measurements. The chemical composition was investigated by energy dispersive x-ray (EDX) spectroscopy, which illustrates the existence of the elements Fe, O and Si.

From analysis of the nitrogen adsorption isotherm, relative narrow mesopore distribution was observed at mean size of 8.2 nm for I-SBA, 2.3-3 nm larger than that reported by Zhao in synthesis of metal oxide in the channels of mesoporous silica monolith (5.2-5.9 nm) (27). In comparison with SBA-15 prepared under the same condition without the presence of $Fe(CO)_5$, I-SBA displays a larger pore size by 2.1 nm, suggesting the swelling property of the micelle in the presence of $Fe(CO)_5$ (Table 2).

**Table 2. Textural parameters and IBU storages of various samples**[19c]

| Sample | $d_{100}$ (nm) | $S_{BET}$ ($m^2g^{-1}$) | Pore size (nm) | $V_P$ ($cm^3g^{-1}$) | PNIPA | Drug storage (wt%) IBU/silica |
|---|---|---|---|---|---|---|
| I-SBA | 9.5 | 586 | 8.2 | 0.95 | | |
| I-SBA-IBU | 9.5 | 122 | 6.2 | 0.25 | | 33.5 |
| I-SBA-PNIPA | 9.6 | 260 | 7.1 | 0.41 | 30 | |
| I-SBA-PNIPA-IBU | 9.6 | 32 | 5.7 | 0.08 | 14.8 | 71.5 |
| SBA-15 | 9.0 | 531 | 6.1 | 0.93 | | |
| SBA-15-PNIPA | 9.1 | 324 | 3.8 | 0.42 | 29.5 | |
| SBA-15-PNIPA-IBU | | | | | | 32 |

It leads to the conclusion that the process here described provides an efficient method to obtain large pore size with magnetic particles formed in situ. A formation mechanism of the mesostructure can be proposed on the basis of the experimental results: $Fe(CO)_5$ as the swelling agent may profoundly penetrate into and swell the core of micelles formed by $EO_{20}PO_{70}EO_{20}$ (P123), leading to a expansion of pore size.

The pore volume is found to be 0.41 $cm^3g^{-1}$ for I-SBA-PNIPA, which is more than three times as much as that reported by Lopez in polymerization of NIPA inside mesoporous MCM-41 (0.133 $cm^3g^{-1}$). The wall thickness of I-SBA-PNIPA can be calculated as 4.2 nm with an average pore diameter of 7.1 nm, which is 1.1 nm smaller than that of the host I-SBA. Together with the analysis of TGA, it reveals a 0.55 nm coating layers of PNIPA through the polymerization of around 30 wt% NIPA formed inside mesopores of I-SBA. The drug delivery system demonstrated a high ibuprofen (IBU) storage of 71.5 wt% (IBU/silica), which is almost twice as much as the highest value based on SBA-15 ever reported (29). As a comparison, the I-SBA system without modification of the pores with PNIPA was found to exhibit only 33.5 wt% storage of IBU. Thus, the magnetic nanoparticles itself could not contribute to the high drug storage.

A comparative stimuli-responsive properties investigation of the effect of drug release on the temperature is conducted. A distinct transition at 32 °C is detected for the I-SBA-PNIPA (Figure 9). On the contrary, no temperature transition is observed for the I-SBA, almost 90 %

drug molecules are released for I-SBA when the temperature reached 24 °C. PNIPA inside I-SBA can respond to the thermal-stimuli from environment.

Further evidence that PNIPA chains can be used to modulate molecular-transport rates within the porous network is provided by data on the release of drug as a function of time at low and high temperatures. Almost complete drugs were entered into solution after 1 h for the sample of I-SBA-PNIPA at 45 °C. In contrast, only 30 wt% of IBU was released after 6 h of release for the same sample at 10 °C. A more rapid increase in UV-vis is observed at 45 °C than that at 10 °C. This further illustrates the PNIPA inside pores can be used for the controlled release.

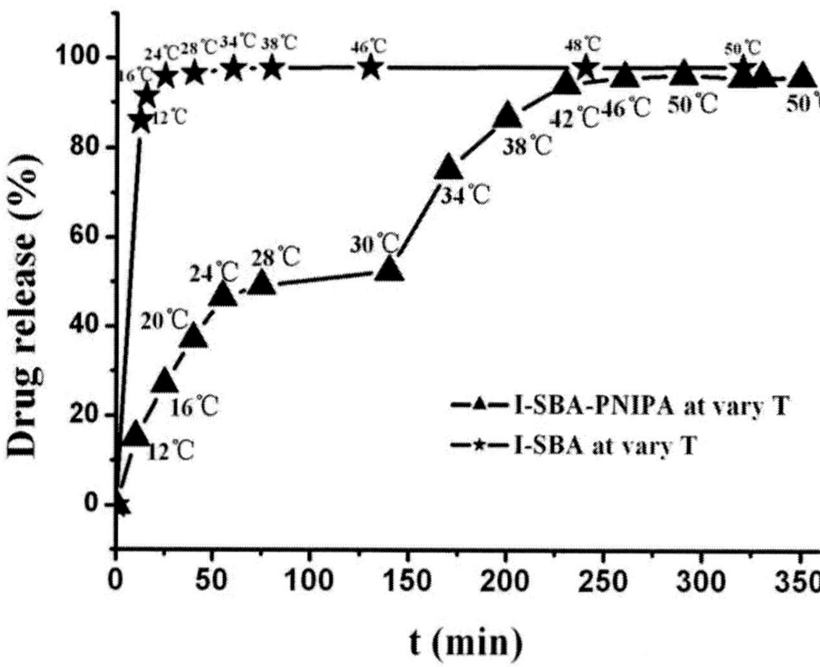


Figure 9. Cumulative drug releases from the system of I-SBA-PNIPA and I-SBA in release media of phosphate buffer on temperature. Reproduced with permission from ref 19b.

The possible mechanism of the drug storage and release profile from the system is schematically exhibited in Figure 10. At low temperature, the drugs are restricted confined in the pores owing to expanding of the PNIPA molecular chain and the formation of hydrogen bonding between PNIPA and IBU. While raising temperature, the polymer-chains are hydrophobic and swell within the pore network resulting in the collapsed of hydrogen-bond, pushing the drug molecules to release from the pores. Consequently, it implies that the swelling of the polymer-chain as a result of temperature can be used for controlled drug release with site label induced by magnetic particles.

Mössbauer spectrum of I-SBA-PNIPA has been measured at 13 K demonstrated the content of $\gamma$-$Fe_2O_3$ in the sample is calculated to be 56 %, whereas that of $Fe_3O_4$ is 39 %. I-SBA display a hysteresis loop with a saturation magnetization (Ms) of 0.81 emug$^{-1}$, indicating a magnetic field can be induced in I-SBA. When an external magnetic field was applied, the

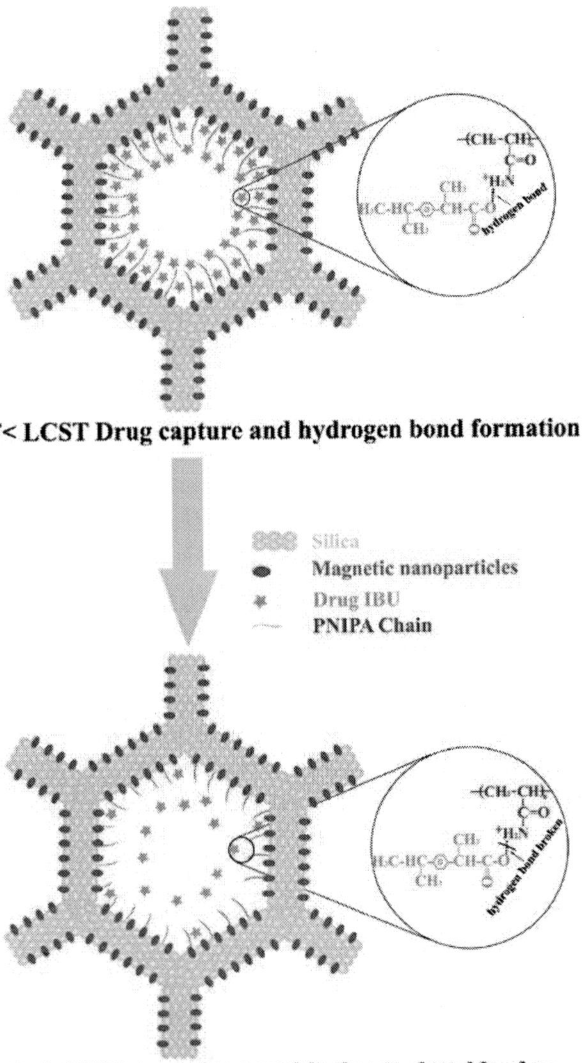


Figure 10. Schematic representation of the stimuli-responsive delivery system based on SBA-15 with magnetic particles formed in situ using $Fe(CO)_5$ as swelling agent. Reproduced with permission from ref 19c.

nanoparticles of I-SBA-PNIPA-IBU were enriched and the dispersion became clear. Thus, magnetic field could be possibly used as traces for target during controlled delivery under the guidance of a magnetic field. In comparison to that described above, a novel controlled-drug release system was successfully synthesized based on stimuli-responsive ordered mesoporous silica with magnetic particles formed in situ. It is encouraging that the modifications of PNIPA inside mesopores not only improves the drug-storage capacity, but also achieves stimuli-responsive release. The aim of this work is to explore a new potential application of highly ordered meoporous silica with large pores and its capability of acting as conventional reservoir for target and controlled drug release, for potential applications in many aspects.

# Conclusion

Drug-delivery systems are currently one of the most evident examples of the transfer of knowledge among science and biomedical materials applications (30). The development of target and controlled release system offers promising possibilities for better medical treatments. Mesoporous materials exhibiting zero premature release seem to be an excellent alternative to conventional systems for the oral administration of highly toxic drugs. Stimuli-responsive materials that have been considered allow the drug-release kinetics to be sped up or slowed down. Other specific long-term treatments could require an increase or decrease in the rate of drug release, depending on the disease evolution. The possibility to control the kinetic release of drug through external stimuli opens a wide field of possibilities for long-term therapies.

The targeted drug delivery systems can improve the efficiency of chemotherapy by allowing the maximum fraction of the delivered drug molecule to react with the disease sites without adverse effects on the normal cells. Magnetic drug targeting allows the concentration of drugs at a defined target site with the aid of a magnetic field.

The use of ordered mesoporous materials with magnetic particles formed in situ and stimuli-responsive polymers inside is a young and growing but still challenging field with still high demand of new and improved materials (31-32). One of the major problems is a real "zero premature release" system to be designed and synthesized. These systems are very desirable when high toxicity drugs are involved. Additionally, the construction of strong magnets should be required for successful site-specific targeting. Most studies are focused on iron oxide or iron oxide based materials. However, it is difficult to synthesize iron oxide particles which possess sufficient magnetic susceptibility to be focused on the targeted site by an appropriate magnetic force. The construction of strong magnets should be required for successful site-specific targeting. Recently, iron was reported to be employed as core for fabricating Fe-$SiO_2$ delivery system with high magnetic efficiency (33). The preparation of $Fe_2O$-$SiO_2$ delivery system with a high surface area is still a challenge. Next, the storage magnesium for the SPR with various drug molecules has to be investigated for reaching a high storage capacity. Moreover, the stability of mesoporous silicas support upon prolonged exposure to aqueous solutions is another important factor for industrial applications. Stimuli-responsive mesoporous silica materials respond to external stimuli (e.g., magnetic fields, thermo-, ion-) or internal pH changes would be very interesting, not only for oral drug release, but also for the implantable purpose.

# Experimental Section

## 1.1. Synthesis of MCF

In a typical preparation, P123 (2.0 g, 0.4 mmol) is dissolved in 1.6 M HCl (75 mL, 120 mmol HCl) at room temperature while being stirred in a beaker covered with a watch glass. 1,3,5-trimethylbenzene (TMB) (0.6 g, 5.1 mmol) is then added, and the mixture is heated to 37-40 °C. Following 1 h of stirring, tetraethyl orthosilicate (TEOS, 4.4 g, 21 mmol) is added. After 20 h at 40 °C, the milky reaction mixture is transferred to an autoclave and aged at 100 °C for 24 h under static conditions. The mixture is then allowed to cool to room temperature, and the white precipitate is isolated by filtration, dried in air, and calcined at 500 °C for 8 h in air to produce the MCF materials.

### 1.2. Surface Modification of MCF

Surface modification of SBA-15 to be hydrophilic was complemented with amino groups by refluxing 0.5 g SBA-15 with 3-aminopropyltriethoxysilane (APTES) in 50 ml of toluene for 12 h. The obtained sample was denoted as m-MCF.

### 1.3. Synthesis of MCF-Supported Initiator

A 100 mL flask containing m-MCF (0.50 g), anhydrous $CHCl_3$ (10.0 ml), N,N-(dimethylamino) pyridine(DMAP, 0.0292 g, 0.2390 mmol), and triethylamine (0.3031 g, 1.667 mmol) was evacuated and thrice filled with Ar. Then 0.3832 g (1.667 mmol) of 2-bromo-2-methylpropionyl bromide dissolved in 5 mL of anhydrous $CHCl_3$ was added dropwise at 0 °C for 60 min. The resulting mixture was stirred for 3 h at 0 °C and then at room temperature for 48 h. The solid was then separated from the mixture by filtration and washed five times with 100 mL of $CHCl_3$. The raw product was dispersed in 20 mL of $CHCl_3$, filtered, and washed three times to remove any adsorbed 2-bromo-2-methylpropionyl bromide. The black solid was collected and dried overnight under vacuum at 40 °C, affording MCF-supported ATRP initiator (Br- MCF).

### 1.4. Synthesis of MCF-PNIPA

Typically 30.0 mg of Br-MCF, 7.2 mg (0.050 mmol) of CuBr, 8.7 mg (0.050 mmol) of N,N,N',N'',N''-pentmethyldiethylenetriamine (PMDETA), and 0.25 mL of ethanol were placed in a 10 mL dry flask, which was then sealed with a rubber plug. The flask was evacuated and filled thrice with Ar. 25.0 mg (0.24 mmol) of NIPA was injected into the flask using a syringe. The flask was immersed in an oil bath at 65 °C, and its contents were stirred for 50 h. By the end of the reaction the viscosity had increased dramatically. The mixture was subsequently diluted with $CHCl_3$ and thrice vacuum-filtered using a 0.22 μm polycarbonate membrane. To ensure that no ungrafted polymer or free reagents were mixed in the product, the filtered mass was dispersed in $CHCl_3$, then filtered, and washed with $CHCl_3$. The resulting solid was predisposed in 5 mL of $CHCl_3$ and precipitated by the addition of 100 mL of methanol. The MCF-PNIPA was obtained by filtration and left drying overnight under vacuum.

### 1.5. Loading of drug IBU

The loading of drug IBU (Aldrich) inside the samples MCF-PNIPA were carried out as follow process: 0.3 g of the samples was added to 20 mL of ibuprofen-hexane solution (35mg/ml) and soaked under stirring at 45 °C for 48 h until the concentration of the solution keeping steady; this was conducted by monitoring the ibuprofen concentration using Lambda

20 UV/Vis spectrometer at a wavelength of 272 nm. Then, the mixture was cooled to ambient temperature soon, washed quickly and thoroughly with hexane, and died under vacuum at 40 °C. Finally, the precipitate was filtered. The sample was denoted as MCF-PNIPA-IBU.

## 1.6. In Vitro Drug Release Study

The release profiles of model drugs were determined by soaking 0.2 g of sample in 100 mL of Phosphate buffer (PH 7.4) under stirring at 100 rpm in a flask. 3.0 mL of the suspension was withdrawn at a predetermined time, replaced with the same volume of fresh medium, and analyzed by using a UV/Vis spectrometer (Lambda 20) at 272 nm. The temperature of the suspension was kept at a certain degree such as 10±0.5 °C and 45±0.5 °C. The release at varied temperatures was controlled with the following methods: the temperature of the suspension was increased to T °C at a rate of 1 °C $min^{-1}$, then keeping at T °C for at least 10 min before the measurement.

## 1.7. Synthesis of Mesoporous SBA-15 Materials

The synthesis process was described as below: First, 4 g tri-block copolymer P123 ($EO_{20}PO_{70}EO_{20}$) was stirred in dilute aqueous hydrochloric acid solution until it completely dissolved at 45 °C. Then, 1.5 g $Fe(CO)_5$ or $FeCl_2$ was added into the solution slowly, stirring for about 1 h. Next, 8.5 g TEOS was added into the mixture under continuous stirring. After that, the mixture was aged at 80 °C for 1 day after stirring at 45 °C for 20 h. The as-prepared sample was recovered by filtration, washed with water, and air-dried at room temperature. Calcinations were carried out at 500 °C for 6 h. Finally, the powder was treated at 450 °C for 4 h in vacuum. The sample thus prepared is denoted to be I-SBA or m-SBA.

## 1.8. Synthesis of Poly(N-isopropylacrylamide) (PNIPA) inside I-SBA

N-isopropylacrylamide (NIPA) solution of alcohol with a free radical initiator 2,2'-azoisobutyronitrile and N,N'-methylene bisacrylamide was added drop wise into 0.5 g I-SBA at ambient temperature. The detailed process was described as follows: 0.25 g of NIPA, 0.003 g of AIBN and 0.02 g of N,N'-methylene bisacrylamide were dissolved together in 1 ml alcohol. After impregnation of the solution about 2 h, the powder was air-dried to remove alcohol. The samples were then subjected to several freeze-vacuum-thaw cycles to remove residual alcohol and oxygen. Polymerization was performed by heating to 80 °C for 24 h. The resultant samples were washed thoroughly with chloroform and ethanol, followed by drying at 80 °C for overnight. The composite powder was designated as sample I-SBA-PNIPA or m-SBA-PNIPA. As a comparison, the sample prepared with the conventional SBA-15 instead of I-SBA and then polymerized the same amount of NIPA inside was denoted as SBA-15-PNIPA.

## References


[1] (a) Torchilin, P. *Nat. Rev. Drug Discov*. 2005, 4, 145-160. (b) Vallet-RegO, M. *Chem. Eur. J.* 2006, 12, 5934-5943.

[2] Lai, C. Y.; Trewyn, B. G.; Jeftinija, D. M.; Jeftinija, K.; Xu, S.; Jeftinija, S.; Lin, V. S.-Y. *J. Am. Chem. Soc*. 2003, 125, 4451-4459.

[3] Slowing, I.; Trewyn, B. G.; Lin, V. S.-Y. *J. Am. Chem. Soc*. 2006, 128, 14792-14793.

[4] Kresge, C. T.; Leonowicz, M. E.; Roth W. J.; Vartuli, J. C.; Beck, J. S. *Nature* 1992, 359, 710-712.

[5] Zhao, D.; Feng, J.; Huo,Q.; Melosh,N.; Fredrickson, G. H.; Chmelka, B. F.; Stucky,G. D. *Science* 1998, 279, 548-552.

[6] Yang, P.; Zhao, D.; Margolese, D.; Chmelka, B.; Stucky, G. D. *Nature* 1998, 396, 152-155.

[7] Vallet-Regi, M.; Rámila, A.; del Real, R. P.; Pérez-Pariente, *J. Chem. Mater*. 2001, 13, 308-311.

[8] Okano, T. *Biorelated Polymers and Gels: Controlled Release and Applications in Biomedical Engineering*; Academic Press: San Diego, 1998.

[9] Vallet-Regi, M.; Balas, F.; Arcos, D. *Angew. Chem. Int. Ed*. 2007, 46, 7548-7558

[10] Hartmann, M. *Chem. Mater*. 2005, 17, 4577-4593.

[11] Takahashi, H.; Li, B.; Sasaki, T.; Miyazaki, C.; Kajino, T.; Inagaki, S. *Micro. Meso. Mater*. 2001, 44–45, 755-762.

[12] (a) Gupta, A. K.; Curtis, A. S. G. *J. Mater. Sci. Mater. Med*. 2004, 15, 493-496. (b) Neuberger, T.; Schopf, B.; Hofmann, H.; Hofmann, M.; von Rechenberg, B. *J. Magn. Magn. Mater*. 2005, 293, 483-496. (c) Lubbe, A. S.; Bergemann, C.; Brock, J.; McClure, D. G. *J. Magn. Magn. Mater*. 1999, 194, 149-155. (d) Schillinger, U.; Brill, T.; Rudolph, C.; Huth, S.; Gersting, S.; Kroetz, F.; Hirschberger, J.; Bergemann, C.; Plank, C. J. *Magn. Magn. Mater*. 2005, 293, 501-508.(e) Scherer, F.; Anton, M.; Schillinger, U.; Henke, J.; Bergemann, C.; Kruger, A.; Gansbacher, B.; Plank, C. *Gene Ther*. 2002, 9, 102-109.

[13] (a) Yoon, T.-J.; Kim, J. S.; Kim, B. G.; Yu, K. N.; Cho, M.-H.; Lee, J.-K. *Angew. Chem. Int. Ed*. 2005, 44, 1068-1071. (b) Chattopadhyay, P.; Gupta, R. B. *Ind. Eng. Chem. Res*. 2002, 41, 6049-6058. (c) Lin, Y.-S.; Wu, S.-H.; Hung, Y.; Chou, Y.-H.; Chang, C.; Lin, M.-L.; Tsai, C.-P.; Mou, C.-Y. *Chem. Mater*. 2006, 18, 5170-5172.

[14] (a) Hernandez, R.; Tseng, H.-R.; Wong, J. W.; Stoddart, J. F.; Zink, J. I. *J. Am. Chem. Soc*. 2004, 126, 3370-3371. (b) Gruenhagen, J. A.; Lai, C.-Y.; Radu,D. R.; Lin,V. S. Y.; Yeung, E. S. *Appl. Spectrosc*. 2005, 59, 424-431. (c) Radu, D. R.; Lai, C.-Y.; Jerzy, W. W.; Pruski, M.; Lin, V. S. Y. *J. Am. Chem. Soc*. 2004, 126, 1640-1641.

[15] Kim, J.; Lee, J.-E; Lee, J.; Yu, J. H.; Kim, B. C.; An, K.; Hwang, Y.; Shin, C.-H.; Park, J.-G.; Kim, J.; Hyeon, T. *J. Am. Chem. Soc*. 2006, 128, 688-689.

[16] Giri, S.; Trewyn, B. G.; Stellmaker, M. P.; Lin, V. S.-Y. *Angew. Chem. Int. Ed*. 2005, 44, 5038- 5044.

[17] Yang, Q.; Wang, S. C.; Fan, P. W.; Wang, L. F.; Di, Y.; Lin, K. F.; Xiao, F. S. *Chem. Mater*. 2005, 17, 5999-6003.

[18] Zhou, Z.; Zhu, S.; Zhang, D. *J. Mater. Chem*. 2007, 17, 2428–2433.

[19] (a) Zhou, Z.; Zhu, S.; Zhang, D. *Chem. Lett.* 2006, 35, 618. (b) Zhu, S.; Zhou, Z.; Zhang, D.; Jin, C.; Li, Z. *Micro. Meso. Mater.* 2007, 106, 56–61. (c) Zhu, S.; Zhou, Z.; Zhang, D. *ChemPhysChem* 2007, 8, 2478 – 2483.
[20] Chang, H.; Shim, C. H.; Kim, B. J.; Shin, Y.; Exarhos, G. J.; Kim, K. J. *Adv. Mater.* 2005, 17, 634-637.
[21] Fu, Q.; Rao, G. V.-R.; Zsta, L. K.; Wu, Y.; Andrzejewski, B. P.; Sklar, L. A.; Ward, T. L.; Lopez, G. P. *Adv. Mater*, 2003, 15, 1262-1266.
[22] Schmidt-Winkel, P.; Lukens, W. W.; Zhao, D.; Yang, P.; Chmelka, B. F.; Stucky, G. D. *J. Am. Chem. Soc.*, 1999, 121, 254-255.
[23] Patten, T. E.; Matyjaszewski, K. *Acc. Chem. Res.*, 1999, 32, 895-903.
[24] Lin, C.; Meng, L. Z.; Lu, X.-J.; Wu, Z. Q.; Zhang, L. F.; He, Y. B. *Macromol. Chem. Phys*. 2005, 206, 1870-1877.
[25] Arruebo, M.; GalPn, M.; NavascuQs, N.; TQllez, C.; Marquina, C.; Ibarra, M. R.; SantamarOa, J. *Chem. Mater*. 2006, 18, 1911-1919.
[26] Radu, R.; Lai, C.-Y.; Jeftinija, K.; Rowe, E.W.; Jeftinija, S.; Lin, V. S.-Y. *J. Am. Chem. Soc*. 2004, 126, 13216-13217.
[27] Yang, H. F.; Lu, Q. Y.; Shi, Q. H.; Yan, Y.; Zhang, F. Q.; Xie, S. H.; Tu, B.; Zhao, D. Y. *Adv. Funct. Mater.* 2005, 15, 1377-1384.
[28] Balamurugan, S.; Mendez, S.; Balamurugan, S. S.; O'Brien II, M. J.; Lopez, G. P. *Langmuir* 2003, 19, 2545-2549.
[29] Song, S.-W.; Hidajat, K.; Kawi, S. *Langmuir* 2005, 21, 9568-9575.
[30] Son, Y.-H.; Park, M.; Choy, Y. B.; Choi, H. R.; Kim, D. S.; Park, K. C.; Choy, J.-H. *Chem. Commun*. 2007, 2799–2801
[31] Fu, Q.; Rama Rao, G. V.; Ward, T. L.; Lu, Y.; Lopez, G. P. *Langmuir* 2007, 23,170-174
[32] Ruiz-Herna´ndez, E.; Lo´pez-Noriega, A.; Arcos, D.; Izquierdo-Barba, I.; Terasaki, O.; Vallet-Regı´, M. *Chem. Mater*. 2007, 19, 3455-3463
[33] Kim, H. J.; Ahn, J. E.; Haam S. J.; Shul, Y. G.; Song, S. Y.; Tatsumi T. *J. Mater. Chem.*, 2006, 16, 1617–1621.

In: New Research on Nanocomposites
Editors: Luis M. Krause et al, pp. 57-78

ISBN: 978-1-60456-799-1


*Chapter 3*

# BONE INSPIRED NANOCOMPOSITES

***Otto C. Wilson***

Jr. the Catholic University of America, Department of Biomedical Engineering, Biomimetics, Orthopedics, and Nanomaterials Education/Composite Research for Advanced Biomaterials (BONE/CRAB) Lab, Washington, DC 20064

## Abstract

Bone is one of the most important and fascinating biomineralized nanocomposite structures in nature. On a macroscopic scale, bone provides structural support for our bodies, protects our internal organs, provides a reservoir for the production and storage of mesenchymal and hematopoietic stem cells, and serves as a storage site for calcium, phosphate, and other biologically relevant ionic species. However, the nanoscale features of bone make it truly remarkable from a material's design viewpoint. Biomineralized nanocomposites share the common design motif of organic matrix mediated nucleation and growth of an inorganic phase with controlled chemistry. The process of mineral phase formation is governed by solubility, surface and complexation chemistry, and bioavailability of important elemental and molecular species such as calcium, carbonate, phosphate, magnesium, and silica. Specific aspects of nanocomposite design related to matrix phase chemistry, degree of crystalline order and orientation, phase distribution and microstructure are uniquely controlled during the birth, healing and remodeling of bone. There are many lessons that we can learn from bone structure and biology to improve our ability to synthesize nanocomposites for various applications. The focus of this chapter is describing unique structural characteristics of biomineralized nanocomposites in nature with an emphasis on bone. Key lessons will be extracted from bone that can be used in the design and synthesis of nanocomposites with tailored functionality. An overview of research efforts that have utilized various aspects of bone inspiration will be provided to highlight current advances in the area of bone inspired nanocomposites and indicate areas for further study.

*As thou knowest not what is the way of the spirit, nor how the bones do grow in the womb of her that is with child; even so thou knowest not the works of God who maketh all.*

***(Ecclesiastes 11:5)***

## Introduction

The treatment of bone disorders is a very important component of medical technology. There is a wide variety of bone and skeletal disorders (Table I) and some treatments require the use of specially tailored implant materials. The development of bone implants that exhibit optimized interactions with bone at the whole tissue, cellular, and sub-cellular levels are needed for people who suffer from medical conditions that influence hard tissue. A series of X-ray images of a unicameral bone cyst (UBC) in the lower left femur region of a seven year old patient is shown in Figure 1. The periosteum in the vicinity of the UBC subsequently broke before treatment. The post break treatment consisted of filling the void region with bone graft and surgical implantation of a titanium plate and screws to fix the broken bone in place during healing. After a 9-12 month healing period, additional surgery will be required to remove the titanium plate and screws because while titanium is durable and strong, it is not capable of growing with the young patient's bone tissue. This is one area that needs improvement for bone fracture fixation. The degree of sophistication in bone implants continues to increase in response to expressed needs from the medical community for multifunctional implants. Bone implants may integrate aspects of structural integrity and support, biodegradability, bioactivity, genetic sequences for cell recognition and recruitment, controlled release and delivery of therapeutic agents such as growth factors, antimicrobial agents to prevent infections, and a host of other functions that are in various stages of vision and development.

**Table 1. Examples of medical conditions that influence hard tissue**

| *Name* | *Description* |
|---|---|
| Unicameral bone cysts | fluid filled void in long bone, may result in thinning of bone diaphysis and subsequent fracture. |
| bone fractures | break in bone ranging from a crack to a complete separation (called a non-union fracture) |
| osteocarcinoma | bone cancer |
| osteoarthritis | condition in which joints undergo degenerative changes. |
| osteoporesis | abnormal loss of bone mass and density |
| osteopetrosis | a genetic condition characterized by abnormal increase in bone density possibly due to deficiency of osteoclasts. |
| osteomalacia | condition caused by inadequate amount of calcium and phosphorous in the blood, resulting in bone softening due to insufficient calcification |
| osteomyelitus | localized infection of bone and bone marrow |
| Pagets Disease | excessive bone destruction and unorganized bone repair |
| Ostemia | abnormal congestion of blood in the bone |
| Ostempysis | accumulation of pus within a bone |
| Osteoaneurysm | an aneurysm within bone tissue |

There is a strong need for the development of innovative design features and processing techniques to exert more control over the microstructure and functionality of nanocomposites for bone tissue engineering. Biological nanocomposites (bone, teeth, nacre, shell) can provide key insights and inspiration to fabricate functionally advanced bioimplants. Nature has an extraordinary way of constructing elaborate composite structures with exceptional properties from relatively simple compounds. Design concepts such as hierarchical nested design and self assembly phenomena are utilized in fascinating ways to synthesize biological nanocomposites by the controlled nucleation of an appropriate inorganic phase in the presence of a functionalized organic matrix. The matrix exerts a high degree of control over the development of the inorganic phase by regulation of ion flux at the matrix interface, growth and morphology modification of the inorganic phase by adsorption/complexation interactions with soluble polymer macromolecules, and functional group mediation of molecular interactions at the polymer/matrix interface (Calvert and Mann, 1988). Biomineralization processes result in elaborate, multifunctional architectures that display an extraordinary degree of control over the phase distribution, size, interfacial structure, morphology, and crystallographic orientation of the inorganic phase.

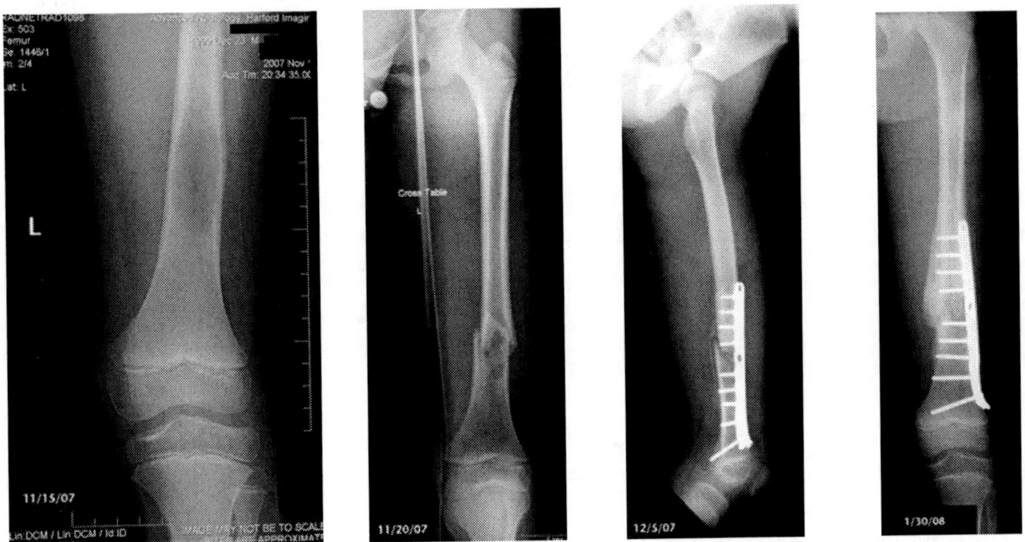


a.) b.) c.) d.)

Figure 1. X-ray images of a bone defect that fractured and was fixed with a titanium plate and screws. a.) Unicameral bone cyst. b.) Unicameral bone cyst after fracture. c.) fracture fixed with titanium plate and screws. d.) Healing region approximately two months post operation.

One approach to achieve the next generation of bone implants involves adopting biomineralization and biomimetic processing strategies. Biomimetic is derived from the root word *bio* (life) and *mimere* (to mimic). The intended goal of biomimetic design and processing is to utilize information gathered from the study of biological systems to improve processing technology. It is a natural extension to utilize biomimetic design principles gathered from bone to enhance the level of control that can be exerted over the development of bone implants. The field of biomimetics can be divided into two focus areas (Sarikaya *et al.*, 1995). Biomimicry is a short term approach that looks for alternative processing routes to achieve the unique structures and functionalities that are exhibited by biological materials such as bone. Bioduplication refers to the direct adoption of biological processes to synthesize unique structures and constitutes a longer term approach to advanced bone implant development. The successful bioduplication of bone combines aspects of tissue engineering with fundamental studies of the structural and cellular biology of bone tissue in order to develop design principles and processing strategies to fabricate bone inspired nanocomposites (BIN).

What does bone inspired mean? Inspiration can be thought of as an intangible element that enables one to envision, achieve, or bring into existence something that was previously unknown or undeveloped. The overall goal of using bone as a biomimetic subject is to draw inspiration from the unique structure and function of hard tissue to make materials to improve bone health. Biomimetic lessons and inspiration from bone can be utilized on three different levels. The first aspect of bone inspiration deals with the development of materials for bone tissue repair based on biomimetic lessons from bone. Another aspect of bone inspiration involves using insights gained from the structure and function of bone and other biological hard tissues to synthesize novel nanocomposites for various technological applications. Bone inspiration can also be utilized for the development of models and strategies to address critical issues in education and mentoring. While all three aspects of bone inspiration are important, the first one related to BIN to enhance the healing and remodeling of bone will serve as the focus for this chapter. The purpose of this chapter is to present a selected review of some of the key features of bone structure, properties, and function that contribute to the design and synthesis of BIN. Key aspects of the structure and properties of bone will be highlighted before providing a view of how researchers are both adopting and adapting these features to design functionally advanced bioimplants to improve bone healing and remodeling.

# I. Bone Structure and Function

Bone is one of the most important and fascinating biomineralized composite structures in nature. On a macroscopic scale, bone provides structural support for our bodies, protects our internal organs, serves as an attachment site for muscle and tendon, and provides a reservoir for the production and storage of hematopoietic and mesenchymal stem cells. On a microscopic scale, bone serves as a storage site for calcium, phosphate, and other biologically relevant ionic species. Bone also provides a framework for the activities of osteoblasts, osteoclasts, and osteocytes to orchestrate the production of extracellular matrix proteins (ecm), resorb excess bone, and help coordinate cellular and tissue interactions. The nanoscale features of bone make this material truly remarkable from a material's design viewpoint. Bone is composed of a cell derived collagen matrix that works in conjunction with ecm and matrix vesicles to influence the in situ nucleation and growth of hydroxyapatite (HAp) crystals (Anderson, 1995). The HAp crystals appear as thin platelets that are intimately incorporated within and on collagen fibrils (Weiner and Wagner, 1998). Bone is a very important tissue in the body due to its structural and metabolic importance in overall body homeostasis. Bone is also a unique tissue because it integrates HAp, a hard calcium phosphate based mineral phase, with a protein based organic matrix to form a strong and tough nanocomposite. Bone derives its properties from the multi-level hierarchy which is built in during the growth, development, and remodeling of this tissue.

The main constituents of bone include HAp, the collagen matrix, various ecm proteins, water, and cells. There is a great deal of interactions between each of these critical building blocks for bone. The major cells in bone include differentiated cells such as osteoblasts, osteoclasts, and osteocytes and progenitor cells including hematopoietic and mesenchymal stem cells. The role of osteoblasts is to build new bone. The osteoblasts are considered specialized fibroblasts (Ducy *et al.*, 2000) that have the ability to express structural proteins such as collagen and ecm components such as osteopontin, osteonectin, alkaline phosphatase, bone sialoprotein, and bone morphogenetic protein that contribute to the organic matrix of bone. The ecm components are glycoproteins which serve various roles in regulating cell activities and biomineralization. Osteoclasts serve an important role in bone remodeling. Osteoclasts are drawn to areas of bone that are no longer needed and remove bone by forming resorption pits. The mechanism of osteoclast bone removal is via enzymatic and pH modulation. Osteocytes are osteoblasts that become entombed in bone tissue and serve a signaling/sensory role in bone tissue health. Osteoblasts and osteoclasts have a very interesting origin and lineage (Ducy *et al.*, 2000; Teitelbaum *et al.*, 2000). The intermedulary cavity of long bone houses the bone marrow which produces the mesenchymal and hematopoietic stem cells that have very important roles in promoting and protecting our skeletal system. Mesenchymal stem cells are pluripotent cells that under appropriate conditions can differentiate into specialized cells to produce connective tissue such as bone, muscle, tendon, cartilage, and adipose (fat). Hematopoietic stem cells can differentiate into cells related to the blood and immune system such as red and white blood cells, macrophages, monocytes, T cells, B cells, and other cells that help protect our bodies from invading pathogens. While the osteoblasts originate from mesenchymal stem cells, the osteoclasts come from the hematopoietic stem cell lineage.

The major acellular components of bone are HAp, collagen, and ecm. HAp makes up approximately 70% of bone by mass. The idealized chemical formula for stoichiometric HAp is $Ca_{10}(PO_4)_6(OH)_2$. Biologically based HAp is liberally substituted by a variety of ions and molecules including Mg, Na, Si, K, and carbonate. Carbonate is a very important molecule in biologically based HAp and it can substitutionally replace either phosphate or hydroxyl groups in the HAp structure (Legeros *et al.*, 1969), resulting in a decrease in degree of crystalline order and morphological changes (Legeros *et al.*, 1967). Bone mineral chemistry plays an important role in influencing the overall bioactivity of nanocomposites (Hing, 2004; Salgado *et al.*, 2004). Some of the unique features of natural bone mineral that have been reported in the literature include low degree of crystalline order (Wopenka and Pasteris, 2005), the absence of OH and the resulting influence on bioactivity (Pasteris *et al.*, 2004), deviation of calcium to phosphate ratio from 1.67 (Liou *et al.*, 2004), and substitutional incorporation of carbonate and its influence on solubility and bioactivity (Cheng *et al.*, 1998).

Collagen is a structural protein that is the main component of the organic matrix of bone. The basic building blocks of proteins are amino acids. There are approximately 20 amino acid types that are routinely encountered in mammalian species. Each amino acid has a similar base structure composed of a carboxyl and an amino end group. Amino acids form peptide bonds in a condensation type polymerization reaction that results in long molecules that are called polypeptide chains. Three polypeptide chains orient in a helical fashion along the axial chain length to form the collagen molecule. The collagen molecule is a cigar shaped structure that is approximately 300 nm in length and about 1-2 nm in diameter. Five collagen molecules assemble in a quarter stagger pattern along their axial length to form the collagen macromolecule. The collagen macromolecule is the basic building block that forms the foundation for tissues in the body including skin, cartilage, bone, blood vessels, muscle, the heart, and the eye.

A number of other structural features need to be considered in bone tissue. Bone tissue is supplied with oxygen and nutrients via numerous blood vessels. Concentric layers of bone matrix (lamella) enclose a central blood vessel within Haversian canals. Osteocytes are interspersed between the lamella in very small spaces known as lacunae. Lacunae and Haversian canals are connected via microscopic channels (canaliculi) that provide a pathway for tissue fluid to reach the bone cells. There are two types of bone that typically occur. Compact bone is characterized by closely spaced Haversian canals and lamellae; hence it is the densest type of bone. Cancellous bone contains many open spaces between trabeculae. The trabeculae are thin processes of bone that join together in different arrangements in different bone types to give structural integrity and strength along the lines of strain in individual bone.

## II. Historical Overview of Key Features in Bone Inspired Nanocomposites

Natural biological materials such as bone, teeth, and ivory have been used in various cosmetic and medical applications since the earliest days of recorded history. Dental bridges were made by fastening carved ivory and/or discarded human teeth with gold as early as 500 BC. Some of the more exotic materials that have been used for dental implants include walrus, hippo, and elephant tusks and seashells. Problems with these implants, such as discoloration and disintegration in the body, motivated people to develop improved biomaterials.

Advances in bone inspired nanocomposites tend to follow technology trends. For the longest time, mechanical properties were considered the most important factor in orthopedic implants because these properties could be monitored and modified as needed. There were many successful implants made of metals that helped improve the quality of life for many with skeletal problems (hips, long bones, etc.). But problems associated with stress shielding (i.e., bone atrophy due to improper stress distribution), corrosion, implant density differences, and the inherent static nature of metallic implants drove efforts to improve on bone implant design. There was a fundamental shift away from monolithic metal implants because of metal's inability to grow, change, and adapt with native bone tissue to something more similar to bone. One of the first considerations in BIN design is that bone tissue is living and vibrant, so an ideal replacement should exhibit these characteristics also.

Autologous bone is the gold standard for bone graft based materials to replace diseased or damaged bone. Autologous bone is harvested from the patient, typically from the hip region. While this material performs on a superior level because it contains all the key factors required for successful integration and stimulation of bone healing and remodeling events with no autoimmune issues, autologous bone harvesting results in a new wound site where risks associated with compromised healing and infection may occur. We are not yet able to generate BIN that rival autologous bone, especially from the aspect of being fully living and vibrant as healthy bone tissue. However, there have been a number of ingenious developments in bone tissue implants that have brought us much closer to being able to produce materials that approach the performance of autologous bone in orchestrating bone healing and remodeling.

Foundational studies on the osteoinductive properties of demineralized bone matrix (DBM) were performed by Urist in the mid 1960's. The implantation of DBM underneath the skin or within muscle tissue initiated a series of reactions that ultimately produced viable bone tissue with marrow. Bioglass was discovered by Larry Hench in the late 1960's. Bioglass revolutionized the field of bioceramics and provided inspiration for a large number of researchers. Bioglass incorporates calcium and phosphorous into a sodium silicate glass to develop a bioactive material. Bioglass develops a tailored surface chemistry to promote bone bonding interactions due to the presence of phosphorous, calcium, and sodium in the glass matrix. In the body, the calcium, phosphorous, and sodium leach out of the glass matrix leaving a silica rich surface layer. The sodium ion regulates the pH in the vicinity of the Bioglass surface and an amorphous calcium phosphate layer forms that subsequently crystallizes and recruits osteoblast cells for bone tissue formation and bonding (Hench *et al.*, 1971, 1973). Bioglass is considered the first synthetic bioceramic that was designed to exhibit tailored bioactivity and introduced the concept of designing nonviable materials so that they recruit appropriate cells for bone tissue healing and remodeling. Since the introduction of Bioglass, the idea of developing bone inspired materials and composites has greatly expanded.

The 1970's witnessed a series of developments that propelled the field of bone tissue regeneration forward by leaps and bounds. There were further developments in Bioglass and supporting studies on the unique role of silica in bone. Carlisle (1971) found that silica was present in the bone tissue of young rats and that the silica concentration in the bone tissue was elevated during the early stages of bone formation. Another big development in BIN during the 1970's occurred with the introduction of coral based implants. Coral has been used as a material to help guide the regeneration of bone tissue since the mid 1970's. The porous $CaCO_3$ coral exoskeleton exhibits a unique ability to bond with bone due to the bioactivity of $CaCO_3$ and the presence of pores (>100 um) which allows vascularized bone tissue to grow into the coral for implant fixation (White *et al.*; 1975, Muller-Mai, *et al.*, 1996). Replamineform techniques were developed to generate the unique coral microstructure with interconnected porosity in other materials to help preserve precious coral reefs (Chiroff *et al.*, 1975). One other development can be highlighted in the 70's. Jarcho *et al.* developed techniques to synthesize nanophase HAp particles and process these powders into bulk shapes via sintering (1976). It is interesting to note that Jarcho achieved sintered HAp microstructures that were nanophase with a large fraction of grains below 100 nm.

Rudimentary efforts to improve BIN by adapting an overall composition more akin to bone were realized in the 1980's. A revolutionary shift in BIN began with Bonfield's work of introducing HAp into the common polymer polyethylene (PE). The PE served as a matrix phase. Matrix comes from the Latin root word for womb. While the PE did not give birth to the HAp or orchestrate its formation like the collagen matrix in bone, this first attempt was groundbreaking in helping researchers realize the potential of polymer/bioceramic BIN. The initial HAp/PE composites of Bonfield *et al.* (1981) exhibited bioactivity at HAp loadings of 20 vol% or more and a ductile to brittle fracture transition near 40 vol% HAp. The mechanical properties of the implants were good, but in the lower range of measured values for the fracture strength of bone. Hence, the commercially available form of this composite material, HAPEX, is used for non load bearing applications such as inner ear reconstructive surgery. The concept of utilizing biomineralization phenomena such as in situ nucleation was adopted more fully in the 1980's on both the theoretical side (Weiner, 1986) and the experimental side (Calvert and Mann, 1988). Just as the collagen-ecm matrix influences the nucleation and growth of HAp crystals (Weiner and Price, 1986; Weiner and Traub, 1986), work began in earnest to utilize biopolymer matrices to orchestrate mineral phase formation via structural, chemical or other signal cues.

Researchers have utilized a wide range of bone inspired concepts in designing BIN since the 1990's. Novel synthesis efforts for nanocomposite development were greatly expanded in the early 1990's (Ozin, 1992; Komarneni, 1992). Researchers from various disciplines and specialties became intrigued by biomimetics and bone structure and joined the quest for BIN. There are a number of different types of nanocomposites that were produced by various methods including in situ nucleation (Bianconi *et al.*, 1991, Calvert and Rieke, 1996) and solution processing (Gonsalves *et al.*, 1997). Inorganic/organic intercalation or layered compounds have been the focus of many studies in recent years. Oriakhi (1998) referred to these materials as nanosandwiches whose synthetic procedures are inspired by antler and bone. Layered nanocomposites can be synthesized via clay exfoliation (sheet separation)-polymer adsorption, in situ polymerization, melt intercalation and template synthesis. The research group of Giannelis (Krishnamoorti *et al.*, 1996; Weimer *et al.*, 1999) used the clay exfoliation technique to synthesize highly dispersed polymer layered silicate nanocomposites by chemically modifying the silicate with organic surfactants to make it less hydrophilic and more compatible with the polymer.

The template synthesis method is most closely related to the biomineralization processes that are utilized in nature to develop nanocomposites because of the similar manner in which the organic phase influences the nucleation and growth of the inorganic phase (Calvert and Rieke, 1996). Ozin reported a template based method to make hierarchical structures reminiscent of diatoms which he referred to as "Skeletons in a Beaker". Stupp *et al.* (1992, 1993a, 1993b 1997) and Muller-Mai *et al.* (1995) synthesized HAp based hybrid inorganic/organic organoapatites and nanoapatites for bioimplant applications. Template synthesis techniques have also been developed to prepare layered double hydroxide nanocomposites (LDHN) based on the coprecipitation of layered double hydroxides (LDH) in the presence of nonionic poly(vinyl alcohol) (Messersmith and Stupp, 1992) and anionic polymer macromolecules such as poly(acrylic acid) and poly(styrene sulfonate) (Oriakhi *et al.*, 1996) in $CO_2$ free water. Layered double hydroxide hybrid nanocomposites (LDHN) are a relatively new class of materials that utilize the permanently charged surface of LDH to serve as a template for the interlayer adsorption of large anionic polymers rather than simple ions such as $Cl^-$, $NO_3^-$, $CO_3^=$ which are usually adsorbed in between the LDH sheets. Whilton *et al.* (1997) synthesized bioinorganic clays by the intercalation of negatively charged aspartic and glutamic acid in the interlayer space of Mg/Al LDH. Ozin *et al.* (1997) used a bone mimetic process to prepare organic-inorganic nanocomposites that consisted of HAp with a calcium dodecylphosphate (CaDDP) lamellar phase. Wilson *et al.* (1999) performed surface and interfacial characterization of Mg/Al LDHN containing interlayer molecules to explore potential biomaterials applications.

Another highlight in BIN during the 90's was the "rediscovery" of nacre as a bioactive material for bone healing and remodeling. The ancient Mayan people used nacre as a dental replacement. Nacre is a layered nanocomposite composed of calcium carbonate, chitin and protein that makes up the inner portion of the shell of deep sea mollusk species such as *Haliotis Rufescens*. Archaeological specimens of Mayan skulls exhibited the unique phenomena of nacre physically bonded to the jaw bone (Westbroek and Marin, 1998). The work of Lopez *et al.* demonstrated that nacre induced osteoblasts to generate new bone. The series of studies included culturing osteoblasts in the presence of nacre chips (Lopez *et al.*, 1992), observing the inducement of osteoblasts to produce a bone "bridge" between neighboring nacre and bone chips with a 1 mm gap (Silve *et al.*, 1992), and the implantation of crushed nacre into periodontal defects in human adult females (Atlan *et al.*, 1997). Westbroek and Marin (1998) pointed out interesting implications regarding common features and chemical factors in the biomineralization processes that developed for the various biomineralized tissues. The idea of a kinship between biomineralized tissues seems plausible due to the observation that the implanted nacre produced no detrimental effects in the female patients and resulted in the growth of new healthy bone tissue in the defect regions (Atlan *et al.*, 1997). Later research on the mechanism for the osteoinductive properties of nacre revealed that nacre contains water soluble matrix proteins that stimulate osteoblast activity and even promote the differentiation of cells to the osteogenic line (Mouries et al., 2002). At least one of the nacre matrix proteins, Perlustrin, has been found to share a number of similarities with human insulin like growth factors (Weiss *et al.*, 2001).

## II.a. Approaches to Bone Inspired Nanocomposites

Some current approaches for bone inspired nanocomposites include aspects of 1.) integration of natural biopolymers in BIN, 2.) incorporating HAp into various polymer matrices, and 3.) enhancement of interfacial bond formation between the HAp reinforcing phase and the polymer matrix.

### II.a.1. Natural Biopolymers

There is a growing trend towards using natural polymers in processing nanocomposites because biological familiarity is a major attraction for natural polymers. It is important to utilize polymers with ecm and collagen-like characteristics. Chitosan and chitin are probably the most popular non collagen based natural polymers that are being used in BIN. Chitosan is a unique polysaccharide that shares a remarkable number of structural and chemical similarities with collagen (Giraud-Guille, 1994). Chitosan has been used as skin grafting template, hemostatic agent, drug and DNA delivery vehicle (Erbacher *et al.*, 1998; Mao *et al.*, 2001), as a wound healing material (Muzzarelli *et al.*, 1999; Lahiji *et al.*, 2000; Khor and Lim, 2003; de Campos *et al.*, 2001), and as a favorable tissue engineering support for osteoblasts and chondrocytes (Lahiji *et al.*, 2000). Chemical functional groups can be grafted onto the surface of chitosan to influence bioactivity. Muzzarelli *et al.* (1993) determined that methylpyrrolidinone chitosan exhibited osteoconductive properties as evidenced by new bone formation in a rat tibia model. Fructose modified chitosan scaffolds and galactosylated chitosan have been used to study hepatocyte cell adhesion in culture (Li *et al.*, 2003; Park *et al.*, 2003). Chitosan has also been mixed with other polymers such as PAA (Hu *et al.*, 2002) and lecitin (Magdassi *et al.*, 1997) to form nanocomposites while chitin whiskers have been used to reinforce rubber (Gopalan-Nair and Dufresne, 2003). The rich body of research regarding biomedical applications for chitin and chitosan has been augmented recently with work highlighting it use as a material for bone healing and regeneration. Chitosan and chitosan/hydroxyapatite composites have been developed to fabricate bone filling pastes (Ito,

1991; Ito *et al.*, 1996) and tissue engineering scaffolds (Viala *et al.*, 1998; Ito *et al.*, 1999; Yamaguchi *et al.*, 2001; Zhao, 2002; Muzzarelli and Muzzarelli, 2002; Tachaboonyakiat *et al.*, 2002; Zhang *et al.*, 2003). Biomimetic inspiration from natural chitin composites has been carried over into studies of biomineralization phenomena using chitosan and chitin (Wan *et al.*, 1998; Muzzarelli and Muzzarelli, 2002; Falini and Fermani, 2004) to form nanocomposites with silica, calcium carbonate, and HAp. Chitosan and Chitin exhibit unique macromolecular conformations and liquid crystalline behavior (Pedroni *et al.*, 2000; Ratto *et al.*, 1995; Girard Guille 1994, 2005). These effects can be utilized to influence biomineralization phenomena in conjunction with soluble polymer molecules (Wilson, Jr, 2002).

There are a vast number of additional natural polymers that can greatly enhance the functionality of BIN. Reis *et al.* (Gomes *et al.*, 2001; Marques *et al.*, 2002) have developed a wide range of starch based biopolymers that exhibit very promising characteristics for BIN. Sodium alginate has been evaluated for use in bone marrow engineering (Wang *et al.*, 2003). In addition, encapsulation of viable cells within chitosan-alginate based fibers has been performed (Wan *et al.*, 2004). Gum Arabic (GA) is another promising natural polymer that can potentially have applications in developing BIN. GA is an aribinogalactan protein (polysaccharide/protein complex) hydrocolloid that is the natural sap-like exudate from the acacia Senegal tree. GA is loosely composed of polysaccharides such as rhamnose, glucuronic acid, and proteins (Phillips and Williams, 1993; Islam et al., 1997). GA has a very colorful history. It has been used as a dispersant for C based inks in ancient Egypt, a paint additive, and it is widely used as an emulsifying agent in the foods and beverages industry. Recently, GA has been used for dispersing C nanotubes (Bandyopadyaya *et al.*, 2002), in ceramic processing (Leong *et al.*, 2001), and similar aribinogalactans have been used for dispersing metal nanoparticles (Mucalo *et al.*, 2002). Gum arabic exhibits strong interactions with $Ca^{2+}$. The water soluble nature of gum arabic makes it a green processing agent. In addition, GA is already FDA approved for use in food products. One of the main attractions of gum arabic involves its similar chemical and structural features with ecm proteins such as decorin, osteoadherin, and bone acidic glycoprotein-75 (Wendel *et al.*, 1998; Midura *et al.*, 2004) which serve important roles in hard tissue biology. Further insights into the relationship between gum arabic, extracellular matrix proteins, and HAp will lead to unique methods for the molecular surface engineering of HAp to optimize biological activity.

### II.a.2. Polymer/HAP Composites

One of the long sought after goals in hard tissue biomaterials is the development of three dimensional (3D) bone analogue materials that function as well as natural bone. Biomimicry approaches which essentially focus on using conventional processing strategies to develop HAP/polymer composites with similar microstructure and mechanical properties as bone have increased in recent years. Most approaches utilize physical mixing strategies to achieve bone analogue materials and have resulted in some impressive implants. Since the pioneering work of Bonfield *et al.* (1981, 1986, 1988), a number of subsequent polymer systems have been investigated for use in HAP/Polymer composite systems including poly(lactic acid) (PLA) (Higashi *et al.*, 1986; Furukawa Kasuga *et al.*, 2001; Rizzi *et al.*, 2001), and PLA/poly(glycolic acids) (PLGA) copolymers (Laurencin et al., 1996; Ambrosio et al., 2001), poly(caprolactone) (Rizzi *et al.*, 2001), starch based polymers (Gomes *et al.*, 2001; Marques *et al.*, 2002), poly(ester) (Calandrelli, 2002), poly(lactones) (Helwig *et al.*, 2001), poly(methyl metacrylate) (Castaldini and Cavallini, 1985; Moursi *et al.*, 2002), poly(ethylmethacrylate) (Deb *et al.*, 1995), poly(propylene fumarate) for injectible composite systems with β-tricalcium phosphate (Mikos *et al.*, 1997; 2000), poly(vinyl phosphonate (Greish and Brown, 2001), and the segmented block copolymer polyethylene glycol/pol(butylene terephthalate) (PEG/PBT Polyactive) (Gaillard and van Blitterswijk, 1994).

There is a growing trend toward adapting in situ nucleation techniques to synthesize BIN. The controlled nucleation and growth of HAp in three dimensional matrices via biomimetic strategies is a challenging proposal due to the complex diffusion and transport issues that are faced, especially in terms of controlling the size, morphology, and crystallographic orientation of the inorganic phase. Rhee *et al.* (2000) reported on the formation of chemically linked HAp in a collagen membrane via aging in a concentrated simulated body fluid solution. Lyn *et al.*, 2005 utilized a co-precipitation process to synthesize collagen, HAp, and silicon nanocomposites. The problem of diffusion can be simplified somewhat by limiting the size of the system. Ambrosio *et al.* (2001) developed a novel method to nucleate an amorphous, carbonated hydroxyapatite within a polymer matrix of bioresorbable pol(lactic acid/glycolic acid) microspheres. Supramolecular assemblies can function in concert with functionalized soluble polymers to control the nucleation and growth of HAp in polymer matrices just as collagen and biopolymers function in the growth and development of bone. Some really interesting developments in the area of matrix assisted nucleation of HAp with nanophase features have been reported. Stupp *et al.* (2001) have synthesized peptide-amphiphile nanofibers that self assemble into matrices at specific pH values and can be crosslinked to nucleate HAP with controlled particle orientation similar to bone.

### II.a.3. Interfacial Bond Enhancement

Some of the issues that arise in comparing synthetic implants with natural bone involve the microstructure, particle size and phase distribution of HAp within the collagen matrix, and the interfacial bond that develops between the HAp and the collagen matrix. In particular, the unique biological function and mechanical properties of bone are closely related to chemical and microstructural features and the intimate interfacial bond that develops between the collagen matrix and nanophase HAp. The enhancement of the interfacial bond that develops between HAp and the polymer matrix is an important issue in BIN, especially when mechanical support is a requirement. It is well known that specific composite properties related to shear and compressive behavior, toughness, fracture modes, and stress transfer phenomena are governed by the interface. While the nature of the HAp/collagen interface is poorly understood at present, the manner in which the collagen matrix controls the crystal growth and orientation of HAp suggest that the HAp interfacial bond is strong (Walsh et al., 1994; Weiner and Wagner, 1998; Ji, 2008).

Improved interfacial bonds were developed between the PE matrix and the HAp filler particles by silanation of the HAp and grafting acrylic acid onto PE. Liu *et al.* (1996) modified the surface of HAp with PAA and poly(ethylene-co maleic acid) prior to mixing with Polyactive (PEG/PBT) and found that the interfacial bond was improved. Organic phosphates including hydroxymethacrylate phosphate (Subirade and Lebugle, 1991), and alkyl phosphates (Tanaka *et al.*, 1997; Ozin *et al.*, 1997) have been utilized to modify the surface of HAp for enhanced interfacial bond formation in composites, to study fundamental adsorption interactions, and in synthesizing novel layered HAp/alkyl phosphate/octacalcium phosphate type nanocomposites (Tanaka *et al.*, 1997). Silane coupling agents have been used with HAp, but the results tend to be mixed in terms of successful interfacial bond enhancement. Deb *et al.* (1995) reported that the use of silane coupling agents with HAp reinforced poly(ethyl methacrylate)-n-butyl methacrylate (PEMA-nBMA) bone cement had mixed results. One main problem has been that the silane-HAp bond is susceptible to hydrolysis (Dupraz *et al.*, 1996). Liu *et al.* (1998) reported that hexamethylene diisocyanate worked effectively as a coupling agent for enhancing bond formation between PEG/PBT and HAp. Bonfield *et al.* have looked at issues involving interfacial bond formation between HAp and PE (Wang and Bonfield, 2001). One promising method for improving the nature and degree of the interfacial bond that develops between HAp and the polymer matrix is to modifiy the surface of the HAp for enhanced bonding with coupling agents. Borum and Wilson (2003a, 2003b) developed a method to modify the surface of HAp with silica for the purpose of modifying the surface chemistry of HAp so that silane coupling agents would adhere better.

Since the turn of the century, there have been further advances in BIN. Researchers have shown that mimicking bone by integrating nanoscale features within BIN, using nanosized HAp and other nanophase reinforcing materials, and providing surface peptide functionality can be used to influence cell activity. Webster *et al.* (1999) showed that osteoblast adhesion was higher on nanograin titania and alumina substrates than on conventional grain sized substrates. The research group of Mikos has investigated the addition of surface modified alumoxane nanoparticles (Horch *et al.*, 2004) and carbon nanotubes (Shi *et al.*, 2006) to polypropylene fumarate (PPF) matrix to improve mechanical properties. Recent work has focused on control of degree of nanosized calcium phosphate dispersion within oligo(poly(ethylene glycol)fumarate) (OPF) polymer matrices (Leeuwenburgh *et al.*, 2007). A high degree of dispersion was achieved by precipitating the calcium phosphate in the presence of OPF monomers. Nanosize HAp particles were incorporated into poly-2-hydroxyethylmethacrylate (PHEMA)/polycaprolactone (PCL) matrix to make new nanocomposites. Porous nanocomposite scaffolds were then produced using a porogen leaching method (Huang *et al.*, 2007). Another method to influence osteoblast cell adhesion involves grafting amino acid sequences onto HAp (Durrieu *et al.*, 1999, 2004). The most common sequence is the RGD peptide sequence which is important for osteoblast cell adhesion. Reyes *et al.*, (2007) investigated the glycine-phenylalanine-hydroxyproline-glycine-glutamate-arginine (GFOGER) collagen-mimetic peptide on titanium implant surfaces. This peptide selectively promotes alpha2-beta1 integrin binding which is important for osteoblastic differentiation. This work and later work (Petrie *et al.*, 2008) demonstrate the important role that peptide functionalization can play in bone healing and opens a wide door of opportunities for enhancing the capabilities of BIN.

## III. Questions at the Interface of Bone and BIN

Our understanding of the intricate processes that orchestrate the healing and remodeling of bone has greatly expanded in recent years. New insights in vital areas related to bone such as the identification of novel ecm proteins and elucidation of their roles in cell signaling and biomineralization (Wendel *et al.*, 1998; Midura *et al.*, 2004) have enhanced our ability to incorporate key aspects of bone architecture and functionality into BIN. However, there are still a number of questions that remain unanswered on both sides of the BIN/bone tissue interface. The questions on the BIN side relate to the optimum bulk and surface chemistries for the mineral phase and the organic matrix, and the overall architecture that best promote seamless integration of BIN into bone tissue. The questions on the bone tissue side involve how growth factors, ecm, cells and biomineralization phenomena work to orchestrate the bone healing and remodeling response. On a more fundamental level, the questions involve the key surface and interfacial characteristics of BINs that best predict and determine optimum hard tissue bioactivity. These key interfacial characteristics relate to factors such as surface charge (Oppermann *et al.*, 1998), chemical composition (Gibson *et al.*, 2002; Hing, 2004), adsorption capacity, matrix orientation and feature size (Fleming *et al.*, 1999; Wilkinson *et al.*, 2002; Andersson *et al.*, 2003), particle and grain size effects (Zhang and Gonsalves, 1997; Wang et al., 1998; Webster *et al.*, 1999; Gao et al, 2003), degree of crystalline order (Wopenka and Pasteris, 2005), porosity (Hing *et al.*, 1999), and solubility effects. Another important question involves how synergistic effects influence materials performance in terms of bioactivity and mechanical performance. There have been a number of fundamental studies that relate mechanical properties of nanocomposites to organic – mineral phase microstructure and length scale (Thompson *et al.*, 2001; Currey *et al.*, 2001; Gao *et al.*, 2003). Sender *et al.* (2007) investigated the mechanical properties of HAp-nylon composites and showed that the important factors in determining nanocomposite strength are related to the filler content, the physical structure of the polymeric matrix, the particles dispersion and the physical interaction strength between organic and inorganic phases. The answers to these questions will help in designing BIN that elicit the optimum bone healing response and expand the range of hard tissue conditions that can be treated. There are a number of excellent review papers that describe additional developments and trends in BIN development (Green *et al.*, 2002; Salgado *et al.*, 2004; Hing, 2004; Chan *et al.*, 2006; Christenson *et al.*, 2007).

## IV. Future Prospects and Vision

The goal of this chapter is to provide an introduction to the field of BIN and to provide an overview of BIN research. Continued research in this field will result in enhanced BIN that will be able to meet the wide range of materials needed for treating bone disorders. Just like there are many types of bone, there will need to be different BIN to help . BIN will require tailoring for specific applications in terms of porosity, strength, bioactivity, and cell recruitment. Projected developments in BIN include better bone graft materials that function on the level of autologous bone, bone implants that can support full body weight, and bone fixation plates and screws that adequately support bone tissue during the healing phase and resorb after their function is complete. One very challenging yet possible goal for BIN is the fabrication of templates that enable the growth and development of fully functioning joints for direct replacement of damaged or malfunctioning joints (hip, knee, elbow, shoulder, finger, toe). This development would require aspects of bioduplication and bone tissue engineering to guide the growth of total replacement joints that contain all the requisite tissue such as cartilage, tendon, and muscle for full, healthy function.

## Acknowledgements

I would like to acknowledge the support of the National Science Foundation (NSF CAREER Award - DMR 0645675).

## References

Ambrosio, A. M., Sahota, J. S., Khan, Y., and Laurencin, C. T. (2001). A novel amorphous calcium phosphate polymer ceramic for bone repair: I. Synthesis and characterization. *J Biomed Mater Res*, **58**(3), 295-301.

Anderson, H. C. (1995). Molecular biology of matrix vesicles. *Clin Ortho Rel Res*, **314**, 266-280.

Andersson, A., Backhed, F., von Euler, A., Richter-Dahlfors, A., Sutherland, D., Kasemo, B. (2003). Nanoscale features influence epithelial cell morphology and cytokine production. *Biomaterials*, **24**, 3427-3436.

Atlan, G., Balmain, N., Berland, S., Vidal, B., and Lopez, E. (1997). Reconstruction of human maxillary defects with nacre powder: Histological evidence for bone regeneration. *C R Acad Sci Paris/Life Sci*, **320**, 253-258.

Bandyopadhyaya, R., Nativ-Roth, E., Regev, O., and Yerushalmi-Rozen, R. (2002). Stabilization of individual carbon nanotubes in aqueous solutions. *Nano Lett,* **2**(1), 25-28.

Bianconi, P. A., Lin, J., and Strzelecki, A. R. (1991). Crystallization of an iorganic phase controlled by a polymer matrix. *Nature*, **349**, 315-317.

Bonfield, W., Grynpas, M. D., Tully, A. E., Bowman, J., and Abram, J. (1981). Hydroxyapatite reinforced polyethylene: A mechanically compatible implant material for bone replacement. *Biomaterials*, **2**(3), 185-186.

Borum, L. and Wilson, Jr., O. C. (2003) Surface modification of hydroxyapatite: I Dodecyl alcohol. *Biomaterials* **24**(21), 3671-3679.

Borum, L. and Wilson, Jr., O. C. (2003). Surface modification of hydroxyapatite: II Silica. *Biomaterials*, **24**(21), 3681-3688.

Calandrelli, L., Immirzi, B., Malinconico, M., Orsello, G., Volpe, M. G., Della Ragione, F., Zappia, V., and Oliva, A. (2002). Biocompatibility studies on biodegradable polyester-based composites of human osteoblasts: a preliminary screening. *J Biomed Mater Res,* **59**(4), 611-617.

Calvert, P. and Mann, S. (1988). Review: Synthetic and Biological Composites formed by in situ Precipitation. *J Mater Sci*, **23**, 3801-3815.

Calvert, P. and Rieke, P. (1996). Biomimetic mineralization in and on polymers. *Chem Mater*, **8**, 1715-1727.

Carlisle, E. M. (1970). Silicon: A possible factor in bone calcification. *Science*, **167**, 279-280.

Castaldini, A. and Cavallini, A. (1985). Setting properties of bone cement with added synthetic hydroxyapatite. *Biomaterials*, **5**, 55-60.

Chan, C. K., Kumar, T. S., Liao, S., Murugan, R., Ngiam, M., Ramakrishnan, S. (2006). Biomimetic nanocomposites for bone graft applications. *Nanomed*, **1**(2), 177-188.

Chiroff, R. T., White, E. W., Weber, J. N., and Roy, D. M. (1975). Tissue ingrowth of replamineform implants. *J Biomed Mater Res*, **9**(4), 29-45.

Christenson, E. M., Anseth, K. S., van den Beucken, J. J., Chan, C. K., Ercan, B., Jansen, J. A., Laurencin, C. T., Li, W. J., Murugan, R., Nair, L. S., Ramakrishna, S., Tuan, R. S., Webster, T. J., Mikos, A. G. (2007). Nanobiomaterial applications in orthopedics. *J Orthop Res,* **25**(1), 11-22.

Currey, J. D., Zioupos, P., Davies, P., Casino, A. (2001). Mechanical properties of nacre and highly mineralized bone. *Proc R Soc Lond B Biol Sci.* **268**(1462), 107-111

Deb, S., Braden, M., Bonfield, W. (1995). Water absorption characteristics of modified hydroxyapatite bone cements. *Biomaterials*, **16**(14), 1095-1100.

de Campos, A. M., Sanchez, A., and Alonso, M. J. (2001). Chitosan nanoparticles: a new vehicle for the improvement of the delivery of drugs to the ocular surface: Application to cyclosporin A, *Int J Pharm,* **224**(1-2), 159-168.

Ducy, P., Schinke, T., and Karsenty, G. (2000). The osteoblast: A sophisticated fibroblast under central surveillance. *Science*, **289**, 1501-1504.

Dupraz, A. M., de Wijn, J. R., v d Meer, S. A., and de Groot, K. (1996). Characterization of silane-treated hydroxyapatite powders for use as filler in biodegradable composites. *J Biomed Mater Res*, **30**(2), 231-238 (1996).

Durrieu, M. C., Pallu, S., Guillemot, F., Breille, R., Amedee, J., Baquey, C. H., Labrugere, C., and Dard, M. (2004). Grafting RGD containing peptides onto hydroxyapatite to promote osteoblast cells adhesion. *J Mater Sci Mater Med.* **15**, 779-786.

Durrieu, M. C., Labrugere, C., Villars, F., Dutoya, S., Guette, A., Bordenave, L., and Baquey, C. (1999). Grafting RGD containing peptides onto hydroxyapatite to promote osteoblast cells adhesion. *J Biomed Mater Res.* **5**, 368-375.

Erbacher, P., Zou, S., Bettinger, T., Steffan, A. M., and Remy, J. S. (1998). Chitosan-based vector/DNA complexes for gene delivery: biophysical characteristics and transfection ability. *Pharm Res,* **15**(9), 1332-1339.

Falini, G. and Fermani, S. (2004). Chitin mineralization. *Tiss Eng,* **10**(1/2), 1-6.

Faatz, M., Gröhn, F., and Wegner, G. (2005). Mineralization of calcium carbonate by controlled release of carbonate in aqueous solution. *Mat Sci Eng C*, **25**(2), 153-159.

Flemming, R. G., Murphy, C. J., Abrams, G. A., Goodman, S. L., and Nealey, P. F. (1999). Effects of synthetic micro and nano-structured surfaces on cell behavior, *Biomaterials*, 20, 573-588.

Gao, H., Ji, B., Jager, I. L., Arzt, E., and Fratzl, P. (2003). Materials become insensitive to flaws at nanoscale: lessons from nature. *Proc Natl Acad Sci USA*, **100**(10), 5597-6000.

Weimer, M. W., Chen, H., Giannelis, E., and Sogah, D., (1999). Direct Synthesis of Dispersed Nanocomposites by in Situ Living Free Radical Polymerization Using a Silicate-Anchored Initiator. *J Am Chem Soc*, **121**, 1615-1616.

Gibson, I. R. and Bonfield W. (2002). Preparation and characterization of magnesium/carbonate co-substituted hydroxyapatite. *J Mater Sci Mater Med*, **13**, 685-693.

Giraud-Guille, M. M. (1994). Liquid crystalline order of biopolymers in cuticles and bones. *Microsc Res Tech*. **27**(5), 420-428.

Glimcher, M. J., Hodge, A. J., and Schmitt F. O. (1957). Macromolecular aggregation states in relation to mineralization: The collagen-hydroxyapatite system as studied in vitro. *Proc Nat Acad Sci:Biochem*, **43**, 860-866.

Gomes, M. E., Reis, R. L., Cunha, A. M. Blitterswijk, C. A., de Bruijn, J. D. (2001) Cytocompatibility and response of osteoblastic-like cells to starch-based polymers: effect of several additives and processing conditions. *Biomaterials*, **22**(13), 1911-1917.

Gopalan Nair, K. and Dufresne, A. (2003). Crab shell chitin whisker reinforced natural rubber nanocomposites. 1. Processing and swelling behavior. *Biomacromol*, **4**(3), 657-665.

Gopalan Nair, K. and Dufresne, A. (2003). Crab shell chitin whisker reinforced natural rubber nanocomposites. 2. Mechanical behavior. *Biomacromol*, **4**(3), 666-674.

Green, D., Walsh, D., Mann, S., and Oreffo, R. O. (2002). The potential of biomimesis in bone tissue engineering: lessons from the design and synthesis of invertebrate skeletons. *Bone*, **30**(6), 810-815.

Greish, Y. E. and Brown, P. W. (2001). Chemically formed HAp-Ca poly(vinyl phosphonate) composites. *Biomaterials*, **22**(8), 807-816.

Helwig, E., Sandner, B., Gopp, U., Vogt, F., Wartewig, S., Henning, S. (2001). Ring-opening polymerization of lactones in the presence of hydroxyapatite. *Biomaterials*, **22**(19), 2695-2702.

Hench, L. L., Splinter, R. J., Allen, W. C., and Greenlee, T. K. (1971). Bonding mechanisms at the interface of ceramic prosthetic materials. *J Biomed Mater Res*, **5**(6), 117-141.

Hench, L. L., Paschall, H. A. (1973). Direct chemical bond of bioactive glass-ceramic materials to bone and muscle. *J Biomed Mater Res*, **7**, 25-42.

Higashi, S., Yamamuro, T., Nakamura, T., Ikada, Y., Hyon, S. H., Jamshidi, K. (1986). Polymer-Hydroxyapatite Composites for Biodegradable Bone Fillers. *Biomaterials* **7**(3), 183-187.

Hing, KA, Best, SM, Bonfield, W. (1999). Characterization of porous hydroxyapatite. *J. Mater Sci Mater Med*, **10**, 135-145.

Hing, K. A. (2004). Bone repair in the twenty-first century: Biology, chemistry or engineering? *Phil Trans R Soc Lond A*, **362**, 2821-2850.

Horch, R. A., Shahid, N., Mistry, A. S., Timmer, M. D., Mikos, A. G., Barron, A. R. (2004). Nanoreinforcement of poly(propylene fumarate)-based networks with surface modified alumoxane nanoparticles for bone tissue engineering. *Biomacromolecules*, **5**(5), 1990-1998.

Huang, J., Lin, Y. W., Fu, X. W., Best, S. M., Brooks, R. A., Rushton, N., Bonfield, W. (2007). Development of nano-sized hydroxyapatite reinforced composites for tissue engineering scaffolds. *J Mater Sci Mater Med*, **18**(11), 2151-2157.

Islam, A. M., Phillips, G. O., Sljivo, A., Snowden, M. J., Williams, P. A. (1997). A review of recent developments on the regulatory, structural, and functional aspects of gum Arabic. *Food Hydrocolloids*, **11**(4), 493-505.

Ito M. (1991). In vitro properties of a chitosan-bonded hydroxyapatite bone-filling paste. *Biomaterials*. **12**(1), 41-45.

Ito, M., Yamagishi, T., Yagasaki, H., Kafrawy, A. H. (1996). In vitro properties of a chitosan-bonded bonefilling paste: studies on solubility of calcium phosphate compounds. *J Biomed Mater Res*, **32**(1), 95-98.

Ito, M., Hidaka, Y., Nakajima, M., Yagasaki, H., Kafrawy, A. H. (1999). Effect of hydroxyapatite content on physical properties and connective tissue reaction to a chitosan-hydroxyapatite composite membrane. *J Biomed Mater Res*, **45**(3), 204-208.

Jarcho, M., Bolen, C. H., Thomas, M. B., Bobick, J., Kay, J. F., Doremus, R. H. (1976). Hydroxylapatite synthesis and characterization in dense polycrystalline form (for bone and tooth implants). *J Mater Sci,* **11**(11), 2027-2035.

Ji, B. (2008). A study of the interface strength between protein and mineral in biological materials. *J Biomech,* **41**(2), 259-66.

Kasuga, T., Ota, Y., Nogami, M., and Abe, Y. (2001). Preparation and mechanical properties of polylactic acid composites containing hydroxyapatite fibers. *Biomaterials*, **22**(1), 19-23.

Khor, E. and Lim, L. Y. (2003). Implantable applications of chitin and chitosan. *Biomaterials*, **24**(13), 2339-2349.

Komarneni, S. (1992). Nanocomposites. *J Mater Chem*, **2**, 1219-1230.

Krishnamoorti, R., Vaia, R., and Giannelis, E.P. (1996). Structure and dynamics of polymer-layered silicate nanocomposites. *Chem Mater*, **8**, 1728-1734.

Lahiji, A., Sohrabi, A., Hungerford, D., and Frondoza, C. (2000). Chitosan supports the expression of extracellular matrix protein in human osteoblasts and chondrocytes. *J Biomed Mater Res*, **5**, 586-595.

Laurencin, C. T., Attawia, M. A., Elgendy, H. E., and Herbert, K. M. (1996). Tissue engineered bone-regeneration using degradable polymers: the formation of mineralized matrices. *Bone*, **19**(1), 93S-99S.

Leeuwenburgh, S. C., Jansen, J. A., and Mikos, A. G. (2007). Functionalization of oligo(poly(ethylene glycol)fumarate) hydrogels with finely dispersed calcium phosphate nanocrystals for bone-substituting purposes. *J Biomater Sci Polym Ed*, **18**(12), 1547-64.

Legeros, R. Z., Trautz, O. R., Legeros, J. P., Klein, E., Shirra, W. P. (1967). Apatite Crystallites: Effects of Carbonate on Morphology. *Science*, **155**(3768), 1409-1411.

LeGeros, R. Z., Trautz, O. R., Klein, E., LeGeros, J. P. (1969). Two types of carbonate substitution in the apatite structure. *Experientia*, **25**(1), 5-7.

Leong, Y. K., Seah, U., Chu, S. Y., Ong, B. C. (2001). Effects of gum arabic macromolecules on surface forces in oxide dispersions. *Coll Surf A: Physicochem Eng Aspects*, **182**, 263-268.

Li, J., Pan, J., Zhang, L., and Yu, Y. (2003). Culture of hepatocytes on fructose modified chitosan scaffolds. *Biomaterials*, **24**(13), 2317-2322.

Liou, S. C., Chen, S. Y., Lee, H. Y., and Bow, J. S. (2004). Structural characterization of nano-sized calsium deficient apatite powders. *Biomaterials*, **25**(2), 189-196.

Liu, Q., de Wijn, J. R., Bakker, D., and van Blitterswijk, C. A. (1996). Surface modification of hydroxyapatite to introduce interfacial bonding with polyactive™ 70/30 in a biodegradeable composite. *J Mater Sci Mater Med*, **7**, 551-557.

Liu, Q., de Wijn, J. R., and van Blitterswijk, C. A. (1998). Composite biomaterials with chemical bonding between hydroxyapatite filler particles and PEG/PBT copolymer matrix. *J Biomed Mater Res*, **40**(3), 490-497.

Lopez, E., Vidal, B., Berland, S., Camprasse, S., Camprasse, G., Silve, C. (1992). Demonstration of the capacity of nacre to induce bone formation by human osteoblasts maintained in vitro. *Tissue Cell*, **24**(5), 667-679.

Lynn, A. K., Nakamura, T., Patel, N., Porter, A. E., Renouf, A. C., Laity, P. R., Best, S. M., Cameron, R. E., Shimizu, Y., and Bonfield, W. (2005). Composition-controlled nanocomposites of apatite and collagen incorporating silicon as an osseopromotive agent. *J Biomed Mater Res A*, **74**(3), 447-453.

Magdassi, S., Bach, U., and Mumcuoglu, K. Y. (1997). Formation of positively charged microcapsules based on chitosan-lecithin interactions. J Microencapsul, 14(2), 189-195.

Marques, A. P., Reis, R. L., and Hunt, J. A. (2002). The biocompatibility of novel starch-based polymers and composites: in vitro studies. *Biomaterials*, **23**(6), 1471-1478.

Mao, H. Q., Roy, K., Troung-Le, V. L., Janes, K. A., Lin, K. Y., Wang, Y., August, J. T., and Leong, K. W. (2001). Chitosan-DNA nanoparticles as gene carriers: synthesis, characterization and transfection efficiency. *J Control Release*, **70**(3), 399-421.

Messersmith, P. B. and Stupp, S. I. (1992). Synthesis of nanocomposites: Organoceramics. *J Mater Res,* **7**(9), 2599-2611.

Midura, R. J., Wang, A., Lovitch, D., Law, D., Powell, K., and Gorski, J. P. (2004). Bone acidic glycoprotein-75 delineates the extracellular sites of future bone sialoprotein accumulation and apatite nucleation in osteoblastic cultures. *J Biol Chem*, **279**(24), 25464-25473.

Mouries, L. P., Almeida, M. J., Milet, C., Berland, S., and Lopez, E. (2002). Bioactivity of nacre water-soluble Matrix from the bivalve mollusk *Pinctada maxima* in three mammalian cell types: fibroblasts, bone marrow stromal cells and osteoblasts. *Comp Biochem Physiol B Biochem Mol Biol*, **132**(1), 217-229.

Moursi, A. M., Winnard, A. V., Winnard, P. L., Lannutti, J. J., and Seghi, R. R. (2002). Enhanced osteoblast response to a polymethylmethacrylate-hydroxyapatite composite. *Biomaterials*, **23**, 133-144.

Mucalo, M. R., Bullen, C. R., Manley-Harris, M., and McIntire, T. M. (2002). Arabinogalactan from the Western larch tree: A new, purified and highly water-soluble polysaccharide-based protecting agent for maintaining precious metal nanoparticles in colloidal suspension. *J Mater Sci*, **37**(3), 493-504.

Muller-Mai, C. M., Stupp, S. I., Voight, C., and Gross, U. (1995). Nanoapatite and organoapatite implants in bone: histology and ultrastructure of the interface, *J Biomed Mater Res*, **29**(1), 9-18.

Muller-Mai, C., Voight, C., De Almeida Reis, S. R., Herbst, H., and Gross, U. M. (1996). Substitution of natural coral by cortical bone and bone marrow in the rat femur. *J Mater Sci Mater Med,* **7**, 479-488.

Muzzarelli, R. A., Zucchini, C., Ilari, P., Pugnaloni, A., Mattioli-Belmonte, M., Biagini, G., and Castaldini, C. (1993). Osteoconductive properties of methylpyrrolidinone chitosan in an animal model. *Biomaterials*, **14**(12), 925-929.

Muzzarelli, R. A., Mattioli-Belmonte, M., Pugnaloni, A., and Biagini, G. (1999) Biochemistry, histology and clinical uses of chitins and chitosans in wound healing. *EXS*, **87**, 251-264.

Muzzarelli, C. and Muzzarelli, R. A. (2002). Natural and artificial chitosan-inorganic composites. *J Inorg Biochem*, **92**(2), 89-94.

Oppermann, D. A., Crimp, M. J., and Bement, D. M. (1998). In vitro stability predictions for the bone/hydroxyapatite composite system. *J Biomed Mater Res*, **42**(3), 412-416.

Oriakhi, C. O., Farr, I. V., and Lerner, M. M. (1996). Incorporation of poly(acrylic acid), poly(vinylsulfonate), and poly(styrenesulfonate) within layered double hydroxides. *J Mater Chem*, **6**(1), 103-107.

Oriakhi, C. (1998). Nanosandwiches. *Chem Brit,* **34**, 59-62.

Ozin, G. A. (1992). Nanochemistry: Synthesis in diminishing dimensions. *Adv Mater*, **4**, 612-649.

Ozin, G. A., Varaska, N., Coombs, N., Davies, J. E., Perovic, D. D., and Ziliox, M. (1997). Bone mimetics: composite of hydroxyapatite and calcium dodecylphosphate lamellar phase. *J Mater Chem*, **7**(8), 1601-1607 (1997).

Park, I. K., Yang, Y., Jeong, H. J., Bom, H. S., Harada, I., Akaike, T., Kim, SI, Cho, C. S. (2003). Galactosylated chitosan as a synthetic extracellular matrix for hepatocytes attachment. *Biomaterials*, **24**(13), 2331-2337.

Pasteris, J. D., Wopenka, B., Freeman, J. J., Rogers, K., Valsami-James, E., van der Houwen, J. A. M., Silva, M. J. (2004). Lack of OH in nanocrystalline apatite as a function of degree of atomic order: Implications for bone and biomaterials. *Biomaterials*, **25**(2), 229-238.

Pedroni, V., Schulz, P. C., Gschaider de Ferreira, M. E., Morini, M. A. (2000). A chitosan-templated monolithic siliceous mesoporous-macroporous material. *Colloid Polym Sci*, **278**, 964-971.

Petrie, T. A., Raynor, J. A., Reyes, C. D., Burns, K. L., Collard, D. M., Garcia, A. J. (2008). The effect of integrin-specific bioactive coatings on tissue healing and implant osseointegration. *Biomaterials*, **29**(19), 2849-2857.

Phillips, G. O., and Williams, P. O. (1993). The specification of the gum arabic of commerce, In K. Nishinari, and E. Doi, (Eds.), *Food Hydrocolloids: Structure, Properties, and Functions* (pp 45-63). NY: Plenum Press.

Ratto, J., Hatakeyama, T., and Blumstein, R. (1995). Differential scanning calorimetry investigation of phase transitions in water/chitosan systems. *Polymer*, **36**(15), 2915-2919.

Reyes, C. D., Petrie, T. A., Burns, K. L., Schwartz, Z., Garcia, A. J. (2007). Biomaterial surface coating to enhance orthopedic tissue healing and integration, *Biomaterials*, **28**(21), 3228-3235.

Rhee, S. H., Lee, J. D., and Tanaka, J. (2000). Nucleation of Hydroxyapatite Crystal through Chemical Interaction with Collagen. *J Am Ceram Soc*, **83**(11), 2890-2892.

Rizzi, S. C., Heath, D. J., Coombes, A. G., Bock, N., Textor, M., and Downes, S. (2001). Biodegradable polymer/hydroxyapatite composites: Surface analysis and initial attachment of human osteoblasts. *J Biomed Mater Res*, **55**(4), 475-486.

Salgado, A. J., Coutinho, O. P., Reis, R. L. (2004). Bone tissue engineering: State of the art and future trends. *Macromol Biosci*, **4**(8), 743-765.

Sanchez, C., Arribart, H., and Guille, M. M. (2005). Biomimetism and bioinspiration as tools for the design of innovative materials and systems. *Nat Mat*, **4**(4), 277-288.

Sarikaya, M., Liu, J., and Aksay, I. A. (1995). Nacre: Properties, crystallography, morphology, and formation. In M. Sarikaya and I. A. Aksay (Eds.), *Biomimetics: Design and Processing of Materials*, (pp. 35-38). Woodbury, NY: *American Institute of Physics* Press.

Sender, C., Dantras, E., Dantras-Laffont, L., Lacoste, M. H., Dandurand, J., Mauzac, M., Lacout, J. L., Lavergne, C., Demont, P., Bernès, A., Lacabanne, C. (2007). Dynamic mechanical properties of a biomimetic hydroxyapatite/polyamide 6,9 nanocomposite. *J Biomed Mater Res B Appl Biomater*, **83**(2), 628-35.

Shi, X., Hudson, J. L., Spicer, P. P., Tour, J. M., Krishnamoorti, R., Mikos A.G. (2006). Injectable nanocomposites of single-walled carbon nanotubes and biodegradable polymers for bone tissue engineering. *Biomacromolecules*, **7**(7), 2237-2242.

Silve, C., Lopez, E., Vidal, B., Smith, D. C., Camprasse, S., Camprasse, G., Couly, G. (1992). Nacre initiates biomineralization by human osteoblasts maintained in vitro. *Calcif Tissue Int*, **51**(5), 363-369.

Sobel, A. E. and Burger, M. (1954). Calcification. XIV. Investigation of the role of chondroitin sulfate in the calcifying mechanism. *Proc Soc Exptl Biol Med*, **87**, 7-13.

Stupp, S. I. and Braun, P. V. (1997). Molecular manipulation of microstructures: biomaterials, ceramics, and semiconductors. *Science*, **277**, 1242-1248.

Stupp, S. I. and Ciegler G. W. (1992). Organoapatites: materials for artificial bone. I. Synthesis and microstructure. *J Biomed Mater Res*, **26**(2), 169-183.

Stupp, S. I., Hanson, J. A., Eurell, J. A., Ciegler, G. W., and Johnson, A. (1993). Organoapatites: Materials for artificial bone. III. Biological testing, *J Biomed Mater Res,* **27**(3), 301-311.

Stupp, S. I., Mejicano, G. C., and Hanson, J. A. (1993). Organoapatites: Materials for artificial bone. II. Hardening reactions and properties, *J Biomed Mater Res*, **27**(3), 289-299.

Subirade, M. and Lebugle, A. (1991). Solubility of an organic phosphate in an apatitic calcium phosphate. *Ann Chim Fr,* **16**, 41-53.

Tachaboonyakiat, W., Serizawa, T., and Akashi, M. (2002). Inorganic-organic polymer hybrid scaffold for tissue engineering--II: partial enzymatic degradation of hydroxyapatite-chitosan hybrid. *J Biomater Sci Polym Ed*, **13**(9), 1021-1032.

Tanaka, H., Yasukawa, A., Kandori, K., and Ishikawa, T. (1997). Surface modification of calcium hydroxyapatite with hexyl and decyl phosphates. *Colloids Surf A: Physicochem Eng Aspects*, **125**, 53-62.

Tanaka, H., Yasukawa, A., Kandori, K., and Ishikawa, T. (1997). Modification of calcium hydroxyapatite using alkyl phosphates. *Langmuir*, **13**, 821-826.

Teitelbaum, S. (2000). Bone resorption by osteoclasts. *Science*, **289**, 1504-1508.

Thompson, J. B., Kindt, J. H., Drake, B., Hansma, H. G., Morse, D. E., Hansma, P. K. (2001). Bone indentation recovery time correlates with bond reforming time., *Nature*. **414**(6865), 773-776.

Urist, M. R. (1965). Bone: Formation by autoinduction. *Science*, **150**(698), 893-899.

Viala, S., Freche, M., and Lacout, J. L. (1998). Preparation of a new organic-mineral composite: chitosan-hydroxyapatite. *Ann Chim Sci Mater*, **23**(1-2), 69-72.

Walsh, W. R., Ohno, M., and Guzelsu, N. (1994). Bone composite behavior: Effects of mineral-organic bonding. *J Mater Sci Mater Med*, **5**, 72-79.

Wan, A. C., Khor, E., and Hastings, G. W. (1998). Preparation of a chitin apatite composite by in situ precipitation onto porous chitin scaffolds. *J Biomed Mater Res*, **41**(4), 541-548.

Wan, A. C. A., Yim, E. K. F., Liao, I. C., Visage, C. L., and Leong, K. W. (2004). Encapsulation of biologics in self-assembled fibers as biostructural units for tissue engineering. *J Biomed Mater Res A,* **71A**(4), 586-595.

Wang, M., Joseph, R., and Bonfield, W. (1998). Hydroxyapatite-Polyethylene Composites for Bone Substitution: Effects of Ceramic Particle Size and Morphology. *Biomaterials*, **19**(24), 2357-2366.

Wang, M., Bonfield, W. (2001). Chemically Coupled Hydroxyapatite-Polyethylene Composites: Structure and Properties. *Biomaterials*, **22**(110), 1311-1320.

Wang, L., Shelton, R. M., Cooper, P. R., Lawson, M., Triffitt, J. T., Barralet, J. E. (2003). Evaluation of sodium alginate for bone marrow cell tissue engineering. *Biomaterials* **24**(20), 3475-3481 (2003).

Webster, T. J., Seigel, R. W., and Bizios, R. (1999). Osteoblast adhesion on nanophase ceramics. *Biomaterials* **20**, 1221-1227.

Weiner, S. (1986). Organization of extracellularly mineralized tissues: a comparative study of biological crystal growth. *CRC Crit Rev Biochem*, **20**(4), 365-408.

Weiner, S. and Price, P. A. (1986). Disaggregation of bone into crystals. *Calcif Tissue Int*, **39**(6), 365-75.

Weiner, S. and Traub, W. (1986). Organization of hydroxyapatite crystals within collagen fibrils. *FEBS Lett*, **206**(2), 262-266.

Weiner, S. and Wagner, H. D. (1998). The Material Bone: Structure-Mechanical Function Relations. *Ann Rev Mater Sci*, **28**, 271-298.

Weiss, I. M., Gohring, W., Fritz, M., and Mann, K. (2001). Perlustrin, a Haliotis laevigata (abalone) nacre protein, is homologous to the insulin-like growth factor binding protein N-terminal module of vertebrates. *Biochem Biophys Res Commun* **285**(2), 244-249.

Wendel, M., Sommarin, Y., Heinegard, D. (1998). Bone matrix proteins: isolation and characterization of a novel cell-binding keratan sulfate proteoglycan (osteoadherin) from bovine bone. *J Cell Biol*, **141**(3), 839-847.

Westbroek, P. and Marin, F. (1998). A marriage of bone and nacre. *Nature*, **392**(6679), 861-862.

Whilton, N. T., Vickers, P. J., Mann, S. (1997). Bioinorganic clays: synthesis and characterization of amino- and polyamino acid intercalated layered double hydroxides. *J Mater Chem*, **7**, 1623-1629.

White, E. W., Weber, J. N., Roy, D. M., Owen, E. L., Chiroff, R. T., White, R. A. (1975). Replamineform porous biomaterials for hard tissue implant applications. *J Biomed Mater Res*, **9**(4), 23-27.

Wilkinson, C. D. W., Riehle, M., Wood, M., Gallagher, J., Curtis, A. S. (2002). The use of materials patterned on a nano- and micro-metric scale in cellular engineering. *Mater Sci Eng C*, **19**, 263-269.

Wilson, Jr., O. C., Olorunyolemi, T., Jaworski, A., Borum, L., Young, D., Siriwat, A., Dickens, E., Oriakhi, C., and Lerner, M. (1999). Surface and Interfacial Properties of Polymer-Intercalated Layered Double Hydroxide Nanocomposites. *Appl Clay Sci*, **15**, 265-279.

Wilson, Jr., O. C. (2002). Soluble Polymers in Bioceramic Systems. In V. A. Hackley, P. Somasundaran, and J. A. Lewis (Eds.) *Polymers in Particulate Systems: Properties and Applications* (Surfactant Science Series, volume 104, pp 325-360). NY: Marcel Dekker, Inc.

Wopenka, B., Pasteris, J. D. (2005). A mineralogical perspective on the apatite in bone. *Mater Sci Eng C*, **25**, 97-104.

Yamaguchi, I., Tokuchi, K., Fukuzaki, H., Koyama, Y., Takakuda, K., Monma, H., and Tanaka, J. (2001). Preparation and microstructure analysis of chitosan/hydroxyapatite nanocomposites. *J Biomed Mater Res*, **55**(1), 20-27.

Zhang, S., Gonsalves, K. E. (1997). Preparation and characterization of thermally stable nanohydroxyapatite. *J Mater Sci Mater Med*, **8**, 25-28.

Zhang, Y., Ni, M., Zhang, M., and Ratner, B. (2003). Calcium phosphate-chitosan composite scaffolds for bone tissue engineering. *Tissue Eng*, **9**(2), 337-345.

Zhao, F., Yin, Y., Lu, W. W., Leong, J. C., Zhang, W., Zhang, J., Zhang, M., and Yao, K. (2002). Preparation and histological evaluation of biomimetic three-dimensional hydroxyapatite/chitosan-gelatin network composite scaffolds. *Biomaterials*, **23**(15), 3227-3234.

In: New Research on Nanocomposites
Editors: Luis M. Krause et al, pp. 79-113

ISBN 978-1-60456-799-1


*Chapter 4*

# STABILITY, ELASTICITY AND STRENGTH OF NANOSTRUCTURED AMORPHOUS CARBON: INSIGHT FROM COMPUTER SIMULATIONS

***Maria Fyta*[*]*, Ioannis N. Remediakis*[†]*, and Pantelis C. Kelires*[‡]**
Physics Department, University of Crete, P.O. Box 2208,
710 03, Heraklion, Crete, Greece

**Abstract**

The present review explores the main characteristics and properties, both structural and mechanical, of nanostructured amorphous carbon; a material with a potential to intermingle properties of carbon nanostructures with those of pure amorphous carbon. Controlling the type and size of the embedded structures opens the possibility to tailor the properties of the amorphous phase. The nanostructures are embedded in an amorphous carbon matrix and are varied both in type and size, ranging from diamond crystallites to nanotubes and concentric-shell graphitic onions bound by dispersion forces, to entirely three-dimensional $sp^2$ covalent conformations, which include porous, open graphene structures with negative curvature. This study is based on classical and semi-empirical simulations, using Monte Carlo schemes within the empirical potential approach and tight-binding molecular dynamics utilizing two different hamiltonians. First, the energetics and stability of the nanocomposites are investigated together with the interaction of the inclusion core with the surrounding matrix and their stress states. Special focus is given to the microscopic character, specifically the behavior of the different carbon hybridization states within these materials, together with the effect of the atomistic features on both the stability and elastic properties. As representative quantities for the latter, the bulk and Young's moduli were calculated for the diamond and nanotube nanocomposites, allowing for comparison to the corresponding values for single phase amorphous carbon. At a final step, the strength under tensile and shear load of dense composites with a diamond inclusion is presented unveiling the fracture mechanism of these highly stable materials.

[*]Present address: Department of Physics, Harvard University, Cambridge, MA, USA
[†]Present address: Department of Materials Science and Technology, University of Crete, Heraklion, Crete, Greece
[‡]Also at: Department of Mechanical Engineering and Materials Science and Engineering, Cyprus University of Technology, Limassol, Cyprus

# 1. Introduction

Interest on nanostructured amorphous carbon (n$a$-C) [1,2] is steadily growing for its properties, both mechanical and electronic, that will supplement those of traditional, single-phase amorphous carbon ($a$-C) [3, 4]. These nanostructured materials are hybrid forms of carbon in which nanocrystallites are embedded in the $a$-C matrix. The variety of embedded nanostructures, and the great number of possible configurations in the matrix, open some new ways to explore the physics. Since some of these nanostructures are proposed to be insulating, while others to be metallic, some of them are rigid, while others are softer, the possibility is opened for tailoring the electronic ane mechanical properties of $a$-C by controlling the type and size of the embedded structures. N$a$-C films can be composed by a variety of carbon nanostructures, ranging from diamond crystallites, to concentric-shell graphitic onions [1, 2] or carbon nanotubes (CNT) [5] bound by dispersion/van der Waals forces, and to entirely three-dimensional sp$^2$ covalent conformations with no van der Waals (vdW) bonding. The latter include porous, open graphene structures with negative curvature (schwarzites) [6–9]. Synthesis of n$a$-C films can be achieved by different methods such as arc plasma methods [10], leading to networks with bucky onions, or by supersonic cluster beam deposition techniques [11, 12], producing schwarzites, or by electron irradiation of $a$-C, producing onions and eventually nucleation of diamond cores [13, 14]. Specifically, the synthesis of nanodiamond composites (nD/$a$-C) and growth mechanisms have been explored in both hydrogenated [15] and pure [16] tetrahedral amorphous carbon (t$a$-C) matrices. The negatively-curved $sp^2$ conformations used here are theoretical structures and correspond to deep metastable phases of carbon. Their occurrence in nature correspond to extreme modes of slow and fast growth, during the accretion growth of graphite. Whereas the negative curvature would derive from seven-membered rings, a positive Gaussian curvature results from the presence of five-membered rings. The positively curved $sp^2$ networks possess vdW forces, examples of which are the carbon nanotubes (both single (SWNT) and multi-wall (MWNT)).

Significant insight was gained through experiments, but several issues about the structure, energetics and stability of n$a$-C are not yet clear. Theoretical work up to now has been mainly focused on the free-standing periodic nanostructures themselves [6–9]. The transformation from graphitic to diamond nanostructures under irradiation conditions is also well studied [13, 14]. However, knowledge about the interaction of the embedded nanostructures with the host amorphous matrix is very limited. This interaction is a crucial factor since it determines the stability of a composite. Variations of the density and coordination of the amorphous matrix are expected to control the stability and the relative energy differences among the carbon phases. From a fundamental point of view, the most important characteristic of n$a$-C is its inhomogeneous nature, characterized by large gradients of density and coordination through the system. From a theoretical point of view, the challenge is to provide a global description of the composite material, extending and/or generalizing concepts and trends which apply to a better understanding of the single-phase system. Here, we present work which aims at such a theoretical description, and is based on Monte Carlo (MC) and tight-binding molecular dynamics (TBMD) simulations.

This review is organized as follows: In section 2. we outline the method we use for our simulations and present some characteristics of single-phase amorphous carbon ($a$-C) which will be the host of the various crystallites also outlined in this part. In sections 3., and 4. we investigate the structural characteristics and the energetics of all composite structures studied here. In section 5. we review their elastic properties, while in section 6. we impose stresses beyond the elastic regime to the diamond composites. We conclude in section 7..

## 2. Methodology

A novel theoretical approach is applied to investigate n$a$-C and to unravel the fundamental principles governing the interaction of nanostructures with the $a$-C matrix [17, 18]. As mentioned previously, we use two different methods including a different description for the interatomic interactions: the Monte Carlo method within the Tersoff empirical potential approach [19], and tight-binding molecular dynamics (TBMD) utilizing separately two hamiltonians, the Mehl & Papaconstantopoulos [20] (TBNRL), and the environment-dependent (EDTB) Wang & Ho [21] models. The empirical Monte Carlo method allows us to simulate large systems, providing greater statistical precision, while with the TBMD schemes, which model the interactions in a quantum-mechanical way, we aim at higher accuracy and thus, bridging the gap between classical and first-principles calculations.

The Tersoff potential is a sum of repulsive and attractive interactions. The coefficient of the attractive term in the pairlike potential (which plays the role of a bond order) depends on the local environment, giving a many-body potential. This potential has proven quite efficient in describing $a$-C [22–24], despite its drawbacks described in the next paragraph. The NRLTB is a non-orthogonal model, using distance- and environment-dependent parameters for transferability between different structures. The parameters for C were obtained by fitting to a first-principles database, which includes the band structure and energies of the diamond, graphite, sc, bcc, and fcc configurations, as well as of the $C_2$ dimer. The fitting was extended to large nearest-neighbor distances, up to 6.6 Å, so the approach properly describes the quantum mechanical effects of bond breaking and microfracture. The calculated phonon frequencies and elastic constants of diamond are reasonably close to experimental values. The EDTB model goes beyond the traditional two-center approximation and allows the TB parameters to change according to the bonding environment. Both accuracy and transferability are improved, as shown from recent successful applications [25]. The TBMD simulations are carried out in the canonical $(N, V, T)$ ensemble, within which the temperature $T$ is controlled via a stochastic temperature control algorithm. The volume $V$ and the number of particles (atoms) $N$ are constant.

A limitation in the Tersoff model is the absence of dispersion (vdW) forces, similar to more refined tight-binding or first-principles methods, which makes the study of structures involving vdW bonding inefficient. There also is an absence of repulsion between nonbonded $\pi$ orbitals [26–28] thus including overly soft bond-bending forces, which favor excessive $sp^2 - sp^3$ mixed hybrids and leat to the overestimation of density for a given $sp^3$ content [24, 29]. This mostly occurs at intermediate $sp^3$ contents, while low-$sp^3$ and high-$sp^3$ (t$a$-C) networks are properly described in this respect. For t$a$-C ($sp^3$ fraction about 80%) the overestimation is $\sim$ 3-4%. This shortcoming is not expected to seriously affect the energetics of the amorphous matrix, because this is determined primarily by the $\sigma$ bond character of the network, which is very well treated by the potential. The absence of $\pi$ repulsion has only a secondary effect on the energetics, through the generation of more dense networks at intermediate contents, not serious enough to change the trends reported here. The $a$-C amorphous networks may not give correct electronic density of states, but they give valuable information on the elastic properties and stress in $a$-C [29, 33], which is relevant to this work.

Here, only carbon nanotubes embedded in $a$-C matrices involve vdW bonding. For the interactions between the CNTs and the host matrix, we use a simple, Lennard-Jones potential, which was shown to succesfully describe the bulk properties of solid $C_{60}$ and multiple-shell carbon fullerenes [30]. The parameters used for this potential are $\epsilon = 2.964\ meV$ and $\sigma = 3.407\ \AA$. We limit the attractive interactions between atoms of the nanotube and atoms in the matrix within a cutoff distance of 0.8 nm. Interactions beyond this distance contribute negligibly to the energy. In the nanotube nanocomposites, we still model the inter-CNT interactions and inter-$a$-C interactions with the Tersoff potential.

## 2.1. Amorphous Carbon

As a step prior to the investigation of the carbon composites, we briefly present the most important features and properties of single-phase $a$-C, which will be crucial for the analysis of the composite material. Amorphous carbon is a non-crystalline and disordered material, whidh does not include crystalline grains. The fact that is has no long-range crystalline order, does not imply that short and medium range order are not preserved. The short range order is directly connected to properties such as optical absorption edges and activated electrical conductivities. Amorphous carbon includes mixtures of hybrid bonding states ($sp^1$, $sp^2$, $sp^3$), thus significant deviations of the interatomic distances and/or interbonding angles with respect to the fully $sp^2$-bonded graphite, as well as to the fully $sp^3$ diamond lattice are evident. This material has been extensively investigated both experimetally and theoretically [23, 31–37]. In high-density (t$a$-C) networks, a clustering of $sp^2$ atoms is evident [35], while in low density networks, which are open structures with voids, long chains and large rings, and with numerous $sp^1$ sites [38]. The variation of $sp^3$ fraction or mean coordination as a function of density is linear over the whole range of possible densities [32,36,37]. Recent tight-binding calculations [39] on $a$-C using the EDTB model, have provided the best up to date agreement with experiment over the whole range of possible densities.

Amorphous carbon networks are mostly isotropic and their bulk moduli follow closely the power-law variation suggested by the constraint-counting model of Phillips and Thorpe [40,41]. According to this model, the bulk modulus vanishes at the floppy transition, having a critical coordination near 2.4, and its variation with the mean coordination has a scaling exponent of 1.5. Specifically, there is considerable softening of all elastic constants as the mean coordination $z$ approaches 3, where the bonding is locally graphitic. As the mean coordination $z$ increases, these as well as the bulk modulus steadily rise and reach high values. In comparison to diamond the moduli of a hypothetically completely tetrahedral $a$-C network, known as the WWW model or amorphous diamond $a$-D [42], remain lower than those of the diamond structure. Especially in the case of t$a$-C, this softening originates mainly from the random orientation of the $sp^3$ hybrids and the accompanying lowering of density and not from the small fraction of $sp^2$ sites [33]. The hardness of $a$-C depends on the $sp^3$ fraction in the sample; the material hardens as it becomes more tetrahedral. Due to the small thickness of these films, though, there are variations in the measured value. For *ta*-C with an $sp^3$ fraction around 85-90 %, a measurement of 80-88 GPa is probably the most accurate [43], because the *a*-C film is the thickest compared to that of other experiments.

Experimentally, *a*-C films are generated from a wide range of deposition processes. Sputtering, evaporation, filtered cathodic arc (FCVA), mass selected ion beams (MSIB), plasma enhanced chemical vapor deposition (PECVD), and laser arcs are a few representative techniques [3, 4]. Theoretically, amorphous films can be generated by quenching a diamond crystal from the melt. This method is not directly related to the kinetics of the growth process of the as-deposited structures, and is a nonequilibrium process, where a metastable state is produced by freezing-in continuously the nonequilibrium local structures. It produces, though, generic structures, which can be definitely associated with the equilibrium ground state of *a*-C films (see relevant discussion in Ref. [29]). The quenching from the melt methodology generates widespread stress distributions, indicating the presence of large internal local stresses, compressive as well as tensile, although the average intrinsic stress of *a*-C films could be zero. Decomposition of the total distribution into contributions of $sp^2$ and $sp^3$ bonded atoms, accredits compressive stress to $sp^3$ sites and tensile to $sp^2$ ones. Therefore, compressive stress generation may promote $sp^3$ bonding [44], while the $sp^2$ sites reduce the total strain. This is related to the fact that the magnitude of the $sp^2$ tensile stresses is on average larger than those of $sp^3$ atoms. This exactly indicates that in mixed phases, the $sp^2$ sites relieve the internal strain energy [33].

## 2.2. Generation of Composite Structures

The quenching from the melt approach is not limited to *a*-C, but can be applied to any nanostructured material [12]. The embedded nanostructures are formed by melting and subsequent quenching at constant volume the corresponding three-dimensional crystal structures, while keeping a certain number of atoms in the central portion of the cells frozen in their ideal crystal positions. In the case of nanotube inclusions, a core is removed from a diamond crystal leaving a void with a shape similar to that of the nanotube. The CNT is then placed into the diamond covering the void. After quenching, which produces amorphization of the surrounding matrix, the cells are thoroughly relaxed with respect to atomic positions and density. Relaxations are particularly important at the interface region, where the crystallites mainly adjust to the host environment. Periodic boundary conditions are applied to the cells. The total number of atoms ranges from $\sim$ 2500 to 4096, depending on the structure, while the number of atoms in the nanocrystals ranges from $\sim$ 300 to 900. The size of the computational cells ranges from about 1.5 to 4.0 nm. For the TBMD relaxed structures, we use smaller cells up to 512 atoms. The diameter (d) of all crystalline inclusions used here, are in the range d=0.6-2.8 nm. For the nanotubes, also structures with the same diameter and different chiralities were used.

Computational cells with various coordinations (densities) of the amorphous matrix can be formed by changing the initial starting density (volume) of the crystal structures. The size (radius) of the nanocrystals is controlled by the number of the atomic shells which are kept frozen during quenching. Their shape depends on the particular crystal symmetry. Diamond inclusions, for example, are almost spherical. On the other hand, the CNTs have open ends and extend through the entire length of the cube. Because periodic boundary conditions (PBC) are applied to the cells, this corresponds to CNTs of infinite length. Due to the PBC, this corresponds to a dense array of CNTs packed in parallel. Although this probably is an idealized model of a CNT nanocomposite, it provides the essential features of the CNT–matrix interaction. The system resembles CNT bundles, but with $a$-C material in between the tubes.

A number of structures that are each one of them *representative of a specific group of conformations*, and which are subsequently embedded in amorphous matrices was selected: the $sp^3$ conformations (a) Diamond (D), and the high-pressure phase (b) BC8 structure for carbon[1], the negatively-curved $sp^2$ conformations with a $D$ periodic minimal surface, that can be described as a three-dimensional linkage of benzenelike rings (c) polybenzene ($6.8^2D$) [7], (d) an analogous network known as $6.8^2P$, that has a $P$ minimal surface and is described as a condensation of truncated-icosahedral $C_{60}$ molecules [7] (e) PCCM (porous conducting carbon modification), an $sp^2$-bonded structure with eightfold rings, reported as a dense packing of compressed nanotubes with shared walls [6], (f) $C_{168}$, a low-density, negative-curvature schwarzite structure ("buckygym") [8], (g) the positively curved $sp^2$ conformations, widely known as carbon nanotubes [5, 46–48] with single (SWNT) and multiple (MWNT) graphitic walls. Structures (c)-(f) do not involve vdW bonding. Polybenzene, $6.8^2P$, and PCCM can be considered as candidate structures resulting from fragmentation and distortion of embedded fullerenes and nanotubes in amorphous matrices. There is extensive experimental research on this method of producing composite carbon films [49–52].

## 3. Structural Characteristics

Six of the generated composite cells with embedded nanostructures are illustrated in Fig.1. In these snapshots, a (a) diamond is embedded in a highly tetrahedral matrix ($z_{am} \simeq 3.8$), (b) polybenzene and (c) PCCM are embedded in a typical $a$-C matrix ($z_{am} \simeq 3.2$), (d) $C_{168}$ schwarzite is embedded in a low-density matrix ($z_{am} \simeq 2.8$), (e) a cross section of a (7,10) SWNT, with a diameter of 1.2 nm, embedded in a matrix with mean coordination $\bar{z} = 3.24$, and (f) a pure $a$-C network with a coordination of 3.69.

In the high density nD/$a$-C, where the diamonds are embedded in a t$a$-C matrix ($sp^2$ fraction less than 30%), the majority of the bonds are $sp^3$-bonded. However, it is essential when investigating the properties of such structures, to take a closer look at the few $sp^2$ atoms. It was found, that these are located in the $a$-C matrix, far from the interface between the matrix and the embedded nanodiamond, and form small clusters with neighboring $sp^2$ atoms, similar to the case of pure t$a$-C networks, where the clustering is also associated to stress relief [29]. In the low coordination/density structures, the amorphous matrices exhibit a characteristic open nanoporous structure resembling random pore schwarzites [12,53]. In the polybenzene and PCCM cells, the matrix is more or less homogeneous, in this sense, but characteristic benzenelike and eightfold rings, respectively, can be easily identified. In the nanodiamond cace, it is interesting to note that there is a remarkable tendency of $sp^3$ atoms in the matrix to gather and enrich the interface region around the nanodiamond. A careful

[1]This structure is fully specified by a single lattice parameter and a single positional parameter $x$=0.0937 in the case of carbon [45].

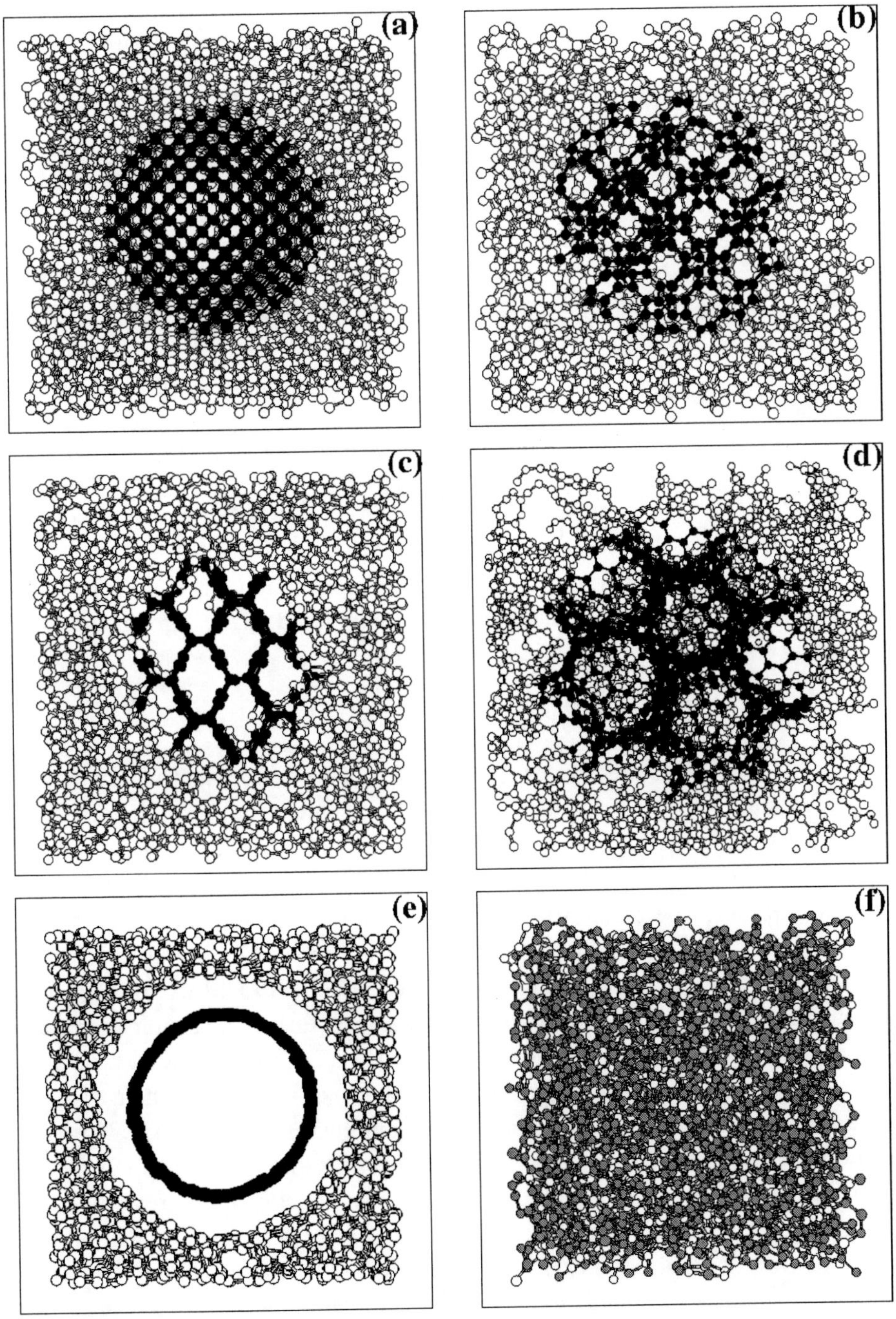


Figure 1. Representative ball and stick models of the carbon nanocomposites studied in this work. Different types of crystallites are embedded in amorphous matrices of different densities: (a) diamond, (b) $6.8^2$D, (c) PCCM, (d) schwarzite $C_{168}$, (e) a (7,10) SWNT, and (f) a single-phase amorphous carbon network. In panels (a)-(e) white and black circles represent the atoms of the matrix and the inclusion respectively, while in (f) in white and black are the $sp^2$ and $sp^3$ atoms of $a$-C.

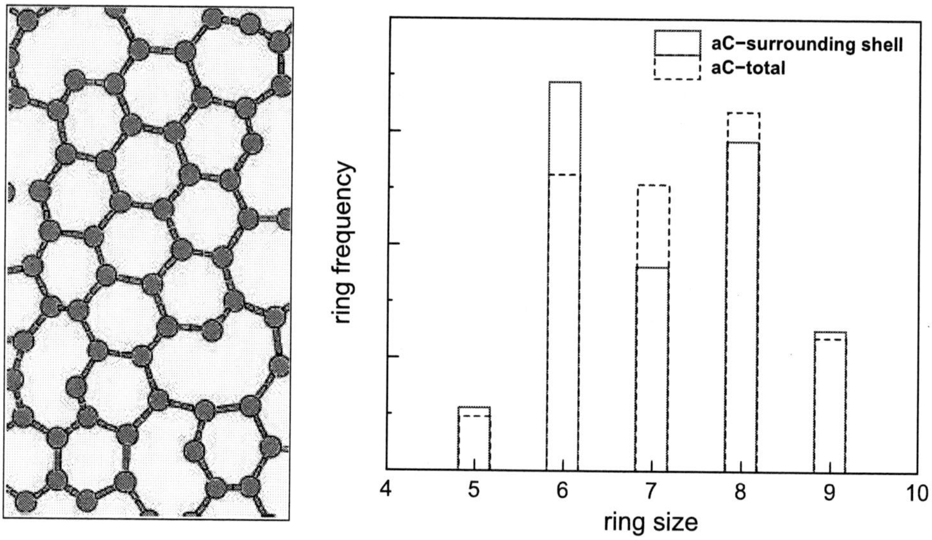


Figure 2. Left panel shows a part of the amorphous shell surrounding a SWNT. Right panel shows the 6-fold ring statistics for the total matrix compared to the corresponding rings in the surrounding shell.

inspection of the local structure reveals an increasing degree of crystallinity, indicating that such tetrahedral atoms may act as possible nucleation centers for expanding the size of the diamond nanocrystal under the appropriate conditions. This will be discussed further in the next sections.

In the case of nanotube composites (CNT/$a$-C), the remarkable feature in the structure is that the atoms of the surrounding matrix reconstruct in such a way as to form an outer wall concentric to the nanotube. A double-wall nanotube is effectively formed. The outer wall is a curved graphitic sheet, but with some degree of disorder, as a close inspection of its structure reveals. Part of this sheet is depicted in Fig. 2 together with a ring statistics analysis, which shows that the sheet mainly consists of six-fold rings, their fraction being considerably higher than the corresponding fraction in the rest of the matrix. However, odd-membered rings do exist, necessitated by the presence of some $sp^3$ atoms in the wall, which serve as bridges with the rest of the matrix. In the lowest-energy structures (see below), the optimum distance between the outer wall and the CNT is close to the graphite interplanar one (0.34 nm). In less stable structures, it deviates from this value depending on the matrix density. This distance is independent of the tube diameter d. We observed a similar pattern in structures with embedded multi-wall carbon nanotubes. The interaction of the MWNT with the matrix forms again a graphitic sheet. The outer interplanar distance is the same with the inner interwall separations in the MWNT. This is independent of the number of concentric shells which compose the MWNT.

The interaction of the matrix graphitic shell with the CNTs does not involve any covalent bonding. In order to unravel the contributions of short- and long-range forces to the formation of this wall, structures were generated by using only the short-ranged Tersoff potential for the interactions through out the system, thus "turning off" the vdW interactions.

**Table 1. Energies $E$ (in eV) and densities $\rho$ (in gr cm$^{-3}$) of bulk carbon crystallites calculated with the VASP ($E^V$, $\rho^V$) and MC/Tersoff methods ($E^T$, $\rho^T$).**

| crystal | $E^V$ | $\rho^V$ | $E^T$ | $\rho^T$ |
|---|---|---|---|---|
| D | 0 | 3.520 | 0 | 3.520 |
| BC8 | 0.7030 | 3.590 | 0.7970 | 3.590 |
| $6.8^2D$ | 0.0920 | 2.130 | 0.0140 | 2.000 |
| $6.8^2P$ | 0.3140 | 2.000 | 0.2240 | 1.860 |
| PCCM | 0.3810 | 2.340 | 0.1020 | 2.230 |
| $C_{168}$ | 0.0893 | 1.189 | 0.1147 | 1.268 |
| G(single-layer) | -0.0200 | 2.040 | -0.1500 | 2.060 |

Again, an outer wall is being formed, but it is defective. Many covalent bridge bonds are generated between the matrix wall and the SWNT, and so the interwall distance necessarily approaches the value of $\sim$ 0.15 nm, much smaller than the interplanar distance in graphite or in MWNTs, and close to the covalent bond length value. This indicates that the repulsive forces of the potential tend to drive the matrix and CNT atoms apart, but its attractive forces impose these artificial bridge bonds, which necessitate the transformation of ideal $sp^2$ sites on the nanotube surface into distorted $sp^3$ sites. Thereby, short-range attractive forces are not involved in the matrix-CNT interaction, and the vdW forces, although weak [54], are driving the perfect reconstruction of the matrix atoms while keeping intact the ideal CNT geometry.

## 4. Energetics and Stability

We begin the analysis with the energetics of $na$-C. The system under study is a composite material, and crystalline nanograins are part of it, thus it is necessary to show that the empirical potential is adequate in describing these phases. In order to establish the reliability of the Tersoff potential in modelling $na$-C, we compare the energetics of various pure crystalline carbon structures to *ab initio* calculations, using the Vienna Ab-initio Simulation Package (VASP) code [55, 56]. We have previously shown [17] that overall, there is very good agreement between the empirical and the *ab initio* calculations both with respect to the energies and the density of the structures, providing confidence in using this potential for the investigation of $na$-C. The agreement is particularly good for the $C_{168}$ schwarzite. Only in the case of PCCM a noticeable relative energy difference exists, but it does not affect the trends in our investigations. The lowest energy structure is polybenzene, followed closely by PCCM and $C_{168}$. The highest energy structure is BC8, as expected, since it is a high-pressure phase. All results are summarized in Table 1.

### 4.1. Formation Energies

A central issue of interest in this work is related to the energetics and stability of the nanostructures. Whether a certain nanocrystal is stable or not can not be told by just comparing the total energy of the composite structure to a reference energy, e.g. diamond. One needs to consider the interaction of the embedded configuration with the host. This is taken care of properly by defining the *formation energy* of a nanocrystal [17], given by

$$E_{form} = E_{total} - N_a E_a - N_c E_c, \tag{1}$$

where $E_{total}$ is the total cohesive energy of the composite system (amorphous matrix plus nanocrystal), calculated directly from the simulation, $E_c$ is the cohesive energy per atom of the respective crystalline phase, $N_c$ is the number of atoms in the nanocrystal, $N_a$ is the number of atoms in the amorphous matrix, and $E_a$ is the cohesive energy per atom of the pure, undistorted amorphous phase (without the nanocrystal) with coordination $z_{am}$. By definition, a negative value of $E_{form}$ denotes stability of the nanostructure, a positive value indicates metastability.

To compute $E_{form}$, we need $E_a$ for an arbitrary coordination of the amorphous matrix, $z_{am}$. To this end, we carried out an energetics study of $a$-C as a function of coordination, or sp$^3$ fraction, over the whole region where $a$-C is a rigid material. Such a study has not been done before. The minimum of energy is, not surprisingly, at $z_{am}$ = 3, where predominance of sp$^2$ bonding is taking place. With increasing coordination, $E_a$ becomes higher because the mixed (sp$^3$/sp$^2$) phase is strained [29, 33]. This is maximized in the tetrahedral region. On the other side, characterized by configurations with many twofold atoms, the variation is more steep with the maximum energy occuring when the networks become floppy ($z_{am} \simeq$ 2.4) [41]. A cubic polynomial best fits the '$z_{am} \geq 3$' region, while a linear fit applies to the '$z_{am} < 3$' region. The functions that were used to fit these two regions, together define the dependence of the energy ($E_a$) of single-phase $a$-C on its coordination ($z_{am}$) are defined through the equations:

$$E_a(z_{am}) = \begin{cases} -1.93678 - 1.6677 z_{am} & z_{am} \leq 3 \\ 0.4206 z_{am}^3 - 4.074 z_{am}^2 + 13.32 z_{am} - 21.61 & z_{am} > 3 \end{cases} \tag{2}$$

We note again, that the absence of $\pi$ repulsion in the potential generates somewhat more dense networks in the intermediate region. This, in turn, overestimates the cohesive energies. Correction of this would slightly shift the energies in this region to higher values, but the trends reported here, based on relative energy differences, remain unaltered.

For each one of the nanostructures considered in this work, we have generated a series of cells with varying $z_{am}$ for the amorphous matrix, but keep the size of the nanocrystal fixed. Then, application of the above methodology yielded $E_{form}$ as a function of $z_{am}$. The energy curves resulting from this analysis are shown in Fig. 3. There are a number of important aspects of these results. The first concerns the question of thermodynamic stability. In principle, astable structure has the potential to expand against the surrounding matrix, provided that the barriers for this transformation can be overcome. Of the six nanostructures considered, only diamond and the $C_{168}$ schwarzite are actually stable ($E_{form} < 0$), the former at high $z_{am}$ (or amorphous density $\rho_{am} \geq 3\ gcm^{-3}$), the latter at low $z_{am}$ ($\rho_{am} \leq 1.2\ gcm^{-3}$). With increasing (decreasing) $z_{am}$ diamonds ($C_{168}$) become more stable, respectively. Overall, diamonds are relatively more stable than schwarzites. The enhanced stability of diamonds in highly tetrahedral networks indicates that once nucleation centers are formed, possibly by intense ion irradiation, large diamond regions can be further grown. The transition of diamonds from metastability to stability at $z_{am} \simeq 3.6$ leads to a quantitative definition of tetrahedral amorphous carbon (t$a$-C), vaguely referred to as the form of $a$-C with a high fraction of sp$^3$ bonding. We define t$a$-C as the form of $a$-C with a fraction of sp$^3$ sites above 60%, in which diamond crystallites are stable. In other words, the predominantly tetrahedral amorphous network of t$a$-C is able to sustain its crystalline counterpart. Networks with sp$^3$ fractions below 60% do not belong to the class of t$a$-C materials, because they can not be transformed into a stable nanocrystalline state.

Other nanostructures studied in this work [(b)-(e)] are metastable ($E_{form} > 0$) through the whole coordination range. However, their curves exhibit a well defined local minimum at which $E_{form}$ is quite small, suggesting that their synthesis is possible, and that they can be maintained in the amorphous matrix under moderate conditions of temperature and pressure. Remarkably, there is a reversal of relative stability between these embedded metastable phases, contrasted to their free-standing energetics. Take for example the case of the BC8 structure, which initially is the highest in energy (see Table 1), but when embedded in a matrix with $z_{am} \simeq 3.4$ becomes only slightly metastable. This demonstrates the significance of the interaction between the crystallite and the host. The interaction becomes favorable when the amorphous environment enforces relaxations in the BC8 nanocrystal that shift the bond lengths and angles to more tetrahedral values. Specifically, there are two sets of bond lengths and angles in the BC8 structure: three bonds are 1.62 Å long and one is 1.44 Å long (relaxed with the present potential). The angle subtended by a long and a short bond is 101.5°, and by two long bonds 116.1°. The embedding induces a shift of the short bond to 1.52 Å, while the longer bond remains unaltered, and of the small angle to 109° (the large angle remains the same). A similar reversal in relative stability occurs between polybenzene and PCCM. The latter becomes more stable in the amorphous matrix.

The formation energies ($E_{form}$) of a number of SWNTs of various diameters versus the matrix coordination are plotted in Fig. 3(b). Well defined minima exist for matrices of $z_{am} \simeq 3.2 - 3.3$, corresponding to densities of about 2.3 – 2.4 gcm$^{-3}$. For all types and sizes of CNTs the formation energy is positive rendering these composites metastable under thermodynamic equilibrium and can be used for practical purposes under normal conditions of temperature and pressure. The more dense (or diluted) the matrix the less stable the system becomes. Indeed, annealing at elevated temperatures results in heavy distortion of CNTs when embedded in dense (diluted) matrices, while nanotubes in matrices with $z_{am}$ near the minima retain their optimal shape. For dense matrices, this result is understood by noting that the more dense the $a$-C matrix the fewer the sp$^2$ sites, and the formation of a graphitic wall to interact *via* vdW forces with the CNT becomes less easy. In fact, the wall in such cases is higly defective, with many sp$^3$ sites, and its distance to the CNT shrinks to $\sim$ 0.24 – 0.28 nm. The incomplete wall – CNT interaction results in CNT distortions. Both deformations in the wall and the CNT increase the energy. Diluted $a$-C matrices, on the other hand, are characterized by a sizeable fraction of sp$^1$ sites, which act as defects on a graphitic plane. We find an increasing number of sp$^1$ sites on the wall with decreasing density. This locally destroys the curved graphitic nature, increasing in this way the energy.

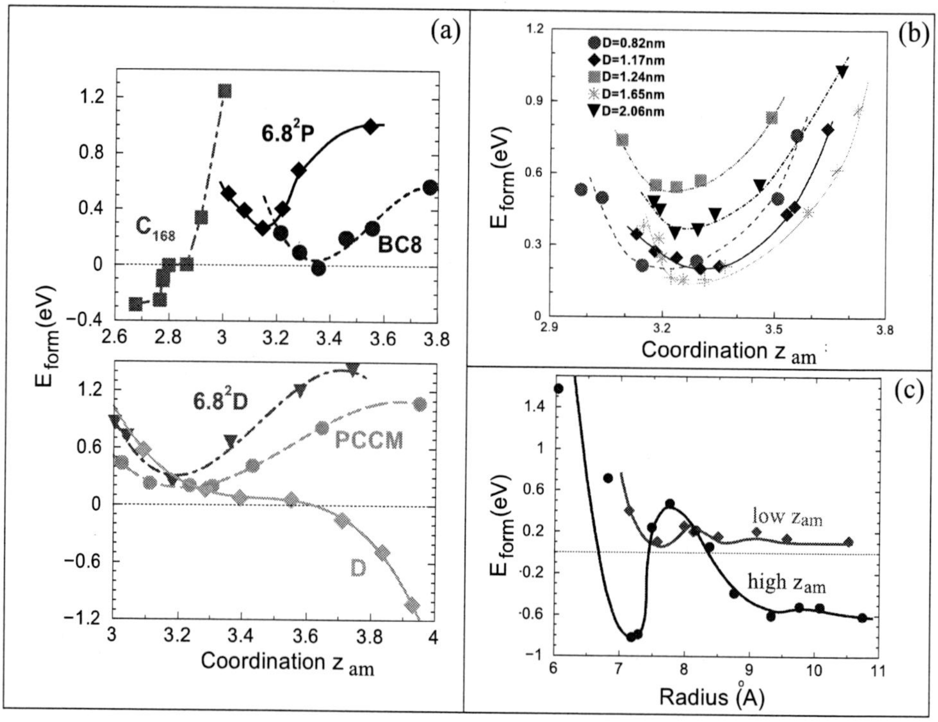


Figure 3. Formation energies for all the carbon composites studied here. Panel (a) shows all the nc/$a$-C, apart from the SWNTs which are shown in (b) for various diameters. Labels are self-explaining according to the text. In (c) the formation energy for diamond crystallites of various diamters embedded in low and high $a$-C matrices, are presented.

## 4.2. Size Dependence

We have checked the stability of the embedded nanocrystals with respect to their size, while keeping the same density (or $z_{am}$) for the amorphous matrix. An example of this analysis is shown in Fig. 3(c) for diamonds. The difference between a stable and a metastable nanocrystal is striking. The former has a deep minimum bounded by a steep branch at small radii, indicating a *critical size* of $\sim$ 6.5 Å, and by a barrier on the right that separates it from other less stable states. Note that, due to the periodic boundary conditions applied to the cells, each crystallite interacts with its neighboring images. This interaction is attractive for certain sizes (distances), e.g. for the minimum and the other low energy states, and repulsive for other (barrier). On the other hand, all states of the second nanocrystal are metastable, and only a shallow local minimum is formed at a *larger* radius than the stable's one. Its critical size is also larger. These findings suggest that by necessity metastable diamonds are on the average larger, *albeit* more deformed, than stable diamonds in a $ta$-C matrix. Note again, that metastable diamonds in a low-$z_{am}$ matrix would be sustained at moderately high temperatures for practical mechanical purposes, but for a better nanostructured material it is preferable to nucleate diamond cores in a $ta$-C matrix. A similar analysis for SWNT composites is also presented in Fig. 3(b). All of them show metastability, without revealing a specific trend with size or chirality. The minima in all cases are

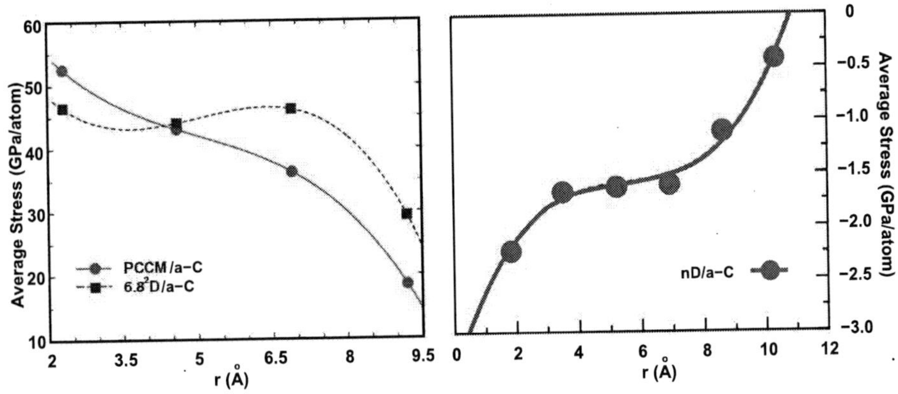


Figure 4. Atomic stresses in PCCM/$a$-C, 6.8$^2$D/$a$-C (left) and diamond (right) nanocomposites, denoted as PCCM/$a$-C, 6.8$^2$D/$a$-C, and nD/$a$-C, respectively. Stresses are averaged over subsequent spherical shells of width 2.1 Å, as a function of the distance from the center. The stresses are calculated at 300 K. Positive values denote compressive stresses. Lines are guides to the eye.

found for low density matrices with a coordination of about 3.2-3.3. This denotes that the interaction of a SWNT with the surrounding matrix is optimum for matrices with a high $sp^2$ components, an environment close to the SWNT structure itself.

### 4.3. Stress States

Inspection of Fig.3(a), reveals another interesting aspect that the formation energies of $6.8^2D$ and PCCM at their minimum are only slightly lower than those of diamond at the same $z_{am}$. This suggests the possibility of transformation of sp$^2$-bonded structures to diamond under the appropriate conditions of pressure. To check whether such conditions exist, the stress fields within the embedded nanocrystals were calculated. High compressive atomic stresses [33] were found in both $6.8^2D$ and PCCM that approach 50 GPa in the center, even at their minimum energy configurations. This is illustrated in Fig. 4(a) for both cases. The compressive stress reduces as we move radially outwards the interface. (For $6.8^2D$, it remains high up to two thirds of the nanocrystal radius). The origin of the compression is attributed to the lower density and the more open configurations of these nanocrystals compared to the amorphous background. On the other hand, diamond nanocrystals in a $ta$-C matrix are somewhat more dense than the tetrahedral amorphous network around them, and they are found to be under slight tensile stress (see Fig. 4(b)).

Such huge compressive stresses in the center of the sp$^2$-bonded structures are proper conditions to locally initiate the sp$^2$ $\rightarrow$ sp$^3$ transition and the nucleation of diamond cores, provided that the barriers for the transformation are overcome by means of external pressure. (Compression is the favorable stress state for the formation of sp$^3$ local geometries) [33]. Let us emphasize that the diamonds produced in this way would be metastable, because their formation energies in this region are positive. Similar compressive conditions are thought to be responsible for the nucleation of diamond cores in the center of spherical

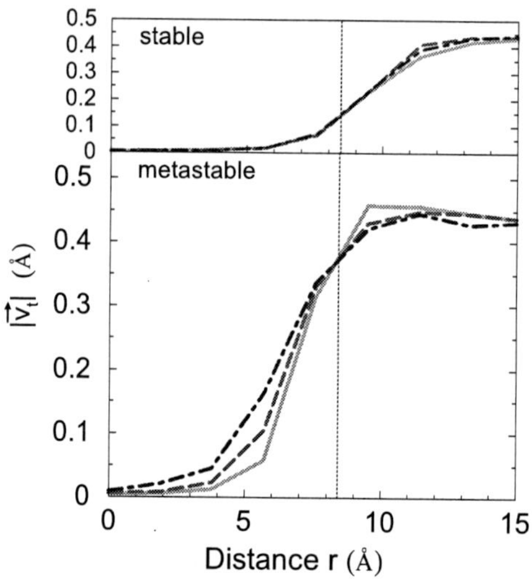


Figure 5. Variation of the magnitude of the tetrahedral vector as a function of distance from the center of diamond nanocrystals. A zero value indicates ideal tetrahedral geometry. Lower panel: metastable case (dilute matrix); upper panel: stable case (dense matrix). The vertical dotted line denotes the position of the interface. Solid (green) line: "as grown"; dashed (red) line: anneal at 1500 K; dash-dotted (blue) line: anneal at 2000 K.

onions under electron irradiation [13].

### 4.4. Annealing Effects

One way of checking the stability of nanocrystals, besides referring to $E_{form}$, is to subject them to thermal annealing. A stable structure should be sustained in the amorphous matrix, while a metastable structure should shrink in favor of the host. We quantitatively demonstrate this by means of the tetrahedral vector [57], which is the sum of the vectors pointing from an atom to its nearest neighbors, and shows the deviation from ideal tetrahedral geometry. The magnitude of this quantity is plotted in Fig. 5 for diamond crystallites, in both the stable and metastable regions, "as-grown" and after annealing. The nanocrystals have a radius of $\sim$ 8.5 Å. Already before annealing, the metastable nanocrystal is heavily deformed in the outer regions near the interface with the amorphous matrix. The stable one is only slightly deformed. Upon annealing, it is evident that the former structure extensively shrinks, and only a small core remains intact, while the stable structure retains its tetrahedral geometry. In the case of nanotube composites, further annealing results in stable structures for an $a$-C coordination close to 3.2-3.3, corresponding to the metastable region. In the case where the $a$-C density is not close to the local minimum for each nanocomposite, the structures tend to collapse after annealing.

In principle, a stable nanocrystal not only retains its shape and structure upon annealing but, as said earlier, should expand against the amorphous matrix. We have not seen this to happen dynamically in the simulations, which were carried out at relatively high temperatures (1500-2000 K) to accelerate the deterioration of the unstable nanocrystals. Actually, such a thermal annealing experiment, either in the laboratory or on the computer, should take place at temperatures below 1200 K, i.e., in the region where the tetrahedral amorphous network is stable against the transformation of $sp^3$ to $sp^2$ bonding [33]. It is doubtful that growth of diamonds will be observed dynamically in simulations at these temperatures because of the enormous computational time needed, but we expect this to happen experimentally.

In the case of CNT/$a$-C, note again (Fig.3(b)), that as the density of the structures rises the formation energy increases significantly. Thus, the more dense (or dilute) the matrix the less stable the system becomes. Indeed, annealing at elevated temperatures results in heavy distortion of CNTs when embedded in dense (diluted) matrices, while nanotubes in matrices with coordinations near the minima retain their optimal shape. For dense matrices, this result is understood by noting that the more dense the a-C matrix the fewer the $sp^2$ sites, and the formation of a graphitic wall to interact *via* vdW forces with the CNT becomes less easy. In fact, the wall in such cases is highly defective, with many $sp^3$ sites, and its distance from the CNT shrinks to $\sim$ 0.24-0.28 nm. The incomplete wall - CNT interaction results in CNT distortions. Both deformations in the wall and the CNT increase the energy. Diluted a-C matrices, on the other hand, are characterized by a sizeable fraction of $sp^1$ sites, which act as defects on a graphitic plane. An increasing number of $sp^1$ sites on the wall with decreasing density was found. This again destroys locally the curved graphitic nature, increasing the energy.

## 5. Elastic Properties

Having established the stability of a wide range of the carbon composites studied here, we turn to their mechanical properties, focusing first on the elastic regime. From this section and on we deal only with composites that contain either diamond or nanotube inclusions. Elasticity [58] is an appropriate starting point for discussing the deformation, since this phenomenon is displayed by all solid materials in at least the initial stage of stressing. For cubic lattices the only three independent elastic constants are $C_{11}$, $C_{12}$, and $C_{44}$. All other vanish, because the energy of a cubic crystal cannot change when the sign of a single component of the displacement field along any of the cubic axes is reversed. By applying the appropriate deformations to the system, the corresponding elastic modulus can be extracted. A hydrostatic deformation, for example, leads to the bulk modulus $B$ which is given from the derivative of the total energy with respect to the volume curve. It is often more convenient, instead of performing the differentiation to determine the minimum energy and the corresponding density and bulk modulus of a structure by fitting the energy versus volume data to Murnaghan's equation of state [59]. This is the procedure we use here. The shear modulus $\mu$ and $C_{44}$ result from orthorhombic (or tetragonal) and monoclinic deformations, respectively. Once the bulk, shear and $C_{44}$ moduli are known, these lead to the remaining elastic moduli of a cubic structure: the Young's modulus $Y$, elastic constants $C_{11}$, $C_{12}$, and Poisson ratio $\nu$. These are given through the following equations:

$$Y = 3B(1 - 2\nu) \tag{3a}$$

$$\nu = \frac{C_{12}}{C_{11} + C_{12}} \tag{3b}$$

$$\mu = \frac{C_{11} - C_{12}}{2} \tag{3c}$$

$$B = \frac{C_{11} + 2C_{12}}{3} \tag{3d}$$

In the case of cylindrical structures, like CNT, the decomposition of their moduli into axial (along the tube axis) and transversal components are of interest. The Young's modulus is defined as the fraction of uniaxial stress over uniaxial strain, and can be defined as: $Y = \frac{1}{V_0}\frac{\partial^2 E}{\partial \epsilon^2}\Big|_{\epsilon=0}$ where $\epsilon$ is the strain, $V_0$ the equilibrium volume and $E$ the strain energy. For cylindrical structures $V_0 = 2\pi LR\delta R$, where L is the length, R the radius and $\delta R$ the shell thickness, taken as the graphite interplane distance ($\delta R = 0.34\ nm$). This is the most common convention in the case of CNTs and and the one that is adopted here. The radial Young's modulus can also be derived by applying a hydrostatic pressure P from the following expression [60]: $Y_\perp = \frac{P}{\epsilon_r} = \frac{PR}{u_r}$ In this case $\epsilon_r$ is the strain in radial direction, $u_r$ the radial displacement, and $R$ is again the tube radius. The Poisson ratio, for an axial strain is defined as [61]: $-\sigma\epsilon = \frac{R(\epsilon) - R_{eq}}{R_{eq}}$, with $R_{eq}$ the equilibrium tube radius and $R(\epsilon)$ the radius at strain $\epsilon$. This ratio is defined as the contraction normal to the applied strain divided by the extension in the direction of the applied strain. According to the theory of isotropic elasticity, its values should be in the range from -1 to 0.5.

### 5.1. Bulk Modulus

We begin our analysis with the calculation of the bulk modulus $B$ of the nanocomposites. A high value of $B$ implies that the structure can not be easily compressed. The bulk modulus for crystalline diamond calculated with EDTB and the Tersoff potential are 428 GPa and 443 GPa, respectively, while the values for the WWW network are 361 GPa and 365 GPa. The results for diamond are in very good agreement with experimental results The WWW model is a theoretical, fully tetrahedral network [42], whose properties are an upper value of the properties of single phase $a$-C and serve as a comparison to those of the composites. Calculations for the Young's modulus with the Tersoff potential give values of 1060 GPa for diamond and 824 GPa for the WWW network [33]. The values for the equilibrium bulk, shear, and Young's moduli and the density of diamond with the TBNRL method are 480 (443) GPa, 494 (476) GPa, 1300 (1145) GPa, and 3.65 (3.51)gcm$^{-3}$, respectively, compared to the experimental values given in parentheses. There is some overestimation in those values, but the trends we are interested in are preserved. The modulus $B$ for a free standing CNT of $\sim$ 1 nm, with the Tersoff potential, was found equal to $\sim$ 200 GPa. These values are in very good agreement with experimental results for the former, and with *ab initio* simulations for the latter [62].

Fig. 6 shows the variation of $B$ for diamond and CNT nanocomposites, as well as for single-phase $a$-C. A composite with a 1.05 nm diamond and matrix with a 30% $sp^2$ fraction has a bulk modulus of 300 GPa. This value for a pure $a$-C network with the same $sp^2$ fraction is about 12% lower. The same composite with a lower $sp^2$ component (11%) in the matrix has an even higher bulk modulus (357 GPa). Comparison with pure $a$-C reveals again that the modulus is about 11% higher. As the size of the diamond increases,

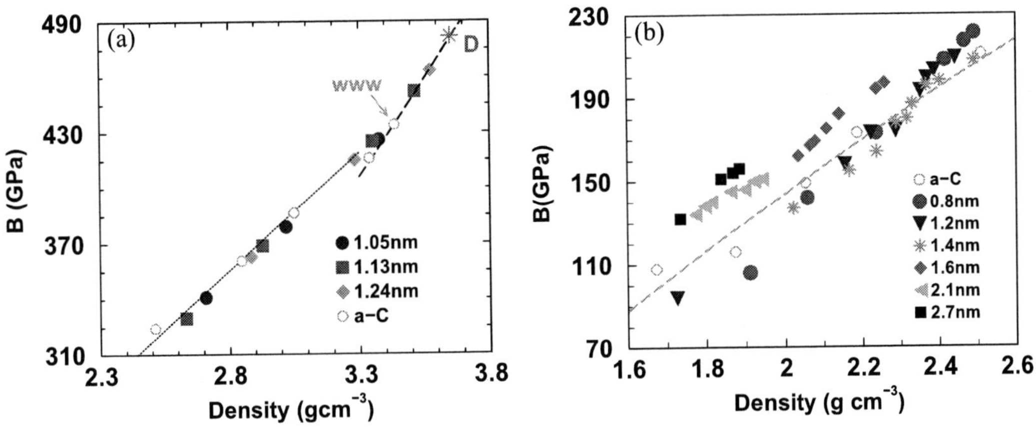


Figure 6. Bulk moduli $B$ for (a) diamond and (b) single-wall carbon nanotube nanocomposites as functions of the density ($\rho$) of the composites. The values for single-phase amorphous carbon (in orange open circles) including the WWW point and the value for bulk diamond (D) in (a) are shown for comparison. In (a) the dotted and dashed lines show the B $\sim \rho$ and B $\sim \rho^{5/3}$ variations, respectively (see text). In (b) the dotted (orange) line is a fit to the $a$-C data.

the bulk modulus of the composite increases and is again enhanced compared to single-phase $a$-C. The modulus is 346 GPa for a 1.24 nm diamond embedded in an $a$-C matrix with 20% threefold atoms, which is increased about 16% over the $a$-C value. It has been predicted that the elastic response of the nD/$a$-C networks to hydrostatic deformation is controlled by the nanoinclusions [63]. Their bulk moduli are enhanced over the values for single phase $a$-C. Diamond is the least compressive material known, and this property is mirrored in nanostructured carbon. Clearly, the moduli of $n$-D/$a$-C networks are considerably higher than those of $a$-C. A mean increase over 10% is evident. The enhancement becomes stronger as the $sp^3$ fraction, or the density, increases. A similar effect is seen as the size, or volume fraction, of the nanocrystals increases. The values for diamond and WWW are shown for comparison. The moduli of $ta$-C networks ($sp^3$ fraction $\geq$ 60% [17]) are quite high, and that of WWW reaches $\sim$ 90% of the diamond value, confirming previous calculations [24, 33]. Impressively, beyond a certain point, $B$'s are shown to exceed the WWW value and approach that of diamond. The enhancement is stronger for dense composites (lower $sp^2$ component) and for larger diamond sizes. Some representative values for $B$ and the Young's modulus $Y$ with respect to the diamond size are shown in Table 2 for composites with similar high density $a$-C matrices ($sp^2$ about 12%), that differ in the size of the diamond they include. As mentioned in previous sections, the $sp^2$ atoms are mainly located far from the interface with the inclusion and are found in clusters with neighboring $sp^2$ sites. These clusters are known to be under tensile stress [33], thereby providing softening mechanisms. Hence, the lower the density of a single-phase $a$-C film, the lower its compressibility. These sites, in the case of nanodiamonds embedded in $a$-C matrices lower the resistance of the whole structure to compression.

The universal dependence of B on density can be understood by considering the micro-

**Table 2. Bulk and Young's moduli for various nD/$a$-C composites embedded in amorphous matrices with an $sp^3$ component of 12%.**

| d (nm) | B (GPa) | Y (GPa) |
|---|---|---|
| 1.9 | 370 | 876 |
| 2.0 | 374 | 884 |
| 2.2 | 375 | 901 |

scopic response of the material to the external pressure. We can think of two regimes: in low-density materials, the pressure is undertaken by appropri- ately adjusting the volume of the void regions of the material. Such voids exist in every low-density material and are a result of induced dipole (van der Waals) interactions. On the other hand, for dense materials, strong covalent bonds have to be deformed, resulting in higher bulk moduli. In this case, the resistance of the electrons to compression fol- lows from their quantum nature and the Pauli principle. In the first case, the scaling of B with respect to can be found by considering a model solid bonded exclusively through van der Waals (vdW) interactions. Using the Lennard- Jones potential we can find that B $\sim \rho$ [64], where $\rho$ is the density of the solid. On the other hand, for high densities, we can get the correct scaling by using the free-electron approximation: in this case, B $\sim \rho^{5/3}$ [64]. This picture is demonstrated in Fig. 6: B$\sim \rho$ for $\rho \lesssim 3.3$ gr/cm$^3$ , while B$\sim \rho^{5/3}$ for $\rho \gtrsim 3.3$ gr/cm$^3$ . These relationships hold with surprisingly good accuracy.

A systematic softening of the materials with respect to $a$-C is observed for CNT inclusions of larger diameters (see Fig. 6(b)). This behavior is the result of a compromise between two competing factors, i.e., the matrix density and the tube diameter. Because we have a high density of tubes in the medium, their size becomes a crucial factor. The larger the tube diameter the smaller the contribution of the matrix to the material's strength, and vice versa. By having larger tubes at low densities, the contribution of the soft matrix declines, and that of the tubes strengthens the material. Smaller embedded tubes operate in the opposite direction, despite the fact that the moduli of isolated CNTs increase as their diameter decreases. In more dense and rigid matrices, inclusion of smaller tubes enhances the strength both because such tubes are very rigid and because the matrix contribution dominates. The rigidity of nanotube nanocomposites is not very high compared to other carbon nanocomposites, such as the nanodiamond ones, due to the fact that the former are more dilute. There is also a region of appreciable volume inside and directly outside the nanotube were no atoms occur. The empty space together with the relative high $sp^2$ ratio, lower the bulk moduli of such materials, making them softer. CNTs are curved graphitic sheets having hexagonal symmetry thus their elastic properties are not affected by the different chiralities. In order to examine this effect, 3-4 different nanotubes were chosen for the analysis. Their diameters were very similar and close to 0.42 nm, which is small but larger than the diameter of the smallest stable nanotube with a diameter of 0.3 nm [66]. For CNTs of similar diameter, but different chiralities, no significant differences were found in their formation energy and rigidity. Improved rigidity is achieved by having the tubes in a "bundle" arrangement without the need to interlink them [10]. In addition, the CNT composite

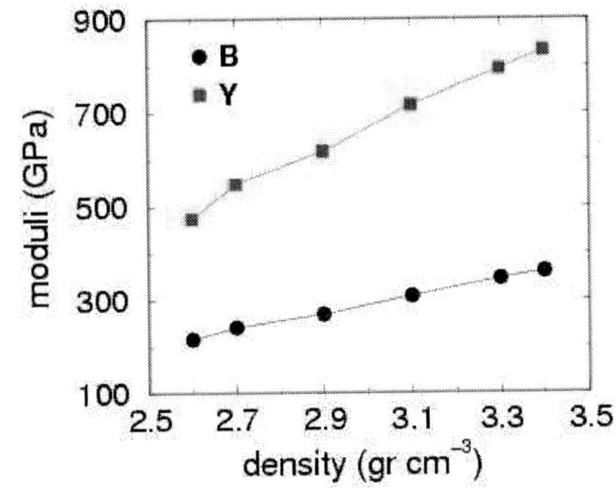


Figure 7. Bulk and Young moduli for nanocomposites of different densities, with diamond inclusions of d=1.7 nm in diameter.

material exhibits much higher bulk moduli than nanotube bundles ($\sim$ 40 GPa) [67].

### 5.2. Young's Modulus

The trends found in the bulk moduli of the diamond and SWNT nanocomposites are similar for the Young's ($Y$) moduli. For nD/$a$-C, $Y$ is found to increase with increasing diamond size or density, as presented in Table 2 and Fig. 7. Note again that $Y$ for the WWW model is 824 GPa, which seems to be overcome for dense composites including large diamonds. In the figure, $Y$ is plotted together with $B$ for a nanodiamond with d=1.7 nm. Again, both $B$ and $Y$ increase for more dense materials, with the slope for the $Y$ variation being higher than the corresponding for $B$. This is a signature of the fact that it is easier to deform the shape of the material (a deformation which leads to $Y$) rather than changing its volume (a deformation which leads to $B$).

In SWNT composites, the increase of the Young's modulus components with the matrix density and the nanotube diameter is again related to the contribution of the surrounding material. For a composite with a SWNT inclusion of d=1.65 nm, we calculate separately the Young's moduli along ($Y_{\|}$) and normal ($Y_{\perp}$) to the tube axis. For these calculations, the empty regions were subtracted from the volume fractions of the constituents. Representative values are given in Table 3. On the whole, SWNTs are stiffer than steel, having a Young's modulus of around 1-1.25 TPa [65,68] and high anisotropy. The increase of the Young's modulus components with the matrix density and the nanotube diameter is again based on the contribution of the surrounding material. For the dense structures, the enhancement originates from the increase of the matrix stiffness, while when comparing structures with different (in diameter) SWNTs, the total density for large nanotubes represents stiffer matrices. The interest, though is in low density matrices, where the nanotubes are stable. Nevertheless, it must be noticed, that also for these densities the composites are stiffer than pure $a$-C materials, as the Young's modulus (the average and it components in transverse directions) is higher than in $a$-C.

**Table 3. Transverse and axial Young's moduli for nanotube composites of various diameters d, coordinations (z) and densities ($\rho$).**

| d (nm) | 0.83 | 1.17 | 1.17 | 1.17 | 1.65 | 1.65 | 2.06 | 2.75 |
|---|---|---|---|---|---|---|---|---|
| z | 3.24 | 3.14 | 3.28 | 3.44 | 3.15 | 3.24 | 3.26 | 3.22 |
| $\rho$ (gcm$^{-3}$) | 2.62 | 2.22 | 2.41 | 2.55 | 1.99 | 2.14 | 1.94 | 1.73 |
| $Y_{\parallel}$ (GPa) | 569 | 476 | 601 | 645 | 478 | 564 | 490 | 404 |
| $Y_{\perp}$ (GPa) | 380 | 290 | 364 | 373 | 231 | 306 | 258 | 216 |

### 5.3. Anisotropy and Rule of Mixtures

In the case of composites, the simplest model for predicting unidirectional continuous fiber composite properties is the *rule of mixtures* [69]. According to this rule, the composite is a simple mixture of two phases with each phase exhibiting mechanical and thermal behaviors, not influenced by their presence in the mixture. As a consequence, the effects of microstructural details, plasticity, and thermal softening are not taken into account [70]. This rule can be applied to mechanical, thermal and electrical properties and has two forms: one for the longitudinal direction and one for the transverse direction. The composite property is estimated as the sum of the responses of the composite components weighted by the component volume fractions. For a composite, where $\lambda_1,\ \lambda_2$ are the volume fractions of the two constituents and $E_1^{\perp},\ E_2^{\perp}, E_1^{||},\ E_2^{||}$, their transversal and longitudinal elastic moduli, respectively, its transverse elastic moduli can be estimated by the rule of mixtures, as follows:

$$E^{||} = \lambda_1 E_1^{||} + \lambda_2 E_2^{||} \tag{4a}$$

$$\frac{1}{E^{\perp}} = \frac{\lambda_1}{E_1^{\perp}} + \frac{\lambda_2}{E_2^{\perp}} \tag{4b}$$

A diamond nanocomposite with a crystalline volume fraction of 33% and a total density $\sim$ 3.3 gcm$^{-3}$ ($sp^2$ component about 25%) has a modulus of 346 GPa. The rule of mixtures for this composite leads to a value of 354 GPa. Similar are the results for all dense structures. For an nD/$a$-C with a total $sp^3$ fraction of 50 % the deviation of the calculated moduli from those given through the rule of mixtures is close to 40 GPa and gets larger with a decreasing $sp^3$ fraction. These features indicate that diamond inclusions are not strongly affected by their presence in high density $a$-C matrices, while this is not the case for dilute $a$-C matrices in which the nanodiamonds are unstable and quite distorted, thus influenced by the embedding effect.

On the other hand, the rule of mixture does not hold for composites with nanotube inclusions. This is a direct evidence that CNTs are strongly influenced by their presence in the mixture and show a stability which is in average lower than that for diamond inclusions. The axial $Y_{||}$ and transversal $Y_{\perp}$ Young's moduli of a composite with a SWNT inclusion of a d=1.65 nm diameter are 564 GPa and 306 GPa, respectively. The values obtained by the rule are $Y_{||}$ = 297 GPa and $Y_{\perp}$ = 323 GPa. In general, the axial moduli compared to the transverse part deviate more from the values calculated through the rule of mixtures and the anisotropy of the elastic moduli of free standing nanotubes is enhanced by their embedding into the matrix. In the axial direction the embedding effect is stronger and leads to a strengthening along this direction, while a slight softening is evident along transverse directions. In transverse directions, the matrix contribution is stronger as it leads to a softening along these directions. It was found that the axial moduli are the ones that differ mostly from the values given by the rule of mixtures. This fact indicates that the embedding effect is higher and leads to a strengthening especially along the axial direction, while a slight softening is evident along the transverse directions. The anisotropy is, in general, enhanced. This argument holds for nanocomposites with the minimum formation energy (most stable and interesting for practical purposes), but also for the remaining (less stable) cases. Nevertheless, the variations from the rule of mixtures are a direct evidence that nanotubes are strongly influenced by their presence in the mixture. The matrix is an isotropic medium and the anisotropy is induced by the CNT. It is stronger for the axial part of the Young's modulus due to the infinite length of the CNT in this direction, along which the CNT actually dominates over the matrix.

Quantitatively the amount of anisotropy in a nanocomposite is defined by the ratio $A = Y_{\perp}/Y_{\parallel}$. Both $a$-C and diamond are isotropic, hence their compound will be isotropic as well. The opposite trend holds for the nanotube composites. A SWNT with a diameter of d=0.8 nm has $Y_{\parallel} = 1100$ GPa and $Y_{\perp} = 620$ GPa, yielding $A = 0.56$, while the nanocomposite with the same CNT and a matrix of density 2.6 gcm$^{-3}$ has $Y_{\parallel} = 570$ GPa, $Y_{\perp} = 380$ GPa, and $A = 0.67$. The anisotropy increases with increasing tube diameter (tube volume fraction), because the tube contribution overwhelms the isotropic matrix part. Another measure of elastic anisotropy for more isotropic structures is provided by the factor $\eta = 2C_{44}/(C_{11} - C_{12})$. A value of unity corresponds to a completely isotropic medium. By calculating the elastic constants for a number of nanocomposites, their high anisotropy was confirmed. For example, the structure shown in Fig. 1(e) a has $\eta = 0.75$. We found for all pure $a$-C networks $\eta \simeq 0.98$.

## 5.4. Elastic Constants

We next calculate the remaining elastic constants. Small strains are applied to the cubic cells, leading to small deviations from cubic symmetry, allowing us again to refer to cubic elastic moduli. We apply the appropriate deformation to the system and compute its total energy as a function of the imposed strain. The curvature of this function at its minimum yields the desired modulus. As mentioned previously, for structures with a cubic symmetry, there are three independent elastic constants, namely $C_{11}$, $C_{12}$ and $C_{44}$, while for an isotropic material, such as $a$-C or n$a$-C, there are only two [71].

For diamond composites, we found that all elastic constants increase with increasing density, similar to the behavior of the bulk modulus as a function of density. For an isotropic material, $2C_{44} = C_{11} - C_{12}$. This relationship holds within 4% or less, for all calculated moduli, demonstrating that nD/$a$-C is a highly isotropic material. The moduli $C_{11}$ and $C_{44}$ are both associated with changes in shape, and this is why their values for nD/$a$-C are much lower than the corresponding values for diamond, where strong directional bonds are bent. On the other hand, $C_{12}$ describes simultaneous elongation along two axes without shearing,

**Table 4. Elastic moduli (in GPa) of diamond composites with a 1.7 nm diamond inclusion and a matrix with density $\rho$ (in ($gcm^{-3}$)).**

| $\rho$ | B | $C_{11}$ | $C_{12}$ | $C_{44}$ | shear | $\nu$ | Y |
|---|---|---|---|---|---|---|---|
| 2.6 | 215 | 493 | 76 | 210 | 208 | 0.13 | 473 |
| 2.7 | 241 | 566 | 78 | 240 | 244 | 0.12 | 547 |
| 2.9 | 268 | 636 | 84 | 289 | 276 | 0.12 | 616 |
| 3.1 | 307 | 735 | 93 | 341 | 321 | 0.11 | 714 |
| 3.3 | 344 | 817 | 107 | 366 | 355 | 0.12 | 792 |
| 3.4 | 359 | 856 | 110 | 387 | 373 | 0.11 | 831 |

and the value of this modulus for diamond follows the trend observed for nanocomposite carbon. The Poisson ratio for diamond and nanotube composites is between 0.1 and 0.3.

In order to elucidate the trend of increasing moduli with increasing density, the calculated elastic constants for a composite with a 1.7 nm diamond inclusion are presented in Table 4. These clearly reveal the hardening of the composite as the $a$-C matrix becomes more dense and the behavior is similar to the case of pure $a$-C [33]. A nanodiamond with d$\sim$ 1.7 nm in diameter embedded in a dense matrix (about 80% of the atoms are four-folded) has a bulk modulus close to 359 GPa and a Young's modulus of 831 GPa. These values for a nanodiamond with a diameter of d=2.1 nm in a similar matrix rise up to 371 and 864 GPa, respectively. The shear modulus for these two composites increases from 373 to 388 GPa. For high densities, the increase over the $a$-C values is $\sim$ 8% for a 1.7 nm diamond inclusion and 13% for a 2.1 nm nanodiamond. The effect of the diamond core on the properties of a composite should be almost the same for all densities (especially for the stable/dense nanocomposites) and the variation of the elastic constants with density should be determined mainly by the matrix, which becomes more incompressible as it gets more dense. In dense composites, both the inclusion and the surrounding matrix are quite incompressible and the whole structure is quite rigid, while in the opposite limit not only the matrix becomes more dilute, but the inclusion becomes more deformed and contributes to the softening of the whole structure. These softening mechanisms are again related to $sp^2$ sites. Therefore, composites with low density (high $sp^2$ content) matrices have elastic constants that are lower than those of single-phase $a$-C. On the other and, the Poisson ratio shows no significant fluctuations, which indicative of the fact that the enhancement in both $C_{11}$ and $C_{12}$ is similar and contraction versus extension is almost the same for all nD/$a$-C structures. Also, the shear value is smaller than the bulk modulus for the same structure, verifying what is known for most materials: they resist a change in volume (determined by the bulk modulus) more than they resist a change in shape. This is also the case for CNT composites.

A first look at the CNT/$a$-C structures reveals that their elastic moduli are in general smaller than those of nD/$a$-C. Two are the main reasons behind this trend. Diamonds are stable in high density, thus more rigid matrices, but CNTs can be found mainly in less dense (more soft) matrices. A second important factor in CNT/$a$-C is related to the empty regions of appreciable volume inside and directly outside the nanotube were no atoms oc-

**Table 5. Elastic moduli (in GPa) for SWNT nanocomposites of different densities $\rho$ (in gcm$^{-3}$) and nanotube diameters d (in nm).**

| d | $\rho$ | $C_{11}$ | $C_{12}$ | $C_{44}$ | Y | poisson | $\eta$ |
|---|---|---|---|---|---|---|---|
| 1.17 | 2.22 | 337 | 121 | 79 | 273 | 0.264 | 0.731 |
| 1.17 | 2.29 | 343 | 89 | 96 | 306 | 0.206 | 0.756 |
| 1.17 | 2.39 | 403 | 103 | 112 | 361 | 0.204 | 0.747 |
| 1.17 | 2.41 | 418 | 104 | 108 | 377 | 0.200 | 0.688 |
| 1.65 | 1.99 | 290 | 78 | 71 | 257 | 0.212 | 0.667 |
| 1.65 | 2.14 | 384 | 81 | 94 | 356 | 0.174 | 0.620 |
| 1.65 | 2.24 | 437 | 73 | 131 | 416 | 0.143 | 0.720 |
| 2.06 | 1.77 | 255 | 73 | 78 | 223 | 0.223 | 0.857 |
| 2.06 | 1.94 | 295 | 79 | 97 | 262 | 0.211 | 0.898 |
| 2.75 | 1.73 | 251 | 72 | 82 | 219 | 0.223 | 0.916 |
| 2.75 | 1.84 | 331 | 61 | 120 | 311 | 0.156 | 0.889 |

cur. As shown in Ref. [18], no covalent bonding between CNT atoms and the matrix exists. Instead, the vdW forces drive a reconstruction of the surrounding medium, producing a graphitic wall at a distance of 3.4 Å from the CNT. Thus, the resulting open space (voids), discussed earlier, together with the relative high $sp^2$ ratio lower the moduli of such materials, rendering them softer. All elastic constants of the dense composites are enhanced compared to more dilute structures, similar to the trends observed for $B$ and $Y$. It is clear that for the same nanotube diameter all moduli increase as the density of the whole material increases. This strenghtening is again necessitated by the more dense (more rigid) $a$-C matrix. For example, a composite with a nanotube d=1.2 nm in diameter has a shear modulus of 79 GPa and a Young's modulus of 273 GPa, while these values rise to 131 and 416 GPa for a 1.7 nm CNT. Both composites have a density of 2 gcm$^{-3}$. The Poisson ratio is slightly higher than for diamond composites and its fluctuations are more significant. This is again a signature of the higher anisotropy that the nanotube composites show. A summary of these points is shown in Table 5 for the elastic moduli of representative SWNT composites. The transverse contraction over the longitudinal extension when strain is applied to CNT/$a$-C is higher than for nD/$a$-C. Note again, that different nanotube chiralities did not affect any of the results, hence the nanotubes are rather anonymous and we refer only to their diameter to characterize them.

## 5.5. Elastic Recovery

In this section we probe the elastic recovery during a compression-decompression cycle of both composite and amorphous structures. Experimentally, this is often performed by atomic force microscopy or standard nanoindentation and nanoscratch experiments. Measurements for $a$-C films show that these recover in a highly elastic manner ($\sim$ 90%) [72], but no variation with the $sp^2$ fraction is known. In our knowledge there are no such exper-

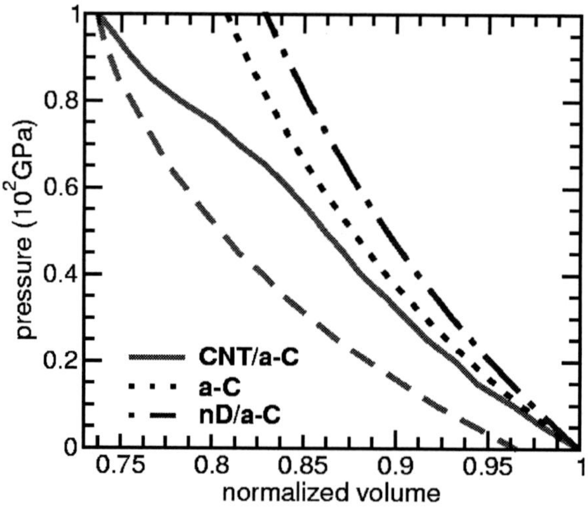


Figure 8. Compression - decompression curves for nD/$a$-C (dotted/blue), CNT/$a$-C (solid/black) and single-phase $a$-C (dashed/red).

iments for the composites studied here and our results serve as predictions for the elastic recovery of these materials. Computationally, by applying hydrostatic pressure on these composite structures their contraction can be measured. A permanent deformation signifies a decrease of the volume expressed as a fraction of its initial value. A perfectly elastic material would have an elastic recovery equal to one, while for a perfect plastic one it would be zero. CNT ropes, as an example, have been found to have high elastic recovery even under high-deformation conditions (over 100 GPa) [73].

In order to reveal the elastic response of the composites, compression of all these stuctures up to 100 GPa took place, followed by full decompression. The behavior is directly dependent on the stability of the composites; the more stable the structure the higher its recovery. We mainly focus on the stable composites, since these are the ones that are of technological interest. It is also more efficient to refer to the total $sp^3$ component for each case. Diamond composites are stable for an $sp^3$ fraction over 60-70%, while for nanotube composites this value is in the range of 20-30%. The elastic recovery of single-phase $a$-C is determined by its density or hybridization. It is almost perfectly elastic in the highly tetrahedral regime ($ta$-C). The recovery is not perfect as the $sp^3$ fraction declines, but it is again high in good aggreement to the experimental results mentioned previously. An almost fully $sp^2$-bonded $a$-C network regains its initial volume at about 95 %. For a much higher compression (up to 500 GPa) the dense networks again recover almost perfectly, but for the dilute ones (fully $sp^2$) the recovery drops to about 80 %. Similarly, in diamond composites, the more stable these are the higher their recovery. It is again almost perfect (even for compression up to 500 GPa), but as their density lowers the recovery drops to $\sim$ 90%. Taking into account the high incompressibility of the diamond crystallites the decrease can only be assigned to the matrix and mainly to its threefold atoms, which are more easily deformed.

The response of nanotube composites is more interesting. Although, they recover their initial volume over 90%, there are essential configurational modifications after decompression. The CNT retains its shape, but a few covalent bridge bonds ($\sim$ 1.6 Å) between the CNT and the matrix have been formed. During decompression the volume is changed (up to $\sim$ 10% depending on density) denoting that the system has been trapped into a local minimum. The bridge bonds, which start to form at a pressure close to 70 GPa, modify the structure and do not allow volume changes during decompression. In the case of unstable nanotube composites the volume change after decompression is again low, but inspection of the nanotube reveals the strong effect inflicted by the pressure. The tubular character is preserved, but the cross section is far from being circular, although no buckling similar to free-standing CNTs undercompression was evident [74]. The matrix and especially its part that immediately surrounds the inclusion is not seriously affected and no bridge bonds were formed.

The elastic recovery for an $a$-C network with $\sim$ 40% threefold atoms, as well as that for diamond and nanotube composites are shown Fig 8. The composites correspond to stable cases, thus they do not all have the same density. The diameters of the inclusions and the $sp^2$ fractions are 1.7 nm and 10% for the diamond composite, and 1.2 nm and $\sim$ 60% for the nanotube composite. The compression curves for $a$-C and nD/$a$-C coincide to those for decompression, indicating the full elastic recovery of these structures. The curves related to CNT/$a$-C show a hysterisis representing the change in the final volume after decompression. This behavior is related to the structural modification in the final configuration mentioned previously. Nanotube composites can be more easily compressed as shown in the figure. At the high pressure of 100 GPa, their volume can be reduced up to 25% from its initial value. The reduction for single-phase $a$-C is 18%, while it is even less for diamond composites. For the latter, the $sp^2$ component is much smaller compared to the CNT/$a$-C and $a$-C networks, but even for a larger component the compressibility is small. (A nD/$a$-C network with 73% of its atoms being $sp^2$ is compressed up to 77% at 100 GPa. This value is again higher than it is for $a$-C and CNT/$a$-C.) Our results indicate that the compressibility from CNT/$a$-C towards $a$-C and nD/$a$-C is decreasing. The crucial quantity for the degree of recovery is the $sp^2$ content. As long as it remains high, there is no perfect elastic recovery.

## 5.6. Single- versus Multi-wall Nanotubes

At this point we should comment on the trends of MWNT composites as compared to those of SWNT composites. MWNTs are also metastable, with well defined minima at low density matrices. The metastable regions coincide with that of SWNTs and the MWNT composites are also more rigid for higher densities and larger nanotube diameters. The surrounding graphite-like wall is again formed. The distance of this wall from the outer wall of the MWNT is, for intermediate densities, close to the graphite interplane value (3.4 $\AA$). Despite the similarities, some essential differences in comparing SWNTs and MWNTs of the same overall diameter, embedded in similar matrices, arise as shown in Fig.9. First, the formation energies of MWNTs are lower than those for SWNT composites, denoting more stable structures. This relies on the fact that SWNTs have, in general, higher cohesive energies than the MWNTs of same diameters, and also because the total system energy lowers as more walls are added to the CNT. A closer look into Fig.9(a) reveals that the minima of the formation energy for MWNTs are slightly shifted towards higher coordinations. This

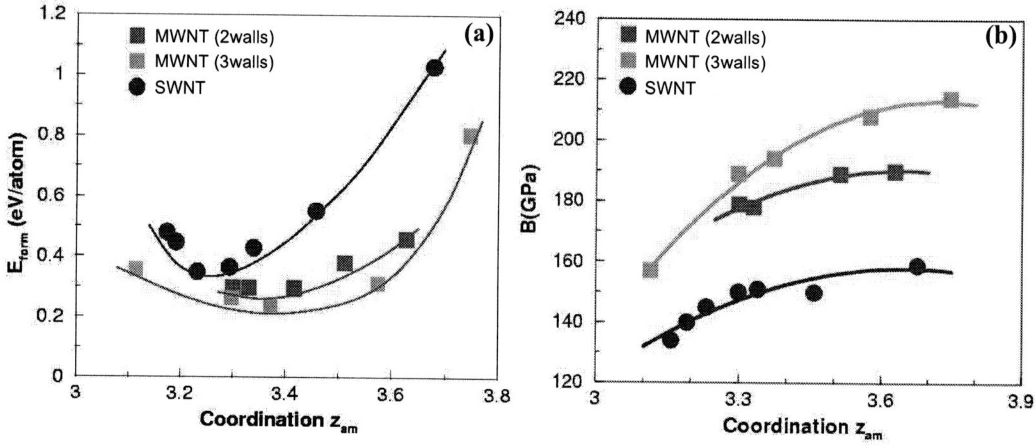


Figure 9. Comparison of the results for SWNT and MWNT composites with the same diameter (d=2.06 nm): (a) formation energies and (b) bulk moduli versus the $a$-C matrix coordination.

feature is probably a consequence of the larger density of a MWNT. Since crystalline inclusions are most stable in matrices with densities (configurations) similar to those of the inclusions, as has already been shown for diamonds, schwarzites and SWNTs, the MWNTs should also be more stable in matrices with intermediate, but slightly higher than that of SWNTs, densities. The number of walls in a MWNT is crucial for the whole composite's rigidity. The more walls a nanotube has, the more rigid the corresponding composite is. The difference lies on the empty space to total volume aspect ratio. The relative volume of these empty regions is smaller in MWNTs. Thereby the corresponding softening mechanism, taking into account that the fractions of the $sp^2$ sites in the matrix are similar, is not so profound, leading to an increase of the total elastic moduli.

## 6. Beyond the Elastic Regime

In this last section we focus only on the diamond nanocomposites and apply stresses beyond the elastic regime, where these materials are expected to break. The response of covalently-bonded materials, such as $a$-C and n$a$-C, under strain can be categorized into three broad regimes: For small strains, the response of the material is elastic, and Hooke's law applies. For a tensile strain applied to an isotropic material, the stress will equal the strain times the Young's modulus. The second regime corresponds to strain beyond the elastic limit, and is usually associated with plastic deformation of the material. The stress experienced by the material increases with increasing strain until a maximum stress (strength) is reached. The third regime is associated with strain beyond that giving the maximum stress. For brittle materials like diamond, the material breaks when the maximum stress is reached, and further increase of the strain results therefore in zero stress. Ductile materials, on the other hand, can be deformed beyond the strain corresponding to the maximum of the stress. Failure occurs when the sum of external and any internal stresses reaches a critical value known as strength. At this point atomic bonds begin to break, elastic energy is released, and new surface is created as the crack propagates in the material. The bonds can break perpendicular to the fracture plane, a process called *cleavage*, or by shearing bonds across the fracture plane, a process called *shear*.

We next investigate the atomistic mechanisms guiding fracture. We calculate the ideal strength of diamond composites, an ultimate test of their hardness. The structures were strained in a series of incremental strains. At each step, atomic positions were fully relaxed with TBMD. The stress was extracted by differentiating the energy with respect to strain, which is simply the ratio of the volume change divided by the initial volume of the sample. The ideal strength, an upper limit to real strength, describes the response of the material to non-hydrostatic deformations. Tensile and shear loads are usually applied to test the strength. In order to check the validity of the method used here, we calculated the ideal tensile and shear strengths for diamond. From the stress–strain curves, and in the case of tensile load along the $\langle 100 \rangle$, $\langle 110 \rangle$, and $\langle 111 \rangle$ directions, the diamond crystal can sustain stresses as large as 209 (0.28), 130 (0.21), and 124 (0.15) GPa. While exploring the case of shearing, a critical stress of 130 GPa (0.23) was found for the $\{111\}\langle 112 \rangle$ slip system. In parentheses, the corresponding strain for the maximum stress is given. These findings are compared to previous *ab initio* results [75, 76], where the tensile strength along the $\langle 100 \rangle$, $\langle 110 \rangle$, and $\langle 111 \rangle$ directions are 225 (0.37), 130 (0.22), and 95 GPa (0.13), respectively. The ideal shear strength is 93 GPa (0.3). Again, the critical strain is given in the parentheses. Deviation between values of *ab initio* and our TBMD results are less than 7-8%, within expexted limits, and the trends are well described (the $\{111\}$ cleavage plane dominates when diamond is fractured).

Having established the reliability of the method [63], we proceed to the study of our amorphous and nanocomposite structures. Tensile load is applied in the $\langle 111 \rangle$ direction and shear load on the $\{111\}$ plane in the $\langle 112 \rangle$ direction. The crystallographic directions are those of the nanodiamond region. In both cases, we choose the easy-slip plane of diamond, so that we can have a direct comparison to diamond, and examine the possibility that such a plane plays a significant role in nanocomposites. For the analysis, we consider three $a$-C samples: a typical $a$-C sample with average coordination z=3.47, a t$a$-C sample with z=3.8 and an "amorphous diamond" (WWW) sample, as well as a nD/$a$-C sample with a diamond with a radius of about 1.2 nm embedded in t$a$-C with a volume fraction of about 30%. The $a$-C phases are highly isotropic, therefore all directions of tensile or shear load are equivalent, and the fracture in the matrix is not expected to be dependent on easy-slip planes. Since, graphite, a fully $sp^2$-like network is known to be much softer than diamond, a fully $sp^3$-like structure, it is reasonable to expect that the strength increases with increasing coordination, as was the case for the elastic constants discussed in Section 5.4..

The resulting stress-strain curves, for the shear load applied here, are shown in Fig. 10. There are a number of important aspects of these results. The first is that all structures have significantly lower ideal strengths and critical strains than diamond, including the WWW model which is 100% $sp^3$-bonded and has no planes of cleavage. This is in contrast to the case of the C-46 carbon clathrates, not synthesized yet, which do not easily offer planes of cleavage and are predicted to have higher strength, but lower elastic moduli than diamond [77]. The maximum stresses are roughly 60, 40 and 30 GPa for z=4.00, 3.78 and 3.50, respectively, so that the strength of $a$-C is roughly proportional to its concentration

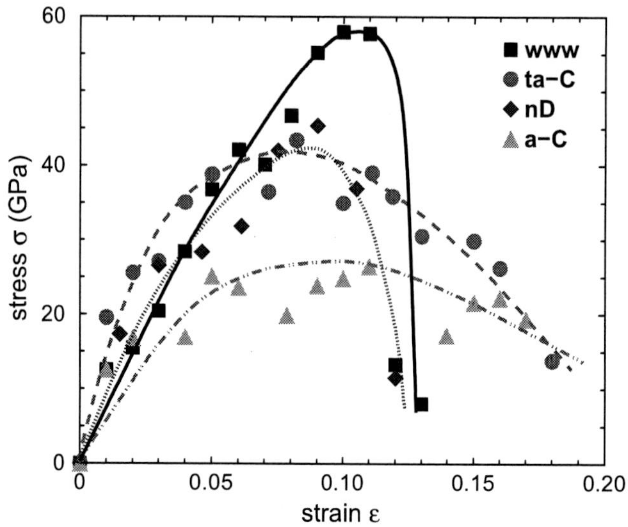


Figure 10. Stress vs strain curves for structures under shear load on the $\{111\}$ plane in the $\langle 112 \rangle$ direction. Results are shown for a WWW network, low and high density single-phase $a$-C, and a stable (negative formation energy) diamond composite (see text for details).

of four-fold atoms. Interestingly, the stress versus strain curve for the low-density $a$-C sample seems to suggest a ductile behavior. We get similar results for tensile strains along diamonds's easy slip plane $\{111\}$.

The difference in the strength stems from the weaker character of the C–C bond in the amorphous network. Note that while the modulus of $a$-D reaches 90% that of diamond, its strength is only about half. This indicates softer angular forces, due to dihedral disorder, which render the $sp^3$ hybrids more easily unstable under non-hydrostatic stresses at the bond-breaking regime, far from equilibrium. On the other hand, radial rigidity near equilibrium is not so weakened, giving rise to high elastic moduli. In the case of $ta$-C, we have, in addition, the presence of $sp^2$ sites, which might contribute to the further decrease of strength (see Fig. 10). We analyze these issues more deeply below. Our most striking finding is that the ideal strength of $n$-D/$a$-C is about equal to that of $ta$-C, for matrices with similar densities, for both tensile and shear loads. This was initially surprising, given the considerable enhancement of elastic moduli. It indicates that the nanoinclusions do not actually contribute to the increase of strength, but that this is rather controlled by the other part of the composite structure, the amorphous matrix.

In the WWW model the number of bonds per unit area for a given direction has to be higher than the number of bonds per unit area on the [111] easy-slip plane of diamond. One could naively expect that the strength of the WWW sample, consisting of four-fold atoms only, could perhaps be higher than that of diamond, due to the lack of such easy slip planes. On the contrary, the calculation reveals that the strength of diamond for tensile load along its easy slip direction, (111), is about twice that of the isotropic WWW. Apparently, the lack of easy slip planes in WWW is compensated by its somehow distorted bonds and the lack of order beyond the first-nearest-neighbor distance. The stress-strain curve for the nanocomposite follows exactly that of the embedding $ta$-C. The crystalline phase remains unaffected by the external load, which is almost completely taken by the surrounding amorphous matrix. The response of n$a$-C to external load beyond the elastic regime is identical to the response of the embedding matrix. As atoms in the amorphous matrix form bonds that are always weaker than the bonds in the crystal, the system prefers to stretch or bend these bonds and keep the strong diamond bonds untouched.

From the stress-strain curves and by extracting the yield stress $E$ (defined as the stress where the strain departs 0.2% from linearity) and the Young's moduli $Y$. Using the empirical relation of Tabor [70] $H/E = 0.07 + 0.6ln(Y/E)$, we estimated the hardness $H$ of our networks. We find $\sim$ 90 GPa for $a$-D (WWW), $\sim$ 70 GPa for $ta$-C and $n$-D/$a$-C, and $\sim$ 40 GPa for 50%-50% $a$-C, compared to 120 GPa for diamond with the present TB method. The value for $ta$-C is in accord with experimentally reported values. The nanocomposites seem to offer additional advantages, because they have superior elastic properties, compared to $ta$-C, and are predicted to also have excellent thermal stability [17]. These properties render nD/$a$-C promising materials for MEMS/NEMS applications.

## 6.1. Fracture Mechanism

In order to understand the fracture mechanisms of the materials we model here, we analyzed the microstructure at the critical strain when bonds start to break. We consider a bond as being broken when its length has become longer than the first minimum in the pair distribution function of the network at equilibrium [78]. The extracted cut-off distances are 1.90 Å, for the amorphous networks, and 1.75 Å for the nanodiamond crystallite. For $n$-D/$a$-C, the partial functions in the two components were calculated.

In Fig. 11, ball and stick models of $ta$-C and nanocomposite carbon under critical tensile strain are presented. Atoms that have lost at least one bond are distinguished, according to their hybridization, from atoms with no broken bonds. It is clear that bond breaking in the nanocomposite takes place in the amorphous matrix and not in the nanodiamond; fracture occurs inter-grain and not intra-grain. Only at the interface some originally elongated nanodiamond bonds are broken. Note, that stress in the nanodiamonds is mostly concentrated towards the interface [17]. This picture explains why the ideal strengths and corresponding strains of $n$-D/$a$-C and $ta$-C are similar. It appears, that the fracture mechanisms in both materials are similar. The vast majority of atoms involving broken bonds are $sp^3$ hybrids, in both $n$-D/$a$-C and $ta$-C materials. This can be understood on the basis of the relative energetics of $sp^3$ and $sp^2$ hybrids in the amorphous phase. It has been shown [22] by decomposing the total energy in atomic contributions, that the average energies of $sp^3$ and $sp^2$ sites in $ta$-C, or in $a$-C in general, are -6.8 and -7.1 eV, respectively, with respect to atomic C. This is a huge energy difference between the two hybrid states compared to the almost degenerate energies in diamond and graphite ($\sim$ -7.4 eV). Therefore, in the presence of critical strain at the onset of fracture, when bonds have to break, it will be energetically favorable to loose bonds belonging to $sp^3$ atoms in the amorphous region. It is clear from Fig. 11 that microfracture originates in $sp^3$-rich clusters.

A close inspection of the $ta$-C network in the same figure shows that the reduction is not directly linked to $sp^2$ sites, as only few of them have broken bonds, but indirectly through their spatial association with $sp^3$ sites. The fracture region involves many clustered

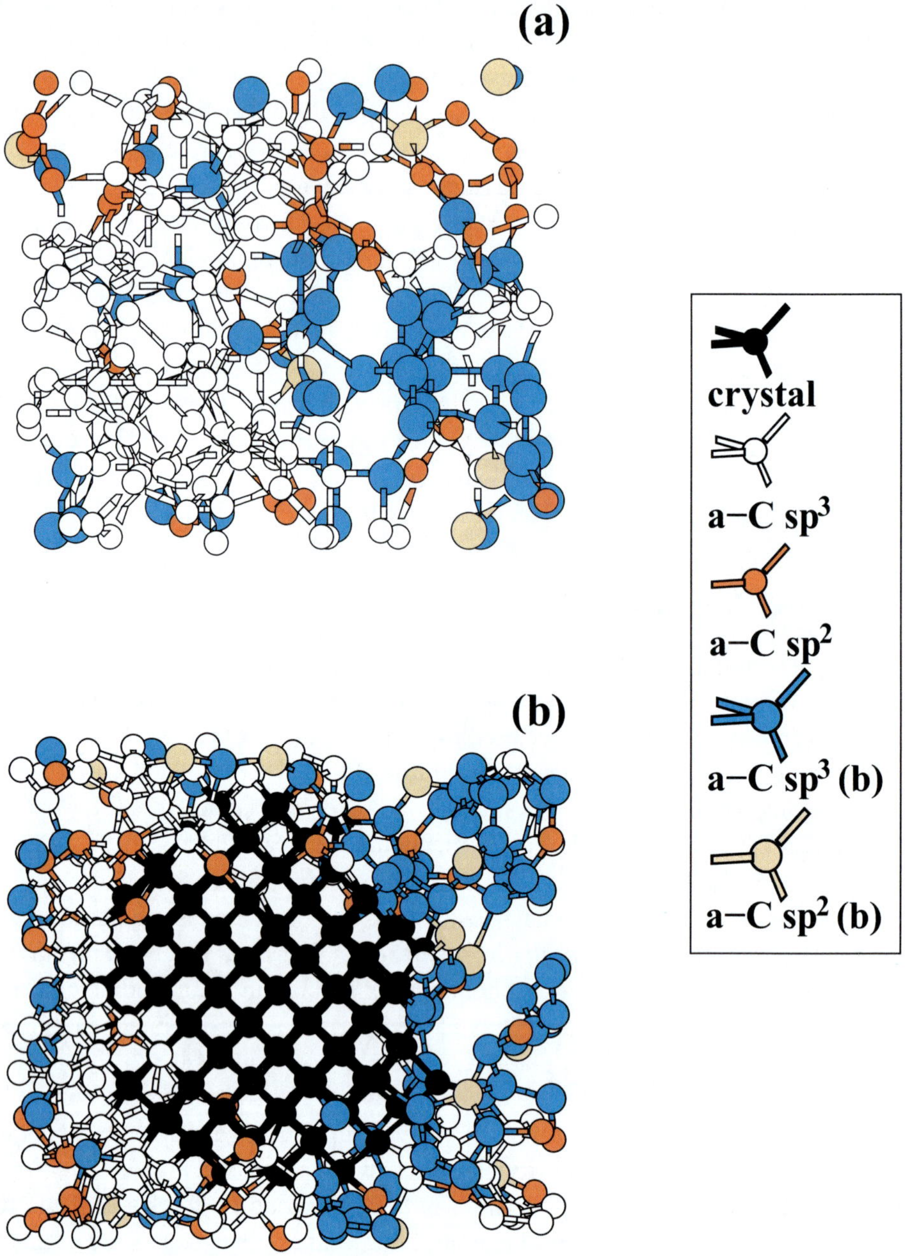


Figure 11. Ball & stick models for fracture in (a) single phase $a$-C and (b) a diamond composite of similar high density. Atoms are shaded according to the legend. Larger spheres (with the label (b) in the legend) denote atoms with broken bonds: in blue are shown atoms with broken $sp^3$ bonds, while in beige are the atoms with broken $sp^2$ bonds. The rest of atoms have no broken bonds: the black atoms of the inclusion, as well as the white $sp^3$, and the redish $sp^2$ atoms of the amorphous matrix. Snapshots are shown at the critical strain, at which bonds have started to break.

$sp^2$ sites, but the weak links are exactly those $sp^3$ sites in their immediate vicinity. This is because these $sp^3$ atoms are mainly under tensile stress [33], contrary to the majority of $sp^3$ sites which are under the more favorable compressive conditions, and so they have higher atomic energies [22]. The unstable sites under strain easily give up a bond to become $sp^2$. This is the favored hybrid state on the surfaces [79] ultimately formed by fracture, when the material breaks apart. The absence of $sp^2$ sites in the WWW model, acting as precursors for the destabilization of $sp^3$ sites, explains its higher strength compared to $ta$-C. the fact that, the $sp^2$ sites are successible to larger plastic deformations is evident in the case of the 50%-50% $a$-C cell (see Fig. 11).

We have identified the fracture mechanisms by adopting the simple, ordered geometry described above, with the strains applied in the weak directions. This model yields a lower bound in the calculated strength. In practice, the nanoinclusions and their easy-slip planes are randomly positioned and oriented. The randomness would help to prevent inter-grain fracture, and thus enhance strength, if the $a$-C material was reduced within narrow grain boundaries. The reinforcement will probably come as a result of the difficulty to propagate microcracks in grain boundaries.

## 7. Summary

In this review, we have reported work on the stability and mechanical properties of carbon composites. These are materials, that consist of a carbon crystallite embedded in an amorphous carbon matrix. Crystallites with sizes in the nanometer region and structures that range from $sp^3$ diamond-like, high density conformations to $sp^2$ graphite-like positively and negatively curved cage-like and open conformations. We have found that diamonds are highly stable in dense amorphous carbon matrices, while nanotubes are metastable in low density matrices. Metastable composites with $sp^2$ inclusions and formation energies close to those of diamond composites, show large compressive stress in their crystalline core compared to the small tensile stresses in the diamond composites' core, suggesting the possibility for transformation of $sp^2$-bonded structures to diamond under the appropriate conditions of pressure. We have focused on the elastic properties of these two composites and investigated the variation of the bulk and Young's moduli with the size of the crystalline inclusions. A noticeable hardening with size was found for the diamond composites, while a softening occured in the case of nanotube inclusions. The application of stress beyond the elastic regime led to the investigation of the fracture mechanism in diamond composites. Under tensile or shear load, fracture occurs inter-grain and thereby the ideal strength of these composite materials is similar to the pure amorphous phase. The onset of fracture takes place at weakly bonded $sp^3$ sites in the amorphous matrix.

The variety of embedded nanostructures used in this work and the large number of possible configurations in the matrix have probed the exploration of the physics of nanostructured carbon. The ability to control the mechanical, but also the -not addressed here-electronic properties of these materials by varying the density of the embedding medium and the type and size of the crystalline inclusion opens up enormous possibilities for potential technological applications of carbon nanocomposites. Still, more work needs to be done along these lines.

**Acknowledgments**

The authors with to thank G. Kopidakis, C. Mathioudakis and G.C. Hadjisavvas for useful discussions. This work was supported by the Ministry of National Education and Religious Affairs of Greece through the action "$E\Pi EAK$" (program "$\Pi Y\Theta A\Gamma OPA\Sigma$".)

## References

[1] Subramoney, S. *Adv. Mater.* 1998, 10, 1157.

[2] Banhart, F. *Rep. Prog. Phys.* 1999, 62, 1182.

[3] Robertson, J. *Mat. Sci. Eng. R* 2002, 37, 129.

[4] Silva, S.R.P. In H*andbook of Thin Film Materials*; Nalwa, H.S.; Ed.; Publisher: Academic Press, 2002; Vol. 4, pp. 403.

[5] Iijima, S. *Nature* 1991, 354, 56.

[6] Winkler, B.; Pickard, C.J.; Milman, V. ; Klee, W.E.; and Thimm, G.; *Chem. Phys. Lett.* 1999, 312, 536.

[7] O' Keeffe, M.; Adams, G.B.; and Sankey, O.F. *Phys. Rev. Lett.* 1992, 68, 2325.

[8] Vanderbilt D.; and Tersoff, J. *Phys. Rev. Lett.* 1992, 68, 511.

[9] Lenosky, T.; Gonze, X.; Teter, M.; and Elser, V. *Phys. Rev. Lett.* 1992, 355, 333.

[10] Amaratunga, G.A.J.; Chhowalla, M.; Kiely, C.J.; Alexandrou, I.; Aharonov, R.; and Devenish, R.M. *Nature*, 1996, 383, 321.

[11] Milani, P.; Ferretti, M.; Piseri, P; Bottani, C.E.; Ferrari, A.; Li Bassi, A.; Guizzetti, G.; and Patrini, M.; *J. Appl. Phys.* 1997; 82, 5793.

[12] Donadio, D.; Colombo, L.; Milani, P.; and Benedek, G. *Phys. Rev. Lett.* 1999, 83, 776.

[13] Zaiser, M.; and Banhart, F. *Phys. Rev. Lett.* 1997, 79, 3680.

[14] Banhart, F.; and Ajayan, P.M., *Nature* 1996, 382, 433.

[15] Lifshitz, Y.; Kohler, T.; Frauenheim, T.; Guzmann, I.; Hoffman, A.; Zhang, R.Q.; Zhou, X.T.; and Lee, S.T. Science 2002, 297, 1531; Zhou, X.T.; Li, Q.; Meng, F.Y.; Bello, I.; Lee, C.S.; Lee, S.T.; and Lifshitz, Y. *Appl. Phys. Lett.* 2002, 80, 3307.

[16] Yao, Y.; Liao, M.Y.; Kohler, T.; Frauenheim, T.; Zhang, R.Q.; Wang, Z.G.; Lifshitz, Y.; and Lee, S.T. *Phys. Rev. B* 2005, 72, 035402.

[17] Fyta, M. G.; Remediakis, I. N.; and Kelires, P.C.; *Phys. Rev. B* 2003, 67, 035423.

[18] Fyta, M.G; and Kelires, P.C. *Appl. Phys. Lett.* 2005, 86, 191916.

[19] Tersoff, J.; *Phys. Rev. Lett.* 1988, 61, 2879.

[20] Mehl, M.; and Papaconstantopoulos, D., *Phys. Rev. B* 1996, 54, 4519; Cohen, R.E.; Mehl, M.; and Papaconstantopoulos, D. *Phys. Rev. B* 1994, 50, 14694.

[21] Tang, M.S.; Wang, C.Z.; Chan, C.T.; and Ho, K.M. *Phys. Rev. B* 1996, 53, 979.

[22] Kelires, P.C. *Phys. Rev. Lett.* 1992, 68, 1854.

[23] Kelires, P.C. *Phys. Rev. B* 1993, 47, 1829.

[24] Mathioudakis, C.; Kelires, P.C.; Panagiotatos, Y.; Patsalas, P.; Charitidis, C.; and Logothetidis, S. *Phys. Rev. B,* 2002 65, 205203.

[25] Raty, J.-Y.; Galli, G.; Bostedt, C.; van Buuren, T.W.; and Terminello, L.J. *Phys. Rev. Lett.* 2003, 90, 037401.

[26] Stephan, U; and Haase, M. *J. Phys.: Condens. Matter* 1993, 5, 9157.

[27] Kaukonen, H.-P.; and Nieminen, R.M. *Phys. Rev. Lett.* 1992, 68, 620.

[28] Jäger, H.U.; and Albe, K. *J. Appl. Phys.* 2000, 88, 1129.

[29] Kelires, P.C. *Phys. Rev. B* 2000, 62, 15686.

[30] Lu, P.; Li, X.P.; and Martin, R.M. *Phys. Rev. Lett.* 1992, 68, 1551; Lu, P.; and Yang, W. *Phys. Rev. B* 1994, 49, 11421.

[31] Li, F.; Lannin, J.S.; *Phys. Rev. Lett.* 1990, 65, 1905; Gao, C.; Wang, Y.Y; Ritter, A.L.; Dennison, J.R. ibid. 1999 62, 945; Comelli, G.; Stohr, J.; Robinson, C.J.; Jark, W. *Phys. Rev. B* 1988, 38, 7511.

[32] Ferrari, A.C.; Libassi, A.; Tanner, B.K.; Stolojan, V.; Yuan, J.; Brown, L.M. ; Rodil, S.E.; Kleinsorge, B.; and Robertson, *J. Phys. Rev. B* 2000, 62, 11089.

[33] Kelires, P.C. *Phys. Rev. Lett.* 1994, 73, 2460.

[34] McCulloch, D.G.; McKenzie, D.R.; Goringe, C.M.; *Phys. Rev. B* 2000, 61, 2349; Marks, N.A.; McKenzie, D.R., and Pailthorpe, B.A. Bernasconi, M.; and Parrinello, M. *Phys. Rev. B* 1996, 54, 9703; Marks, N.A.; McKenzie, D.R.; and Pailthorpe, B.A.; Bernasconi, M.; and Parrinello, M. *Phys. Rev. Lett.* 1996, 76, 768.

[35] Frauenheim, T.; Blaudeck, P.; Stephan, U.; and Jungnickel, G.; *Phys. Rev. B* (1993), 48, 4823; Drabold, D.A.; Fedders, P.A.; and Stumm, P. *Phys. Rev. B* 1994, 49, 16 415.

[36] Wang, C.Z.; Ho, K.M.; and Chan, C.T. *Phys. Rev. Lett.* 1993 70, 611.

[37] Galli, G.; Martin, R.M.; Car, R.; and Parrinello, M., *Phys. Rev. Lett,* 1989, 62 555 (1989); Galli, G.; Martin, R.M.; Car, R.; and Parrinello, M. *Phys. Rev. B* 1990, 42, 7470.

[38] Barborini, E.; Piseri, P.; Li Bassi, A.; Ferrari, A.C.; Bottani, C.E.; and Milani, P. *Chem. Phys. Lett.* 1999, 300, 633; Ravagnan, L.; Siviero, F.; Lenardi, C.; Piseri, P.; Barborini, E.; Milani, P. Casari, C.S.; Li Bassi, A.; and Bottani, C.E. *Phys. Rev. Lett.* 2002, 89, 285506.

[39] Mathioudakis, C.; Kopidakis, G.; Kelires, P.C.; Wang, C.Z.; and Ho, K.M. *Phys. Rev. B* 2004, 70, 125202.

[40] Phillips, J.C. *J. Non-Cryst. Sol.* 1979, 34, 153.

[41] He, H.; and Thorpe, M.F. *Phys. Rev. Lett.* 1995, 54, 2107.

[42] Wooten, F.; Winer, K.; and Weaire, D.; *Phys. Rev. Lett.* 1985; 54, 1392.

[43] Friedmann, T.A.; McCarty, K.F.; Barbour, J.C.; Siegal, M.P.; and Dibble, D.C. *Appl. Phys. Lett.* 1996, 68, 1643.

[44] McKenzie, D.R. *J. Vac. Sci. Technol. B* 1993, 11, 1928.

[45] Crain, J.; Clark, S.J.; Ackland, G.J.; Payne, M.C.; Milman, V. Hatton, P.D.; and Reid, B.J.; *Phys. Rev. B.* 1994, 49, 5329.

[46] Saito, R.; Dresselhaus, G.; and Dresselhaus, M.S. P*hysical Properties of Carbon Nanotubes,* Imperial College Press, 1998.

[47] Iijima, S.; and Ichihashi, T. *Nature* 1993, 363, 603.

[48] Bethune, D.S.; Klang, C.H.; de Vries, M.S.; Gorman, G.; Savoy, R.; Vazquez, J.; and Beyers, R. *Nature* 1993, 363, 605.

[49] Alexandrou, I.; Scheibe, H.-J.; Kiely, C.J.; Papworth, A.J.; Amaratunga, G.A.J.; and Schultrich, *B. Phys. Rev. B* 1999, 60, 10903.

[50] Hirono, S.; Umemura, S.; Tomita, M.; and Kaneko, R. Appl. Phys. Lett. 2002, 80, 425.

[51] Kozlov, M.E.; Hirabayashi, M.; Nozaki, K.; Tokumoto, M.;and Ihara, H. *Appl. Phys. Lett.* 1995 66, 1199.

[52] Lyapin, A.G.; Brazhkin, V.V.; Gromnitskaya, E.L.; Popova, S.V.; Stal'gorova, O.V.; Bayliss, S.C.; and Sapelkin, A.V.; *Appl. Phys. Lett.* 2000, 76, 712.

[53] Townsend, S.J.; Lenosky, T.J.; Muller, D.A.; Nichols, C.S.; and Elser, V. *Phys. Rev. Lett.* 1992, 69, 921.

[54] We estimate the contribution of VdW forces to the cohesive energy of the system to be $\sim 0.2\%$.

[55] Kresse, G.; and Hafner, *J. Phys. Rev. B* 1993, 47, RC558; Kresse, G.; and Furthmüller, *J. Comput. Mat. Sci.* 1996, 6, 15.

[56] The generalized fradient approximation (GGA) was used for the exchange-correlation potential and Vanderbilt-type ultrasoft pseudopotentials. The numerical parameters (energy cut-off and number of k-points) were chosen so that error in the calculated lattice constants and energies are of the order of 0.1%.

[57] Bernstein, N.; Aziz, M.J.; and Kaxiras, E. *Phys. Rev. B* 1998, 58, 4579.

[58] Landau, L.D.; and Lifshitz, E.M.; Theory of Elasticity; Course of Theoretical Physics; Publisher: Pergamon Press, 1986; Vol. 7.

[59] Murnaghan, F.D. *Proc. Natl. Acad. Sci. U.S.A.* 1944, 30, 244.

[60] Li, C.; and Chou, T.-W. *Phys. Rev. B* 2004, 69, 073401.

[61] Hernández, E.; Goze, C.; Bernier, P.; and Rubio, A. *Phys. Rev. Lett.* 1998, 80, 4502.

[62] Reich, S.; Thomsen, C.; and Ordejon, P. *Phys. Rev. B* 2002, 65, 153047.

[63] Fyta, M.G.; Remediakis, I.N.; Kelires, P.C.; Papaconstantopoulos, D.A. *Phys. Rev. Lett.* 2006, 96, 185503.

[64] Ashcroft, N.W.; Mermin, N.D.; *Solid State Physics,* Publisher: Solid State Physics, 1976.

[65] Treacy, M.M.J.; Ebbesen, T.W.; and Gibson, J.M. *Nature* 1996, 381, 680.

[66] Zhao, X.; Liu, Y.; Inoue, S.; Suzuki, T.; Jones, R.O.; and Ando, Y. *Phys. Rev. Lett.* 2004, 92, 125502.

[67] Reich, S.; Thomsen, C. ; and Ordejón, P. *Phys. Rev. B* 2002; 65, 153407.

[68] Wong, E.W.; Sheehan, P.E.; Lieber, C.M. *Science* 1997, 297, 1971.

[69] Paul, B. *Transactions of the Metallurgical Society of AIME* 1960, 36-41.

[70] D. Tabor, *Rev. Phys. Technol.* 1970, 1, 145.

[71] Kaxiras, E.; *Atomic and Electronic Structure of Solids*; Publisher: Cambridge University Press, 2003.

[72] Charitidis, C.; Logothetidis, S. *Thin Solid Films*, 2005, 482, 120.

[73] Liang, H.; and Upmanyu, M. *Phys. Rev. Lett.* 2005, 94, 065502.

[74] Yakobson, B.I.; Brabec, C.J.; and Bernholc, *J. Phys. Rev. Lett.* 1996, 76, 2511.

[75] Roundy, D. and Cohen, M.L. *Phys. Rev. B* 2001, 64, 212103.

[76] Chacham, H.; and Kleinman, L. *Phys. Rev. Lett.* 2000, 85, 4904.

[77] Blase, X.; Gillet, P.; San Miguel, A.; and Mélinon, P. *Phys. Rev. Lett.* 2004, 92, 215505.

[78] Galli, G.; Gygi, F. and Castellani, A.; *Phys. Rev. Lett.* 1999, 82, 3476.

[79] Kelires, P.C.; *J. Non-Cryst. Solids* 1998, 227-230, 597; Davis, C.A.; Knowles, K.M.; and Amaratunga, G.A.J. *Phys. Rev. Lett.* 1998, 80, 3280.

In: New Research on Nanocomposites
Editors: Luis M. Krause et al, pp. 115-148
ISBN 978-1-60456-799-1


*Chapter 5*

# ULTRANANOCRYSTALLINE DIAMOND/AMORPHOUS CARBON NANOCOMPOSITE FILMS

***Wilhelm Kulisch*[1]*and *Cyril Popov*[2]†**
[1]European Commission Joint Research Centre
Institute for Health and Consumer Protection
Via Enrico Fermi 1
21020 Ispra (VA), Italy
[2]Institute of Nanostructure Technologies and Analytics
University of Kassel
Heinrich-Plett-Strasse 40
34132 Kassel, Germany

## ABSTRACT

Ultrananocrystalline diamond/amorphous carbon (UNCD/a-C) films have been deposited by microwave plasma chemical vapour deposition from 17% $CH_4/N_2$ mixtures at temperatures between 450 and 770°C. This process results in a nanocomposite type of films consisting of diamond nanocrystallites of 3-5 nm diameter, which are embedded in an amorphous carbon matrix. The first part of this contribution is devoted to the basic film properties such as crystallinity, composition, and the bonding environment, while the second addresses the nucleation and growth processes and the underlying mechanisms, discussing among others the role of parameters such as the substrate pre-treatment and the substrate temperature.

The final and major part of the contribution addresses properties relevant for future applications of this type of nanocomposite films, e.g. electrical, optical, mechanical, tribological, and bio(techno)logical. Emphasize is laid on a discussion of the influence on the nanocomposite character of these films, i.e. the existence of an amorphous carbon matrix, on these properties. It will be shown that the matrix renders the optical properties for the worse as compared to pure UNCD films but is of advantage e.g. for the mechanical and electrical properties.

*E-mail address: wilhelm.kulisch@jrc.it
†E-mail address: popov@ina.uni-kassel.de

# 1. Introduction

Diamond is a material with many extreme properties; among others it is the hardest known single phase material, possesses the highest Young's modulus, the highest thermal conductivity and the highest Debye temperature [1, 2]. In addition, it is a semiconductor with a large band gap and chemically very stable. Beside natural occuring crystals (i.e. gemstones), diamond is nowaday available in form of crystals fabricated by high pressure/high temperature processes (HP/HT) but also in form of thin polycrystalline CVD films (PCD) [3, 4]. For many applications of diamond in such diverse fields as tribology, electronics, optics, biotechnology, etc. of course CVD thin films would be the solution of choice. However, owing to their polycrystalline nature with crystallite sizes on the order of micrometers or even larger, such PCD films are usually rather rough, thus preventing their applications in fields where smooth films are required [5, 6]. Although techniques have been developed to overcome this problem [6, 7], this has to be paid off by rather high film thicknesses and also by a degradation of film properties. In order to overcome this problem, recently nanocrystalline diamond and also ultrananocrystalline diamond films (UNCD)[1] have been developed [10, 11] which retain to a large extent the extreme properties of diamond but bear the advantage of much smoother surfaces.

Diamond has the disadvantage of being extremely brittle [12], which can lead to fatal brittle failure [13]. It is generally agreed upon that for many applications not hardness is the decisive property but rather toughness [14–16]. Recently, approaches of modern nanotechnology have been used to design and prepare films which are not only superhard but also supertough [13]. Among others, multilayer [17–19] and nanocomposite systems [20–22] have been developed. The latter usually consist of nanocrystallites of some nm diameter of a hard or superhard material which are embedded in an appropriate amorphous matrix.

Thus, the question arises whether nanocomposite films containing diamond nanocrystallites (i.e. the hardest material) with superior properties can be developed. Amorphous carbon offers itself as a matrix material as it would simplify the deposition process. But of cource the presence and the nature of an amorphous matrix will not only influence the mechanical properties but also affect the optical, electrical, thermal, etc. characteristics of the nanocomposite.

In this contribution, we will first describe the experimental details of the deposition of UNCD/a-C nanocomposite films by microwave plasma chemical vapour deposition (MWCVD) from $CH_4/N_2$ gas mixtures. In the following section, we will present the basic properties of these films, thereby showing that they are indeed nanocomposites, consisting of diamond nanocrystallites embedded in an amorphous carbon matrix. Section 4. is devoted to a discussion of the nucleation and growth mechanisms; among others, the influence of the substrate pre-treatment on the nucleation and that of the substrate temperature on the growth are addressed. The final section deals with application relevant properties of the films (e.g. electrical, optical and mechanical). In this section, also the question is

[1] Please note: In all our previous publications we had labeled our films NCD and NCD/a-C. In recent times, after the introduction by Gruen [8] and the definition by Butler [9] it has become common to label this type of film UNCD. In order to avoid confusions we decided to follow this definition. The differences between UNCD and NCD will become apparent in Sec. 4.3. below.

discussed to what extent these properties are influenced by the nanocomposite nature of the films.

# 2. Deposition of UNCD/a-C Films

## 2.1. Substrate Pre-treatment

**Table 1. Parameters of the sustrate pre-treatment procedures used in this study.**

| process | amount NCD [mg] | size NCD [nm] | amount UDD [mg] | $n_{nuc}$ [cm$^{-2}$] |
|---|---|---|---|---|
| standard I | 50 | 250 | — | $2\text{–}4 \cdot 10^8$ |
| series I | 50 | 250 – 1000 | — | $1\text{–}4 \cdot 10^8$ |
| standard II | 50 | 250 | 80 | $1 \cdot 10^{10}$ |
| series II | 50 | 250 | 0 – 80 | $1 \cdot 10^8 - 1 \cdot 10^{10}$ |

Most of the experiments described in this contribution have been carried out with (100) Si substrates, but the deposition of UNCD/a-C was also successfully accomplished on a variety of materials such as polycrystalline diamond films, cubic boron nitride films, TiN, Ti-6Al-4V alloy, $Si_3N_4$ and GaAs. The influence of the substrate material on nucleation, growth and properties has been discussed in [23,24]. The results presented in this contribution mainly concentrate on films on Si substrates. They have been pre-treated ultrasonically with a suspension of diamond powder in an organic solvent in order to increase the nucleation density. This is a common technique applied not only in the deposition of NCD and UNCD films [25,26] but also for polycrystalline diamond films [27,28]. The organic solvent used in our study was n-pentane (75 ml) in all cases. As diamond powders, either nanocrystalline diamond with various mean grain sizes (from 250 nm to 1 $\mu$m) or a mixture of NCD powder with ultradisperse diamond powder (UDD, 3-5 nm) was used. The different procedures employed in this investigation and the nucleation densities $n_{nuc}$ achieved are summarized in Table 1. The depositions on PCD and c-BN substrates also discussed in this contribution have been carried out without any pretreatment.

## 2.2. Deposition Process

For the deposition of the UNCD/a-C films discussed in this contribution, a microwave plasma enhance chemical vapour (MWCVD) deposition set-up (Plasma Technology) has been used which is schematically shown in Fig. 1. Details of this set-up and the deposition procedure have been published previously [29,30]. In brief, the signal of a 1.5 keV magnetron microwave source (Astex) is coupled via a waveguide through a quartz window into the stainless steel deposition chamber. The substrates are placed on a graphite holder which can be heated inductively by means of a copper coil inside the holder. The temperature is measured by means of a thermocouple placed directly below the substrate. The process gas

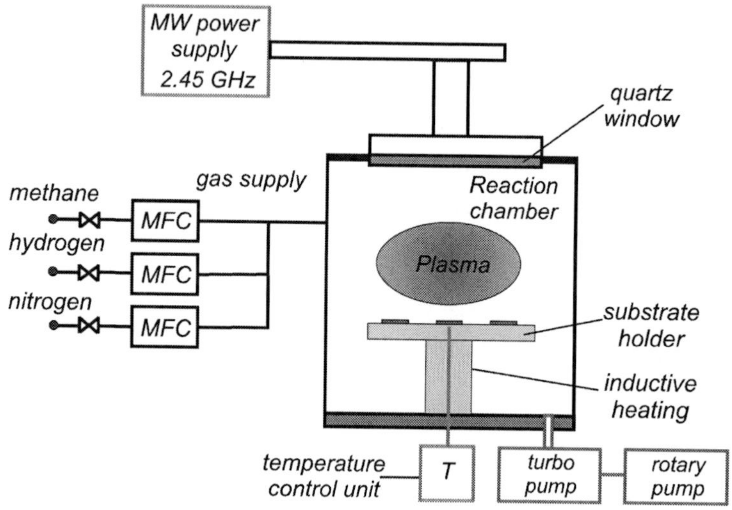


Figure 1. Schematic diagram of the deposition set-up used in this study.

consists of a mixture of $CH_4$ and $N_2$ with a total flow of 300 sccm, leading to a pressure of 2.2 kPa.

**Table 2. Parameters of the standard deposition process used in this study. Also given are the ranges of parameter variations.**

| parameter | | standard | variation |
|---|---|---|---|
| microwave power | [W] | 800 | |
| substrate temperature | [°C] | 600 | 450 – 770 |
| pressure | [kPa] | 2.2 | |
| total gas flow | [sccm] | 300 | |
| $CH_4$ fraction | [%] | 17 | 9 – 17 |
| duration | [min] | 390 | 45 – 720 |

The parameters used in our standard process are summarized in Table 2. In the studies reported here, besides the substrate pre-treatment procedure (see Table 1) mainly the substrate temperature, the methane fraction of the gas mixture, and the deposition time have been varied. The ranges of these variations are also listed in Table 2.

## 3. Basic Film Properties

A variety of techniques has been applied to establish the basic properties of these UNCD/a-C films, i.e. composition, morphology and structure, crystallinity, and bonding environment. In the following, we will give a short summary of these investigations. Especially, we will show that the films possess indeed a nanocomposite nature. For details of the experimental techniques applied in these studies, we refer to the papers cited in this section.

## 3.1. Crystallinity

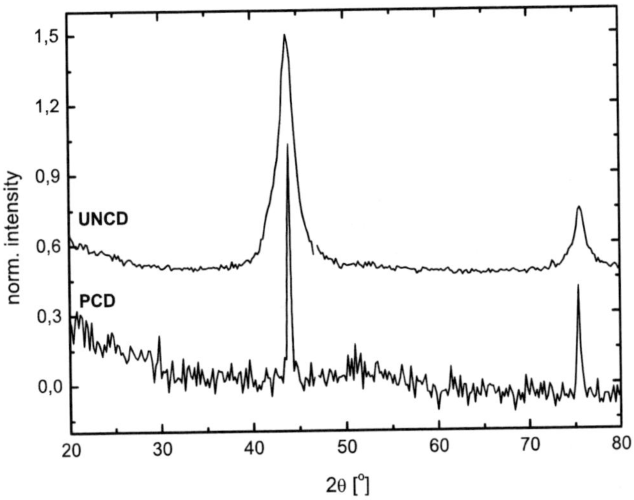


Figure 2. Glancing angle XRD measurements ($\phi = 5°$) of a poly- and a ultrananocrystalline diamond film.

Fig. 2 shows grazing angle ($\phi = 5°$) scans of a polycrystalline diamond film (deposited in the same set-up at 770°C but from a 1% $CH_4/H_2$ mixture), and a UNCD/a-C film obtained at the same temperature from a 17% $CH_4/N_2$ mixture [31]. Both measurements show two peaks at $2\theta = 43.8$ and 75.6°, which can, according to JCPDS card 6-675, be assigned to the (111) and (220) reflections of diamond. However, whereas the peaks for the PCD film are rather sharp (here, the width is given by the step width of the measurement of 0.2°), the peaks for the UNCD film are rather broad, indicating a limited crystallite size. From a modified Debye-Scherrer formula [32], the crystallite size could be determined to 3-4 nm. Thus, two conclusions can be drawn immediately from the diagram in Fig. 2: i) the UNCD film contains indeed crystalline diamond and ii) this diamond is nanocrystalline with very small grain sizes.

## 3.2. Properties of the Matrix

Auger electron spectroscopy (AES) measurements have been performed in order to establish the $sp^2$ content of the surface region [33]. Fig. 3 presents derivative spectra in the region of the $C_{kvv}$ peak of two samples grown at 600°C but with different deposition times. In the inset, spectra of a polycrystalline diamond film and of highly oriented pyrolytic graphite are shown. It has been demonstrated previously in the literature [34–36] that the so-called "D-parameter" $\Delta E_D$, the energetical difference between the position of the minimum of such derivative spectra around 270 eV, and that of the highest maximum can be taken as a

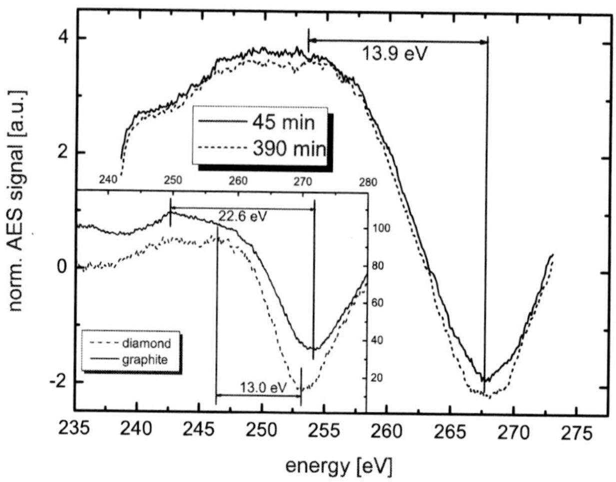


Figure 3. Derivative AES spectra in the region of the $C_{kvv}$ transition for UNCD/a-C films deposited at 600°C for 45 and 390 min, respectively. Spectra of a polycrystalline diamond film and graphite are shown in the inset.

quantitative measure of the $sp^2$ content of a carbon-based material. For both UNCD films shown in Fig. 3, $\Delta E_D$ is about 13.9 eV. From this and the values of $\Delta E_D$ for diamond and graphite also given in the figure, it can be concluded that the surface region of the UNCD films contains about 10% of $sp^2$ hybridized carbon, independent of the deposition time.

AES is a surface sensitive method with an information depth of a few nm only. In contrast, Raman spectroscoy is a bulk method, although in the case of carbon based films the information depth and the sensitivity depend on the nature of the films and also on the excitation wavelength [37–39]. Fig. 4 shows a Raman spectrum of a 600°C UNCD film recorded at 514.5 nm. It consists of a broad band between 1100 and 1700 $cm^{-1}$ which can be deconvoluted into four contributions:

i) The graphite-related G band at 1570 $cm^{-1}$.
ii) The graphite-related D band at 1363 $cm^{-1}$.
iii) The NCD-related band at 1175 $cm^{-1}$.
iv) The NCD-related band at 1482 $cm^{-1}$.

While the first two are common to all films containing $sp^2$ hybridized carbon, the assignment of the latter two has been controversely discussed in the literature. The 1175 $cm^{-1}$ peak (literature values reach from 1140 to 1190 $cm^{-1}$) was first observed by Nemanich et al. [40] and assigned to nanocrystalline diamond. According to this interpretation it is caused by the small crystallite size and related to the main diamond peak at 1332 $cm^{-1}$ in a similar way as the graphite D band is related to the G band. Prawer et al. [41] observed it also in the Raman spectra of nanocrystalline diamond powder, which was free of any $sp^2$ carbon, and identified it with surface plasmons. These assignments have been critized

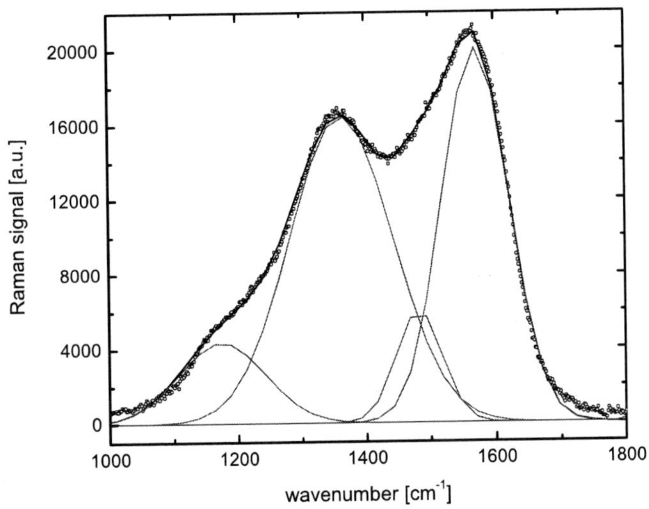


Figure 4. Raman spectrum of a UNCD/a-C film deposited at 600°C.

recently by Ferrari and Robertson [42] and subsequently also by other authors [43]. They claim that this peak is caused by trans-polyacethylene groups $(C_2H_2)_n$ situated at the grain boundaries of NCD and UNCD films. The present authors are no experts in the field of Raman spectroscopy; thus we have no intention to decide this discussion. However, our following findings should be included in such a discussion:

i) To our knowledge, this peak is observed only for films containing nano- or ultrananocrystalline diamond.
ii) This peak is observed in all films containing nano- or ultrananocrystalline diamond.
iii) The assigment to trans-polyacethylene would require the presence of hydrogen bonded to $sp^2$ hybridized carbon. Thus the intensity of this peak should correlate to the concentration of such groups in the films.
iv) As discussed below in Sec. 4.3. the nature of the grain boundaries can vary considerably in UNCD films. It is difficult to imagine that in all cases trans-polyacethylene plays a similar role.

Thus, at the time being the presence of the 1140 $cm^{-1}$ peak can only be taken as a fingerprint for the presence of (ultra)nanocrystalline diamond. Owing to the different sensitivities for $sp^2$ and $sp^3$ carbon (see below) [37,38], it is anyway not possible to derive quantitative informations from such Raman spectra. It is, however, clear that $sp^2$ hybridized carbon is present in the films.

Fig. 5 shows Raman spectra of the same UNCD/a-C sample recorded with four different excition wavelengths reaching from the infrared (785 nm) to the ultraviolett (244 nm). It can be seen that — although all spectra stem from the same sample — they are extremely different. This has also been reported by other authors [38, 39, 42]. The reason can be found in the fact that in the visible range the energy of the exciting light is close to the band gap of $sp^2$ carbon (ca. 2 eV) which leads to resonance effects with electronic $\pi - \pi^*$ transitions which enhance the Raman cross section. This is the reason for the high sensitivity of standard Raman for $sp^2$ carbon mentioned above. In the UV, the energy is close to that of $\sigma - \sigma^*$ transitions in both, $sp^2$ and $sp^3$ carbon [38,39].

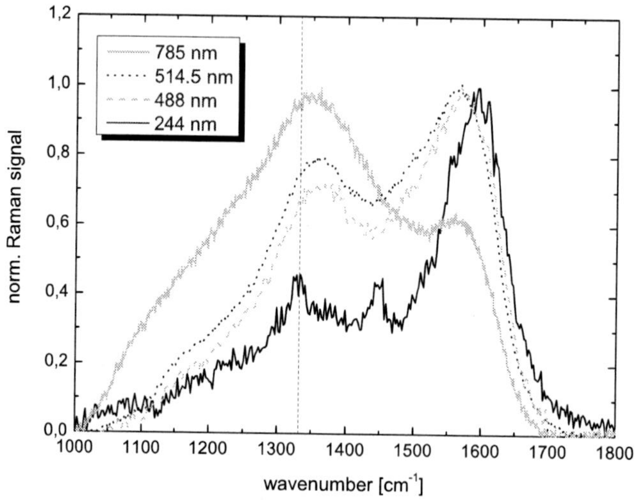


Figure 5. Raman spectra of a UNCD/a-C film deposited at 600°C obtained at four different excitation wavelengths. The vertical line shows the position of the diamond peak at 1332 $cm^{-1}$.

As a consequence, in the 244 nm spectrum even the diamond line at 1332 $cm^{-1}$ is present. Its large width is caused by the nanocrystalline nature of the films. Nevertheless, its existence is an independent proof of the presence of diamond within the films. All other tendencies observed in Fig. 5, e.g. the strong reduction of the ratio of the intensities of the D and the G band with increasing energy of the excitation light, or the vanishing of the D band and also the 1175 $cm^{-1}$ peak in the 244 nm spectrum are in agreement with the papers cited above.

The results presented above have been confirmed by complementary methods. X-ray photoelectron spectroscopy (XPS) core level and valence spectra prove the existence of 10 - 15% $sp^2$ hybridized carbon in the surface region [31]. The same result was obtained by electron energy loss spectroscopy (EELS) for the film bulk [44]. Finally, infrared spectra have shown that the hydrogen which is present in our films (see the next section) is almost exclusively bonded to $sp^3$ hybridized carbon. The spectra revealed that a network of CH, $CH_2$ and $CH_3$ groups with varying bond lengths and angles exist which can be assigned to the amorphous carbon matrix [31].

### 3.3. Composition

The surface composition of the NCD/a-C films has been investigated by XPS [31,45], the bulk composition by elastic recoil detection (ERD) [31]. In addition, for some selected

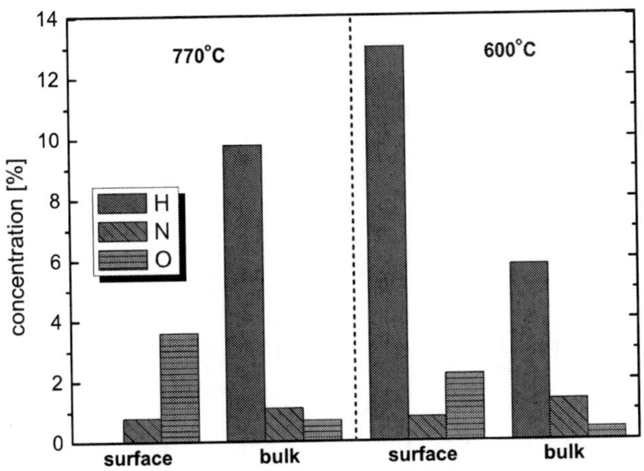


Figure 6. Surface (XPS) and bulk (ERD, NRA) composition of UNCD/a-C films deposited at 770 and 600°C, respectively. The surface hydrogen content of the 770°C sample has not been established.

samples nuclear reaction analysis (NRA) measurements were performed in order to record hydrogen depths profiles [46, 47]. Fig. 6 summarizes data on the surface and bulk composition for films deposited at 770 and 600°C, respectively.

There are several aspects of the data in Fig. 6 worth to be discussed. First of all, it has to be noted that the films contain considerable amounts of hydrogen. This is in contrast to polycrystalline diamond films for which rather low H contents have been reported (e.g. 0.001 - 0.2% reported in Ref. [48]). On the other hand, values for (U)NCD films are much higher, reaching from 3 to 15% [49–52]. Moreover, it has been shown that the hydrogen content for PCD [48] as well as for (U)NCD films [50] scales inversely with the grain size. These observations suggest that the hydrogen is predominantly incorporated at the grain boundaries and not within the diamond (nano)crystals. In our case this means that the hydrogen is located within the amorphous matrix.

Second, the nitrogen content of the films has to be addressed. In view of the fact that the gas mixture used for the deposition of these films contained 83% $N_2$, it is surprisingly low as can be seen in Fig. 6. UNCD films are usually grown from hydrogen poor $Ar/CH_4/H_2$ mixtures (for a detailed discussion of the conditions leading to UNCD instead of PCD growth see [10, 11]). However, there are several papers in the literature reporting the use of $N_2$ containing mixtures. As early as 1993, Badzian et al. [53] used $CH_4/N_2$ to deposit diamond containing films but this approach was almost neglected at that time. More recently, several groups used either pure $CH_4/N_2$ mixtures [54, 55] or nitrogen containing $CH_4/N_2/H_2$ mixtures [52, 56–58]. Despite the considerable amount of nitrogen in the gas phases used in these experiments, the concentrations of N incorporated within these films never exceeded 1% [52, 54, 55, 57, 59]. From this it can be concluded that the nitrogen in the gas mixtures used for the deposition of these various UNCD films must play a major — but yet unknown — role in the formation of the films but is not incorporated within the films in larger amounts.

Irrespective of the substrate temperature, the oxygen content of the film bulk of 0.4 – 0.7% is very low. The oxygen surface concentration of the 600°C film of 2.2% is also very low. It should be noted that this sample was sealed in vacuum after deposition until shortly before its introduction into the XPS system [45]. Thus it was exposed to the laboratory atmosphere only for a short time (a couple of hours). In contrast, the 770°C film was exposed to air several months before the XPS measurements have been performed. Nevertheless, the surface concentration of only 3.6% is surprisingly low. This aspect will be discussed further in Sec. 5.4.1.

## 3.4. Summary: Basic Properties of UNCD/a-C Films

From the results presented in this section a clear picture of this type of films emerges. From Sec. 3.1. it is evident that diamond nanocrystallites of 3-5 nm diameter are present. It is also clear that the films contain in addition some $sp^2$ hybridized carbon material and also $sp^3$ $CH_x$ groups. It is suggesting to ascribe the latter two to the grain boundaries rather than to the diamond nanocrystals, in other words to amorphous material surrounding the nanocrystals.

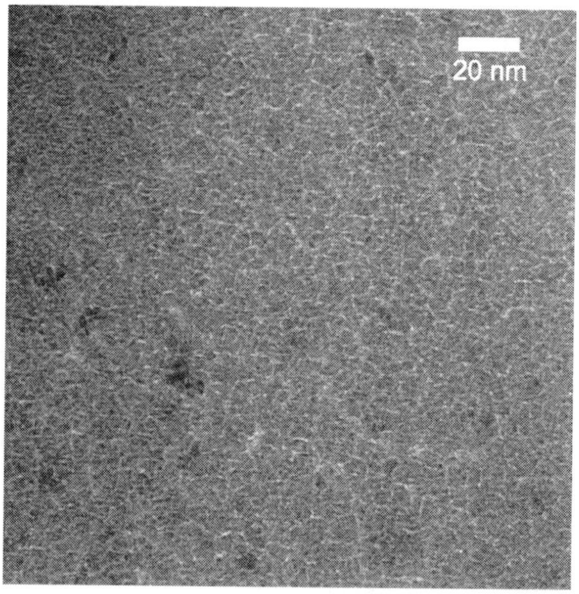


Figure 7. TEM image of a UNCD/a-C film.

Fig. 7 shows a transmission electron microscopy picture of a film deposited at 770°C. Here, clearly nanocrystallites, but also agglomerates of nanocrystallites are visible which are surrounded by an amorphous matrix. The width of the matrix could be derived from this and similar images to 1 - 1.5 nm. X-ray reflection measurements [31] yielded a density of 2.75 $gcm^{-3}$. The density of diamond is 3.5 $gcm^{-3}$, thus it is obvious that also a phase with a lower density must exist. Taking all these informations together it turned out that the volume ratio of nanocrystallites and matrix is about unity, with a density of the matrix of about 2.0 $gcm^{-3}$.

Thus, most importantly, the results presented in this section show that the films have a nanocomposite nature. They consist of diamond nanocrystallites embedded in an amorphous carbon matrix.

# 4. Nucleation and Growth Mechanisms

## 4.1. Influence of the Substrate Pretreatment

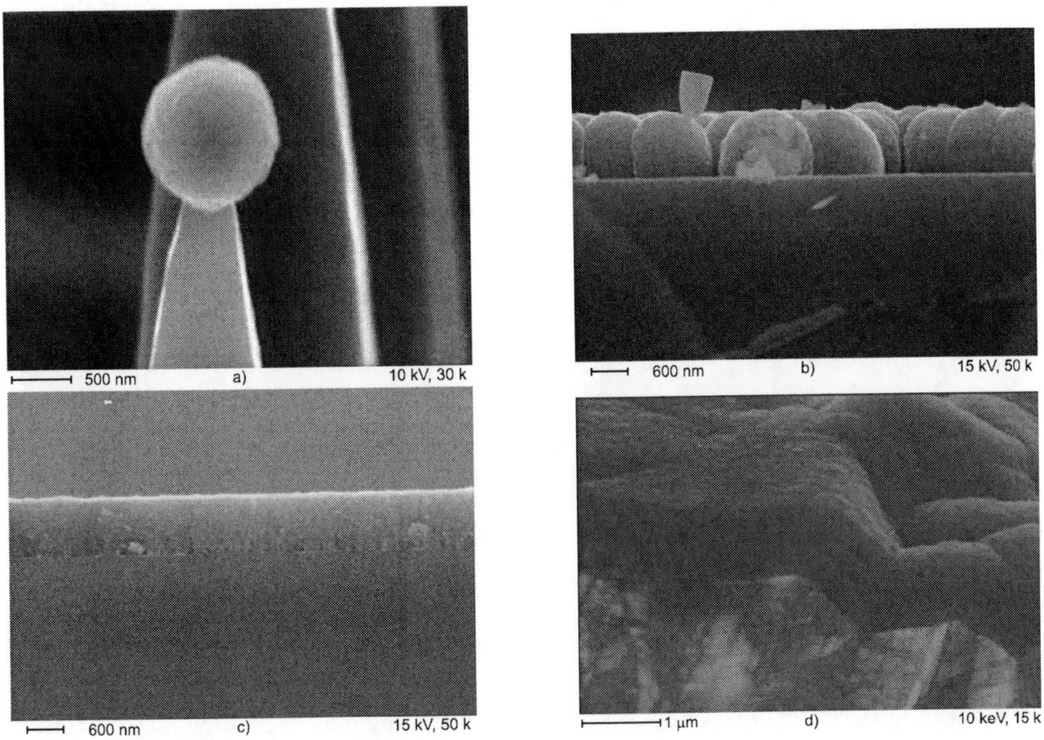


Figure 8. SEM images of a) a UNCD/a-C ball deposited on top of a micromachined silicon needle; b) a film deposited on silicon with a nucleation density of $1 \cdot 10^8$ cm$^{-2}$; c) a film deposited on silicon with a nucleation density of $1 \cdot 10^{10}$ cm$^{-2}$; d) a film deposited on a HFCVD PCD film.

Fig. 8 shows four SEM pictures of samples, the nucleation densities of which vary from the lowest possible to the highest possible values. The deposition conditions were the same for all samples with the exception of the UNCD ball shown in Fig. 8a), which had been obtained at 770°C (one hour deposition); for the other three samples, $T_s$ was 600°C and $t$ = 390 min:

i) Fig. 8a) shows an UNCD ball deposited on a sharp silicon needle fabricated by standard micromachining techniques [60]. No pretreatment has been applied; the nucleation density is very low (in a sense one per needle). The picture shows that — without any geometrical constraints — an almost perfect ball has developed which is, however, not perfectly smooth.

ii) Fig. 8b) stems from a sample which has been pretreated by the Standard I procedure listed in Table 1. The nucleation density is about $1 \cdot 10^8$ cm$^{-2}$. As a consequence, the film — which is about 1.1 $\mu$m thick — is not closed but consist of individual nodules only. It is important to note that these nodules show the same surface morphology as the single UNCD ball shown in Fig. 8a).

iii) The film shown in Fig. 8c) has been deposited in the same run, but this time the pretreatment procedure Standard II was applied which led to a nucleation density of about $1 \cdot 10^{10}$ cm$^{-2}$ [61]. As a consequence the film is closed instead of consisting of individual nodules. However, a closer inspection shows that there are voids present at the interface which are a consequence of a nucleation from individual but denser nodules.

iv) Fig. 8d), finally, stems from a film which has been deposited under exactly the same conditions on a polycrystalline diamond film (obtained by hot filament CVD from a 0.4% $CH_4/H_2$ mixture at 850°C). No pretreatment of the substrate has been applied. It can be seen that the PCD film is covered by a compact UNCD film without any structures or voids at the interface. Similar films have been obtained on cubic boron nitride films [23]. Evidently, the nucleation density must have been extremely high in this case; as discussed in [11], every point of the substrate surface serves as a nucleation site for the subsequent UNCD growth. The same holds for deposition on cubic boron nitride, which is regarding structure, lattice constant and properties very close to diamond [2].

We have already pointed out in [11] that the nucleation behaviour of NCD films is exactly the same as that of polycrystalline CVD diamond films. On smooth untreated surfaces the nucleation density is very low; nucleation takes place only at prominent surface sites (e.g. defects and scratches, but also on sharp needles as in Fig. 8a) or — after pretreatment — on nucleation sites created by this process. This means that the differences between nanocrystalline and polycrystalline films are solely based on the growth process, not on the nucleation step [11]. This conclusion is corroborated by the observation that for closed films such as shown in Fig. 8c,d) the rms surface roughness is independent of the nucleation density. As discussed in detail in [46, 61], the films "loose their memory" of the nucleation step once they are closed. Thus the nucleation density determines only the minimum thickness of closed films, and the density, width and height of voids at the interface (Fig. 8c)). As a consequence, the properties of closed UNCD/a-C films are influenced by the primary nucleation density only in cases in which these voids play a role.

## 4.2. Influence of the Substrate Temperature

The deposition of NCD/a-C films was investigated in the temperature range between 450 and 770°C. These experiments have been carried out to meet two aims: On the one hand it should be explored whether there is a low temperature limit for the deposition of this type of films. Well-facetted polycrystalline diamond films require under standard conditions a substrate temperature of at least 700°C [2, 27]. In fact, too low temperatures can lead to the formation of NCD instead of PCD [62]. For the deposition of PCD films at low temperatures gas phase chemistries utilizing halogens have to be used [63, 64]. On the other hand, the deposition of NCD films at temperatures below 500°C under conditions leading to UNCD also at higher temperatures has been reported by [65,66].

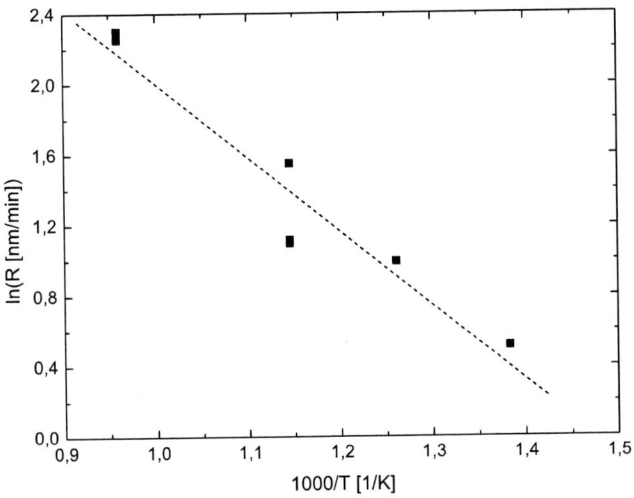


Figure 9. Deposition rates of UNCD/a-C films as a function of the substrate temperature in an Arrhenius presentation.

The second aim of this temperature series concerns the growth mechanisms, as the temperature dependence of e.g. the growth rate or of major film properties may give some insight as to what extent temperature dependent processes are involved in the growth. Fig. 9 shows the growth rate of UNCD/a-C films as a function of the substrate temperature in an Arrhenius presentation. The data lie fairly well on a straight line from which an activation energy of 0.36 eV can be derived. This is again in agreement with the growth of polycrystalline diamond for which, however, activation energies of 0.5 – 1.6 eV have been observed (see [2,67,68] and the references cited therein). On the other hand, McCauley et al. [68] reported an activation energy of 0.25 eV for the growth of UNCD films, which is quite close to the value found in the present study.

Fig. 10 shows grazing angle XRD measurements of UNCD/a-C films as a function of the substrate temperature. It can be seen that neither the peaks observed (the diamond (111) and (220) peaks) nor their positions, their widths and their intensity ratio vary with temperature. This means that in the temperature range covered by these experiments the properties of the nanocrystallites are not affected by the substrate temperature. This regards the crystallite size (peak width), the absence of any texture (peak ratio) and the absence of noticeable stresses (peak positions) likewise.

It has been mentioned above in Sec. 3.4. that the density of the films deposited at 770°C has been determined by XRR measurements to 2.75 $gcm^{-3}$. Nuclear reaction analysis studies yielded for a 600°C film a density of 2.65 $gcm^{-3}$ which is slightly lower. From the XRD results presented above it is clear that the properties of the crystalline fraction of the films are independent of the substrate temperature. This leaves two possible explanations

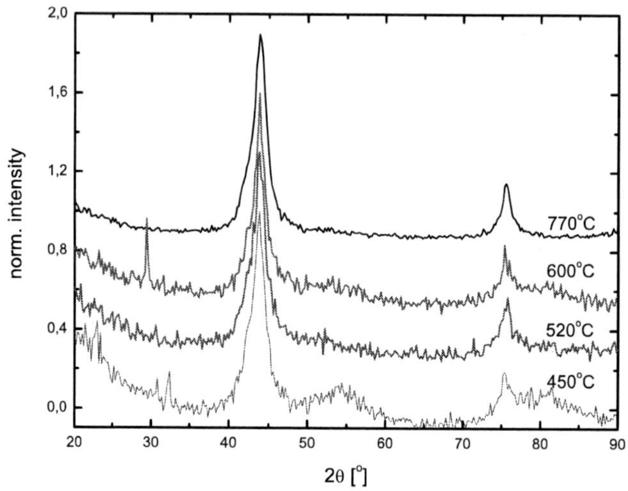


Figure 10. Glancing angle XRD spectra as a function of the substrate temperature in the range from 770 to 450°C.

for the reduced density of the 600°C film: i) either the volume ratio of crystallites and matrix has changed or ii) the properties of the matrix are different at lower temperatures. Although Raman and FTIR measurements indicate subtle changes of the matrix at lower temperatures as discussed in detail in [67], at the present time we can not decide this question definitively. Investigations to give more insight are presently under way.

### 4.3. Discussion

**Table 3. Nature of diamond films obtained by CVD processes as a function of the primary nucleation density $n_{nuc}$ and the rate of secondary nucleation $R_{sec}$.**

| $R_{sec}$ | $n_{nuc}$ | nucleation phase | thin films | thick films |
|---|---|---|---|---|
| low | low | few crystallites | growing crystallites | thick PCD film |
| low | high | dense crystallites | thin NCD film | thick PCD film |
| high | low | few nodules | growing nodules | thick UNCD film |
| high | high | dense nodules | thin UNCD film | thick UNCD film |

From the results presented in Sec. 4.1. it is clear that the morphology of UNCD films, once they are closed, is almost independent of the primary nucleation density $n_{nuc}$. There is a general agreement in the literature that the very mechanism leading to the growth of ultrananocrystalline diamond with all its properties (that means including the morphology) is a very high rate of secondary nucleation $R_{sec}$ [9–11,69]. The difference between nanocrystalline and ultrananocrystalline diamond, as defined by Butler [9], can in fact be found in the frequent re-nucleation which is unique to the latter. As shown schematically in Table 3, nanocrystalline films can also be obtained under PCD conditions (i.e. with a low rate of secondary nucleation), if the density of primary nucleation is very high ($10^{11}$ – $10^{12}$ $cm^{-2}$) [9]. But owing to the evolutionary growth of polycrystalline diamond [70, 71] leading to a survival of the fastest growing crystals while suppressing slowly growing ones, the crystallite size will increase with increasing thickness.

In contrast, if the rate of secondary nucleation is high, the films will stay (ultra)nanocrystrystalline. Once these films are closed their morphology and properties are solely determined by the secondary nucleation. For example, the mean crystallite size of a 4 $\mu$m thick film deposited under our conditions was 3-5 nm, the same value as observed for films of 1 $\mu$m thickness. The most impressing data stem from Hogmark et al. [72] reporting of a 140 $\mu$m thick UNCD film.

This discussion leaves two questions open. The first regards the conditions leading to such high rates of secondary nucleation. Although we are not able to give a final answer it is interesting to note that — as discussed in more detail in [11] — high values of $R_{\mathrm{sec}}$ are observed in many cases in which the rather narrow parameter window of PCD deposition [2, 73] is left. This can be done by either using hydrogen poor gas mixture instead of the $\approx$ 1% $CH_4/H_2$ mixture usually used in diamond CVD (our case of 17% $CH_4/N_2$ is an extreme example) or by lowering the pressure, by using low temperatures or by applying a bias voltage throughout the deposition process (see Ref. [11] and the references cited therein). Nevertheless, the question is not solved why all these measures lead to UNCD films with unique but similar properties.

The other open question in this context is why under our special conditions UNCD/a-C nanocomposite films are obtained, in contrast to many papers in the literature reporting on UNCD films with only low amounts of grain boundary material. Again, the question cannot be answered finally, but there may be a rather gradual transition between grain boundaries in the strict sense of the term, on the one hand, and an extended matrix as in our case, on the other hand. For example, the group at the Argonne National Laboratory observed an increase of the grain boundary width from 0.3 to 1.5 nm when up to 20% nitrogen was added to their 1% $CH_4/Ar$ gas mixture [57, 58].

# 5. Application Relevant Properties

In this section we will briefly summarize some application relevant properties of our NCD/a-C nanocomposite films. In each case the question is discussed to which extent they are influenced by the nanocrystalline nature of the films.

## 5.1. Mechanical Properties

The mechanical and tribological properties of UNCD/a-C films have been investigated by nanoindentation, nano scratch test and nano tribo test measurements [23, 74]. In the following we will briefly summarize the most important results.

Diamond is the hardest of all known single phase materials, with hardness values around 100 GPa, depending on the crystalline direction. The hardness values reported for poly-

**Table 4. Typical values of hardness $H$ and Young's modulus $E$ of single crystalline diamond, polycrystalline diamond films, nanocrystalline diamond films, and our UNCD/a-C nanocomposite films.**

| material | Ref. | $H$ [GPa] | $E$ [GPa] | remark |
|---|---|---|---|---|
| diamond | [1] | 100 | 1005 – 1210 | depends on orientation |
| PCD | [75] | 80 – 100 | 500 – 533 | MWCVD, $CH_4/H_2$ |
| PCD | [76] | — | 925 – 1080 | HFCVD, $CH_4/H_2$, 710°C |
| NCD | [77] | — | 517 – 1120 | MWCVD, $CH_4/H_2$, 720°C |
| UNCD | [56] | 87 – 92 | 750 | MWVCD, $CH_4/H_2$, 850°C |
| UNCD | [78] | 65 – 95 | — | MWVCD, $CH_4/H_2$, 500°C |
| UNCD/a-C | this work | 40 ± 2 | 390 ± 20 | MWVCD, $CH_4/N_2$, 770°C |
| UNCD/a-C | this work | 27 ± 2 | 265 ± 5 | MWVCD, $CH_4/N_2$, 600°C |

crystalline diamond films are in the range of 80 – 100 GPa, i.e. rather close to the single crystalline values (Table 4). The same is true even for nanocrystalline diamond films, for which hardnesses up to 95 GPa [56] have been reported. Thus on a first view the hardness values found for our 770°C UNCD/a-C films of 40 ± 2 GPa [23, 74] seem to be rather low. The same is true for the Young's modulus, as can be seen from the data also listed in Table 4.

These results can only be explained by the presence of the amorphous carbon matrix, which forms roughly one half of the material. On the other hand, as has already been pointed out in the introduction, for many mechanical or tribological applications not the hardness but rather the toughness is decisive. Diamond is an extremely brittle material [12], which means that it is prone to fatal brittle failure [13]. It has been shown that nanocomposites consisting of a hard or superhard nanocrystallites embedded in an amorphous matrix can add considerable toughness without loosing too much of the hard character.

From Table 4 it can be seen that the values of hardness and Young's modulus are even lower for films grown at 600°C. This may be caused by the slightly reduced density obtained at this temperature as discussed above in Sec. 4.2.. But the reduced film thickness (1 $\mu$m instead of 4 $\mu$m) may also have contributed to this decrease of $H$ and $E$. In this context it should be noted that for a 1 $\mu$m thick UNCD/a-C film deposited onto a polycrystalline diamond film at 600°C values of $H$ = 33.7 ± 4.1 GPa and $E$ = 362 ± 49 GPa have been observed.

Fig. 11 shows a 3-D image of a scratch with a length of 1 mm obtained during the adhesion investigation with a Nano Scratch Tester. The tests were performed with a Rockwell diamond indenter, applying a progressive load with a rate of 600 mN/min starting from 3 mN. Four different parts can be distinguished alongside the scratch: (i) no penetration of the indenter in the UNCD/a-C film, (ii) increasing of the penetration depth and beginning of the delamination, (iii) full delamination of the films, (iv) damage of the underlaying silicon substrate in those areas where full delamination had occurred, proving the protecting nature

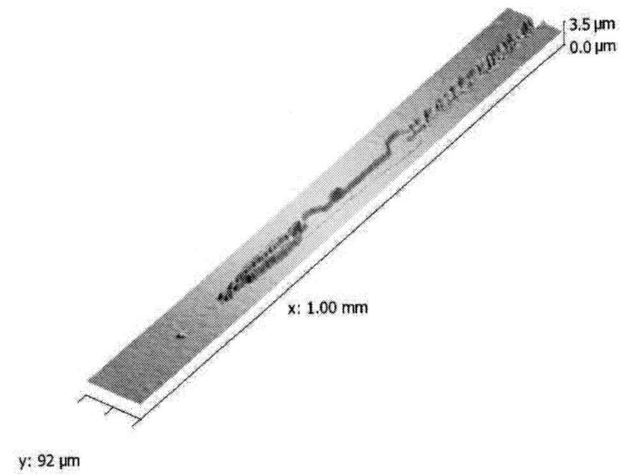


Figure 11. 3-D image of a 1 mm long scratch on a UNCD/a-C film on a silicon substrate created with a nano scratch tester.

of the UNCD/a-C coating [23].

The results of such nano scratch tests, performed with 600 and 770°C samples [23, 74], and of additional nano tribo tests (770°C only) can be summarized as follows:

i) Irrespective of the substrate temperature, the films possess a protective effect on Si substrates.
ii) Very good adhesion was observed for UNCD/a-C films on Si substrates, but also on cubic boron nitride and TiN.
iii) The friction coefficient, measured in ball-on-disk configuration with an $Al_2O_3$ ball as counterpart, was initially about 0.5 but decreased soon to values below 0.1, which are in agreement with literature values for UNCD films [79]. The high starting value is probably due to the surface roughness of 9 – 13 nm of these films which is quickly worn off.
iv) No delamination of the films was observed during these friction tests even after 10,000 laps although the $Al_2O_3$ counterpart was considerably worn.

## 5.2. Electrical Properties

Two completely different sets of measurements have been performed to establish the electrical properties of the UNCD/a-C films. The first consists of $I/V$ and Hall measurements performed in van der Pauw geometry as shown schematically in Fig. 12a). The second series was performed in vertical geometry as shown in Fig. 12b). It consists of impedance measurements carried out in the frequence range from 1 Hz to 1 MHz and the temperature range from 216 to 295 K.

### 5.2.1. Van der Pauw-Measurements

Results of the measurements performed in van der Pauw geometry are shown in Fig. 13. From the I/V curve presented in the figure, and from Hall measurements performed with

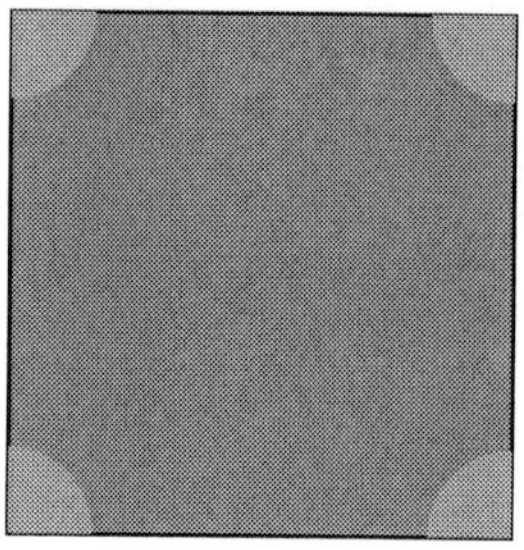

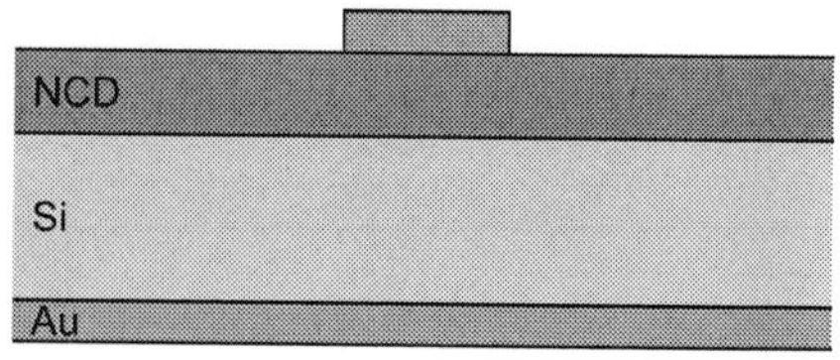


Figure 12. Geometries applied for the measurement of the electrical surface (left) and bulk (right) properties of UNCD/a-C films.

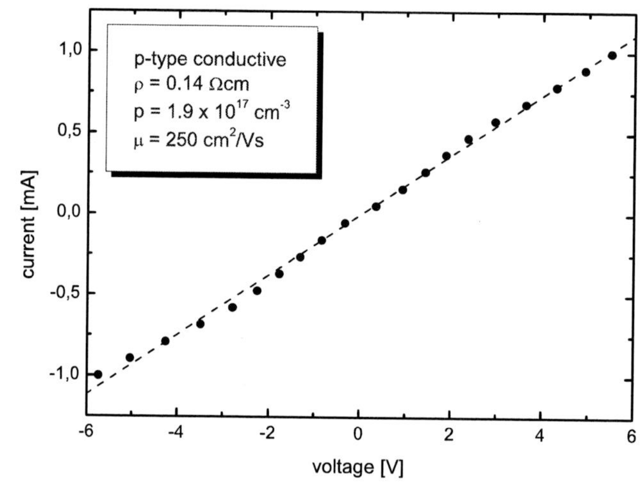


Figure 13. Current/voltage curve of a UNCD/a-C film recorded in van der Pauw geometry. Results of Hall measurements performed with the same geometry are given in the inset.

the same geometry (for details see [80]) the following results have been obtained:

i) The I/V curve is almost linear, indicating an ohmic behaviour.
ii) The film is p–type conductive with a resistivity of 0.14 Ωcm.
iii) The carrier concentration and the carrier mobility are $1.9 \cdot 10^{17}$ $cm^{-3}$ and 250 $cm^2/Vs$, respectively.

### 5.2.2. Impedance Measurements

Results of the impedance measurements carried out to establish the bulk electrical properties are presented in Fig. 14. At room temperature, the conductivity is $7.7 \cdot 10^{-7}$ S/cm, corresponding to a resitivity $\rho = 1.3 \cdot 10^6$ Ωcm. It can be seen from the diagram that at lower temperatures the conductivity decreases; the Arrhenius plot of these data gives a straight line from which an activation energy of 0.17 eV can be derived. The inset of Fig. 14

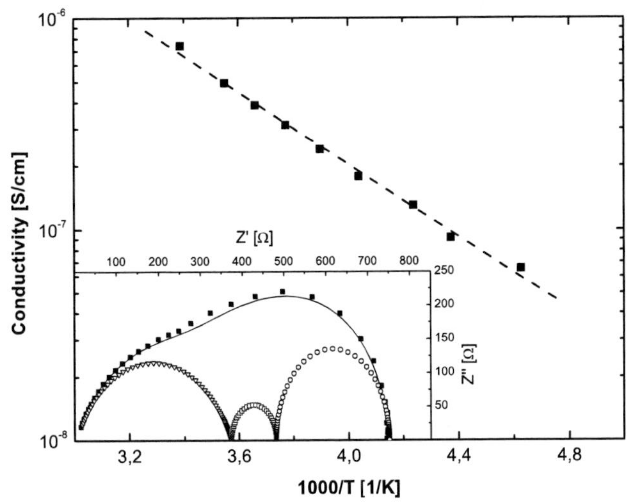


Figure 14. Bulk conductivity of a UNCD/a-C film as a function of the temperature in the range from 210 to 295 K. The inset shows a Cole-Cole plot for 295 K.

shows a Cole-Cole plot of the frequency depending complex impedance (i.e. the imaginary part Z"($\omega$) vs the real part Z'($\omega$)). The best fit was obtained by assuming three contributions to the conductivity of the film, two RC elements and one constant phase element.

### 5.2.3. Discussion of the Electrical Properties

On a first view, there seems to be a strong contradiction between the van der Pauw measurements of the electrical surface properties ($\rho = 0.14$ Ωcm) and those of the impedance measurements of the bulk properties ($\rho = 1.3 \cdot 10^6$ Ωcm). However, this apparent contradiction can be explained rather easily. It is well-known that all hydrogen-terminated diamond surfaces are p-type conductive with sheet carrier concentrations on the order of $10^{13} - 10^{14}$ $cm^{-2}$ [81, 82]. This holds for single- [83], poly- [82, 84] and nanocrystalline diamond [85] likewise. The reason for this surface conductivity can be found in the electronic band structure in the surface region, which not only causes this p-type conductivity but also a negative electron affinity of H-terminated diamond surfaces [81].

From TOF-SIMS measurements and other investigations we know that the as-grown surfaces of our UNCD/a-C films are also hydrogen-terminated [45, 86]. This means that irrespective of the nanocomposite nature of these films, i.e. the presence of an amorphous matrix, the electrical surface properties are very similar to those of PCD and pure (U)NCD films (the sheet hole concentration of our films is $7.6 \cdot 10^{13}$ $cm^{-2}$).

Whereas thus the electrical surface properties of our UNCD films are in agreement with literature data, this is not the case for the bulk conductivity. Bhattacharyya et al. [57] observed that on addition of up to 20% $N_2$ to a 1% $CH_4$/Ar mixture the bulk conductivity increased from 0.016 S/cm (for 1% $N_2$) to 143 S/cm (for 20% $N_2$). This has been ascribed to grain boundary conductivity caused by nitrogen related defects in the grain boundaries [58]. On the other hand, the nitrogen content of these films ($2 \cdot 10^{20}$ cm$^{-3}$ [57]) is even somewhat lower than that of ours (ca. $10^{21}$ cm$^{-3}$). It must thus be concluded that our matrix must possess a structure different from these grain boundaries, as we have observed a much higher resistivity.

On the other hand, Feliciangeli et al. [87] and Ye et al. [88] found in impedance spectroscopy measurements with NCD films only two contributions to the conductivity, one of which becomes apparent only at higher temperatures (450°C [87] and 250 °C [88], respectively). They have been accociated with the grain interior and the grain boundaries, respectively. The fact, that we observed three contributions, all of them fully developed even at 216 K, is another indication that there are fundamental differences concerning the structure of the non-crystalline material between "pure" UNCD and UNCD/a-C films. However, at the present time we are not able to ascribe these three contributions visible in the impedance spectra to possible conduction paths within the films.

## 5.3. Optical Properties

The optical properties of the UNCD/a-C nanocomposite films have been investigated by reflectance and scattering measurements in the UV/VIS/NIR range from 200 to 1000 nm, by spectral ellipsometry [89] and by variable angle ellipsometry [46]. Details of these measurements can be founds in the references given. The major result of these studies is that the optical properties of our nanocomposite films are markedly inferior to those of PCD or pure UNCD films.

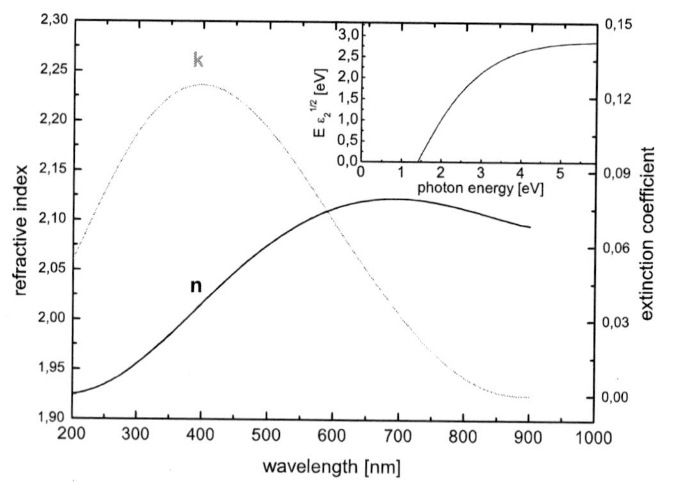


Figure 15. Wavelength dependence of the refractive index $n$ and extinction coefficient $k$ of a UNCD/a-C film deposited at 770°C. The inset shows a Tauc plot of these data.

An example is given in Fig. 15, showing the wavelength dependence of the refractive index $n$ and the extinction coefficient $k$ for a 4 $\mu$m thick film deposited at 770°C. The data represent a fit on the basis of the Tauc-Lorentz dispersion model to both, reflectance and spectroscopic ellipsometry measurements. It can be seen that the refractive index increases from 1.93 at 200 nm to ca. 2.12 at 600–700 nm. This has to be compared to the value of 2.41 of single crystalline diamond [2], but also values of 2.34 [90] for NCD films. It can also be seen from Fig. 15 that there is considerable absorption for wavelengths $\leq$ 800 nm. The extinction coefficient $k$ possesses a maximum of 0.12 at ca. 400 nm. Considerable absorption especially at low wavelengths has also been reported for nanocrystalline diamond by other auhors [90–92].

The inset of Fig. 15 presents a Tauc plot derived from these data. It shows that the optical band gap of the film is ca. 1.4 eV, which is by far lower than that of diamond (5.4 eV [1, 5]) but on the order of values reported for amorphous carbon films. Thus from these results it must be concluded that the optical properties of our nanocomposite films are dominated by the amorphous carbon matrix. Especially the presence of $sp^2$ bonded carbon in the matrix (see Sec. 3.2.) may be the reason for the high absorption within the films.

## 5.4. Biological Properties

In the rapidly developing field of biotechnology, there are at least two applications currently envisioned for (ultra)nanocrystalline diamond films. Both of them make use of the intrinsic properties of diamond and the smoothness of NCD/UNCD surfaces alike. The first regards tribological and protective coating for implants, e.g. temperomandibular joint replacements [93, 94] or retinal microchips [26], utilizing the mechanical and chemical properties and the biocompatibility of diamond, while in the second they are used as a template for the immobilization of biomolecules, e.g. for biosensors and DNA-chips [95–98], utilizing the extreme chemical stability of diamond. In the following we report on preliminary experiments carried out in order to investigate the suitability of our UNCD/a-C nanocomposite films for these two purposes.

### 5.4.1. Surface Characterization and Modification

In order to use UNCD/a-C films for any biotechnological application, it is necessary to know the chemical nature of their surface and to be able to modify it if required for a given application. For this reason, five different surfaces have been prepared by plasma and chemical treatments:

i) An as-grown UNCD/a-C surface (AG in the following);
ii) A surface subjected to a $H_2$ MW plasma (HP);
iii) A surface subjected to a $O_2$ MW plasma (OP);
iv) A surface subjected to a $CHF_3$ rf plasma (FP);
v) A surface subjected to a chemical treatment with aqua regia ($HCl/HNO_3$ with a ratio of 3:1) for 90 min at room temperature (CT).

The conditions of these treatments are summarized in Table 5. The resulting surfaces have been characterized by XPS and time-of-flight secondary ion mass spectrometry (TOF-SIMS), AFM and contact angle measurements with respect to their composition, roughness and hydrophobicity. Table 5 also contains the compositions of the different surfaces as revealed by XPS [45]. It can be seen that the AG and CT surfaces are extremely clean, especially if one takes into account that the UNCD film bulk contains already 0.5 - 1% oxygen

**Table 5. Parameters used for the preparation of differently treated surfaces. The results of the characterization of the surfaces by XPS, AFM and contacts angle measurements are given also.**

| Parameter | | AG | HP | OP | FP | CT |
|---|---|---|---|---|---|---|
| Plasma | | 2.45 GHz | 2.45 GHz | 2.45 GHz | 13.56 MHz | — |
| Power | [W] | 800 | 800 | 200 | 70 | — |
| Gas | | 17%$CH_4/N_2$ | $H_2$ | $O_2$ | $CHF_3$ | — |
| Gas flow | [sccm] | 300 | 250 | — | 5 | — |
| Pressure | [kPa] | 2.2 | 6 | 0.2 | 0.066 | — |
| Temperature | [°C] | 600 | 400 | rt | rt | rt |
| Duration | [min] | 390 | 30 | 10 | 10 | 90 |
| C | [%] | 96.9±0.6 | 93.3±0.2 | 87.4±0.1 | 81.9±2.3 | 97.6±0.5 |
| O | [%] | 2.2±0.4 | 4.1±0.3 | 12.0±0.4 | 4.8±0.2 | 1.8±0.1 |
| N | [%] | 0.8±0.1 | 1.2±0.3 | 0.7±0.1 | 0.8±0.2 | 0.7±0.1 |
| Si | [%] | — | 1.4±0.1 | — | — | — |
| F | [%] | — | — | — | 12.5±2.4 | — |
| rms | [nm] | 10.0 | 12.1 | 12.1 | — | 10.5 |
| $\theta_c$ | [°] | 86±2 | 88±2 | 7±1 | 100±3 | 67±2 |

and nitrogen (see Section 3.3.). The slight oxygen contamination of the HP surface stems very probably from $SiO_2$ species from the quartz window of the chamber in which the $H_2$ plasma treatment has been performed. More important is the observation that the surfaces exposed to $O_2$ and $CHF_3$ plasmas contain ca. 12% oxygen and fluorine, respectively. Thus it seems that the UNCD surfaces can be functionalized with O and F.

Further insight was brought by TOF-SIMS investigations. Fig. 16 show count rate ratios $C_xH/C_x$ for $x = 2, 4, 6$, which have been normalized with respect to the values of the AG surface. These ratios can be taken as a measure of the hydrogen termination of the surfaces [45]. The hydrogen plasma treatment leads to a slight increase of the ratios which means that already the as-grown surface is hydrogen terminated, in agreement with recent nuclear reaction analysis results [47]. On the other hand, for the $O_2$ and $CHF_3$ plasma treatments the values are low, indicating that surface C-H bonds have been replaced by oxygen and fluorine containing species, respectively. The value for the aqua regia treated sample is somewhat in between, which means that the hydrogen termination has been removed partially although according to the XPS data in Table 5 it was obviously not replaced by another element.

The TOF-SIMS measurements had revealed further that in the case of the $O_2$ plasma treatment the surface is terminated by C-O and C-OH species rather than by carboxylic acid COOH groups [45]. Likewise it could be shown that the $CHF_3$ plasma did not lead to the deposition of a $C_xF_yH_z$ polymer-like films but just to the formation of $CF_x$ groups at the surface [86].

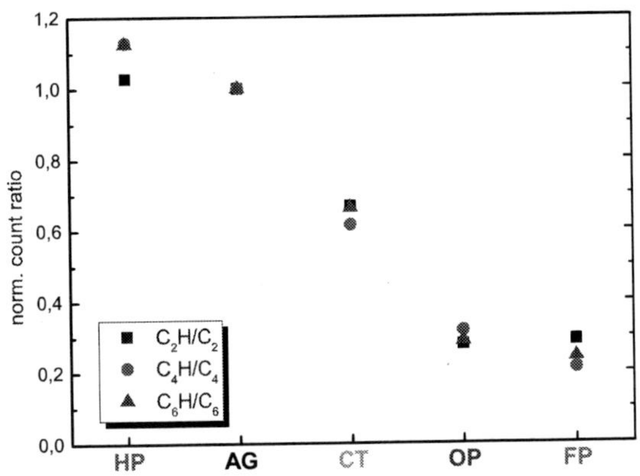


Figure 16. TOF-SIMS count ratios $C_xH/C_x$ for $x$ = 2, 4, 6, normalized with respect to the AG values, for the differently treated surfaces.

The values of the surface roughness also given in Table 5 do not vary significantly. They are in the close range of 9 – 14 nm found for all our samples [46, 61]. It can thus be concluded that none of the treatments has a significant effect on the morphology of the films. Especially, grain boundary etching can be excluded.

The contact angle $\theta_c$ against purified water for as-grown surfaces was 86 $\pm$ 2°. After hydrogen plasma treatment it increases slightly which is in agreement with the slight improvement of the hydrogen termination evident from Fig. 16. For the fluorine treated surface $\theta_c$ is even higher (100 $\pm$ 3°). One the other hand, for the $O_2$ plasma treated surface very low values of $\theta_c$ were observed. These contact angle measurements thus confirm the successful functionalization of UNCD/a-C surfaces with oxygen and fluorine, respectively. The intermediate value of $\theta_c$ observed for the aqua regia treated surface is in agreement with the partial removal of the H-termination revealed by the TOF-SIMS investigations.

In order to investigate the stability of these different UNCD surfaces, contact angle measurements have been performed from time to time over the period of one year, during which the samples have been stored in air. The results are shown in Fig. 17. They can be summarized as follows:

i) The most stable surfaces are the as-grown and the $H_2$ plasma treated ones, i.e. the hydrogen terminated surfaces. In both cases, the contact angles were > 80° and > 85°, respectively, over the period of a year.
ii) For the fluorine terminated surface a slight decrease of the contact angle from 111000 $\pm$ 3° to 95 $\pm$ 2° was observed. Whether this decrease is significant still has to be established.
iii) For the aqua regia treated sample an increase of the contact angle is observed. Here again the reasons are not clear, but eventually the surface recovered somewhat from the partial removal of the hydrogen termination caused by the treatment.
iv) The oxygen plasma treated sample shows a pronounced increase of the contact angle with time. In addition, there is a large scattering of the data especially for interme-

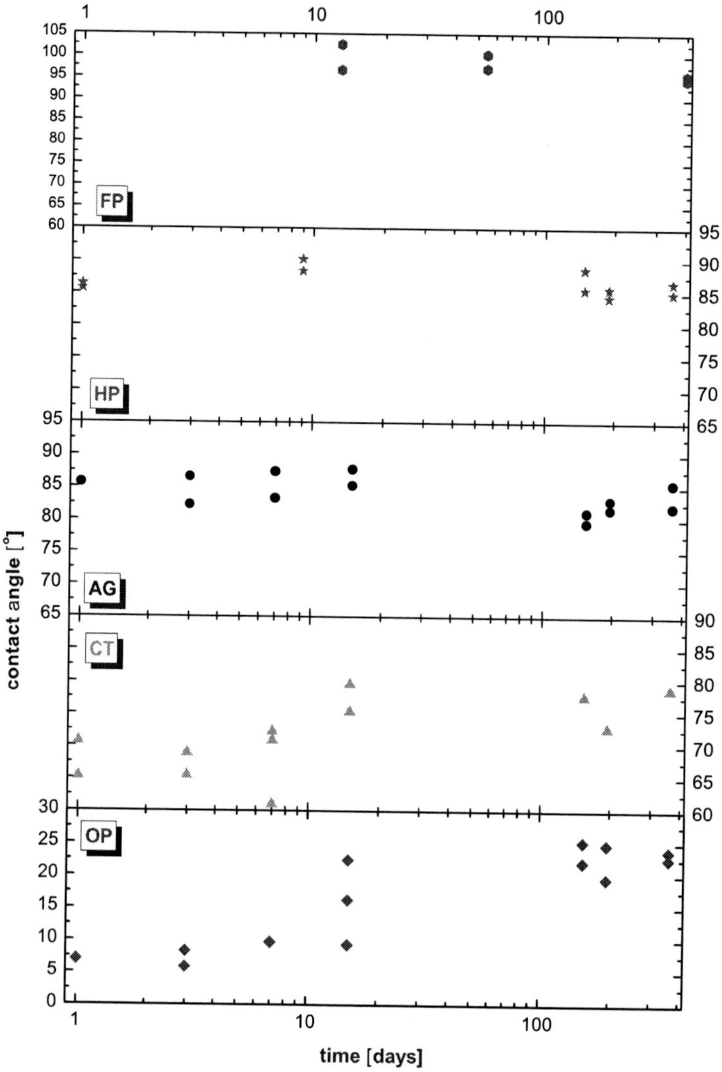


Figure 17. Contact angles as a function of time for the five different surfaces of this study.

diate times. It has been shown by TOF-SIMS [45] that this surface is very prone to contaminations e.g. by amine ($C_xH_yN$) molecules. Very probably, such contaminations lead to a degradation of the oxygen functionalized surface.

### 5.4.2. Biocompatibility and Bioinertness

The biocompatibility of the UNCD/a-C films has been investigated by direct contact tests with four different types of cells:

i) Osteoblast-like SaOS-2 cells [80].

ii) Human embryonal lung fibroblasts $LEP_{19}$ [99].
iii) Human endothelial cells EA.hy 926 [100].
iv) Mouse embryonic fibroblasts Balb/3T3.

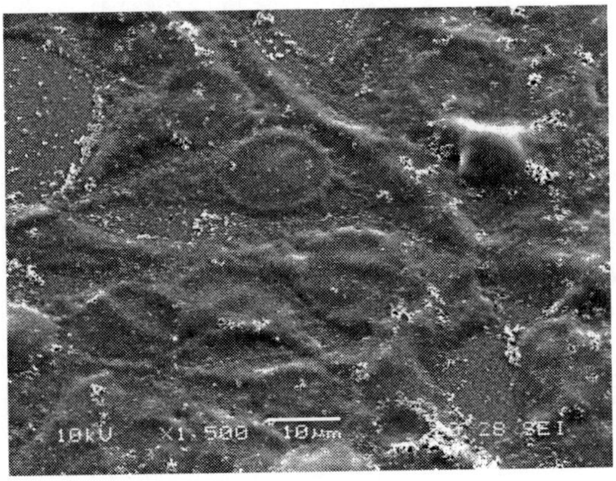


Figure 18. Human endothelial cells EA. hy 926 in contact with a UNCD/a-C surface after 5 days of growth.

It turned out in all cases that continuous hydrogen terminated UNCD/a-C films are biocompatible; they are not cytotoxic and do not affect the health of the cells. An example is shown in Fig. 18 showing human endothelial EA.hy 926 cells in contact with such a surface. The cells show good attachment and growth; similar to control cells they have developed elongated extensions.

For the Balb/3T3 cells, similar results were not only obtained for as-grown, H-terminated but also for fluorine and oxygen terminated surfaces. It turned out, however, that not-closed, nodule-type surfaces such as shown in Fig. 8b) are detrimental for the health of the cells. This was observed for mouse embryonic fibroblasts as well as for human endothelial cells. For the latter, a reduced size of the cells and irregular cell borders were observed [100]. This is in agreement with the general observation that the topography of a surface has a great influence on the viability and proliferation of cells [101, 102].

In another series of experiments we were able to show that the UNCD/a-C films are bioinert [80]. This means that in the presence of a simulated body fluid (SBF), a solution very close to the composition of blood plasma, no hydroxy apatite (HA, $Ca_5(PO_4)_3OH$) is formed on them even after exposure for ten days. In the case of bioactive surfaces the formation of HA takes place immediately. It can be observed by changes of the SBF but also directly by XRD, FTIR or X-ray fluorescence. In our case, neither a significant change of the fluid (by atomic absorption spectrocopy) nor the presence of HA (by SEM and XRD) could be observed which means that UNCD/a-C films are bioinert.

### 5.4.3. Non-Specific Adsorption of Biomolecules

Another study has been carried out to investigate whether as-grown UNCD/a-C surfaces are prone to unspecific interactions with biomolecules. To this end double-stranded RNA (with a length of $118 \pm 4$ nm) and proteins which can bind to RNA (20 nm diameter) have been deposited on UNCD surfaces but also on mica and glass for control purposes [99]. Thereafter, the samples were carefully cleaned and characterized by AFM and scanning force spectroscopy.

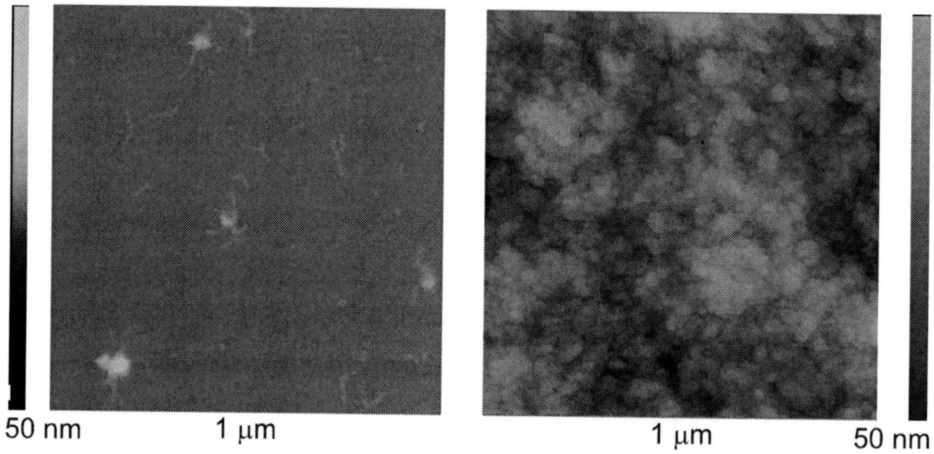


Figure 19. AFM images of a mica (left) and a UNCD/a-C surface (right) which have been exposed to dsRNA and proteins.

Fig. 19 shows AFM topography images of a mica sample and a UNCD surface after this treatment. On the mica surface, clearly RNA strands but also rounded proteins are discernable. They are completely absent on the UNCD surface, which shows only the typical morphology of ultrananocrystalline diamond [46]. These observations were confirmed by the phase images of these measurements (not shown graphically) [99]. To further corroborate these results, scannning force spectroscopy measurements have been carried out. A silicon cantilever which was functionalized with bovine serum albumin has been used for this purpose. 120 force/distance curves have been recorded at random positions on a glass control sample and an UNCD/a-C surface. Whereas in the former case in 38% of the measurements a clear interaction of the cantilever with the surface took place, this was observed in none of the measurements with the UNCD surface [99]. Thus, these results prove that as-grown (i.e. H-terminated) UNCD/a-C surfaces are not prone to unspecific interactions with biomolecules which is of great advantage for the application of these films in biosensors.

### 5.4.4. Attachment of RNA

Finally, first experiments have been carried out to functionalize UNCD/a-C surfaces with RNA molecules. A photochemical approach similar to those reported recently in the literature [95–98, 103–105] has been used [100], which is show schematically in Fig. 20. It consists of the following steps:

(1) Photochemical attachment of 1-amino-3-cyclopentene hydrochloride ($C_5H_{10}ClN$) by exposure to UV light (244 nm) for 195 min.
(2) Deprotection of the amino groups in 10% $NH_4OH$.
(3) Attachment of sulfosuccinimidyl 4-(N-maleimi-domethyl)cyclohexane-1-carboxylate (SSMCC) which serves as a linker molecule.

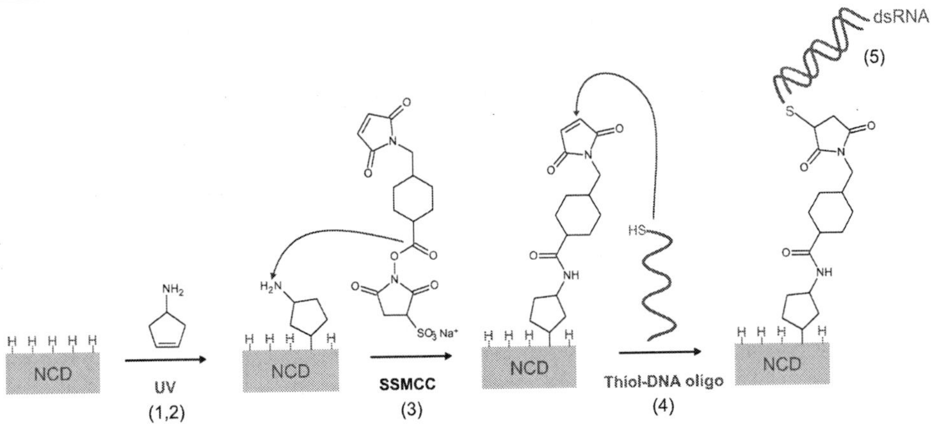


Figure 20. Photochemical approach to functionalize UNCD surfaces with RNA.

(4) Functionalization with a thiolated DNA oligonucleotide (5' GATCCCCGGTA-CCGAGCTCGAATTCGCCC-SH 3').
(5) Attachment of a 374 base pair long double RNA strand with a 30 nucleotide single stranded overhang at the 5' end, which can form Watson-Crick base pairs with the thiolated DNA oligonucleotide.

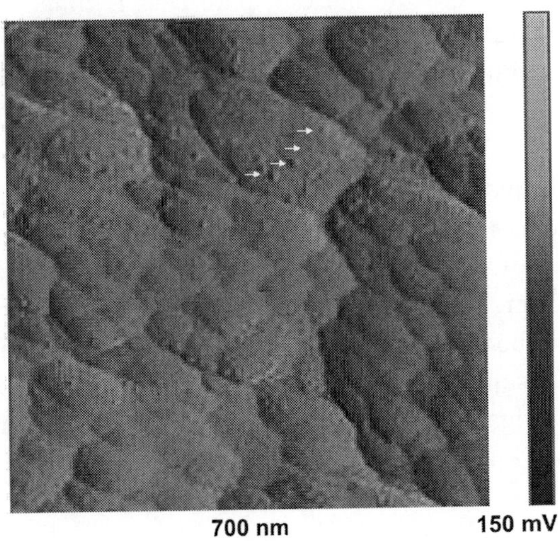


Figure 21. AFM phase image of an UNCD/a-C surface functionalized according to the scheme shown in Fig. 20.

The results of this procedure has been examined by AFM. In the phase image shown in Fig. 21 a number of objects (ca. 20 nm) can be seen (some of them are indicated by arrows). Such objects were not observed on control samples, neither on a UNCD/a-C surface which was exposed to the RNA but not functionalized, nor on a functionalized sample which was not exposed to RNA. It is therefore tempting to identify these objects with RNA molecules which have been attached to the functionalized surface [100]. Further experiments to verify this identification by other detection methods are presently under way.

# 6. Conclusion

Ultrananocrystalline diamond/amorphous carbon (UNCD/a-C) nanocomposite films have been deposited by microwave plasma chemical vapour deposition from 17% $CH_4/N_2$ mixtures in the temperature range from 450 to 770°C. They consist of diamond nanocrystallites of 3 - 5 nm diameter which are embedded in an amorphous carbon matrix. A pretreatment of the substrate with diamond powder is necessary to achieve a high nucleation density and an early coalescence to continuous films. However, once the films are closed they loose their memory of this nucleation step; the morphology is solely determined by the high rate of secondary nucleation, which is unique to UNCD deposition processes and the reason for the very small crystallite size of 3 - 5 nm. It could be shown that the growth process is thermally activated, which means that the growth rate increases with temperature. The crystallite size it not affected by the temperature, but there are subtle changes of the nature of the matrix and/or the crystallite/matrix volume ratio.

These UNCD films have been charactarized with respect to their mechanical, electrical, optical, and bio(techno)logical properties. In all cases where the bulk of the films is decisive, the presence of the amorphous carbon matrix has a pronounced influence on these properties. For example, the hardness and the Young's modulus are lower than those of diamond and also "pure" UNCD films, but on the other hand nano scratch tests and nano tribo tests have revealed a pronounced protective effect e.g. on silicon substrates. The optical properties are clearly inferior to those of pure UNCD films. The refractive index is that of an effective medium consisting of diamond nanocrystals and an amorphous carbon matrix, while the absorption is solely defined by the matrix which contains $sp^2$ hybridized carbon. Surprisingly, the surface properties of the UNCD/a-C films are not affected by the presence of the amorphous matrix. An example is the p-type conductivity of the surface, which is common to all hydrogen terminated diamond surfaces. But it also holds for the chemical behaviour of our UNCD/a-C films and for their interaction with biomolecules and larger biological entities. Like all diamond surfaces, they are biocompatible, i.e. not cytotoxic, bioinert and not prone to unspecific interactions with biomolecules such as RNA. Finally, the chemistry to functionalize UNCD/a-C surfaces and to immobilize biomolecules on them seems to be the same than that required for PCD and pure UNCD films although here clearly further investigations are necessary which are currently under way.

## Acknowledgments

A work such as presented above could of course not have been carried out without the cooperation with many coworkers at our home institutes, the JRC in Ispra and the INA in Kassel, but also with collegues from all over the world. We would therefore like to thank at the JRC: G. Ceccone (XPS), L. Cereotti (cell tests), P.N. Gibson (XRD), D. Gilliland (TOF-SIMS), H. Rauscher (Ellipsometry), T. Sasaki (SEM), L. Sirghi (AFM), F. Rossi (Discussion); at the INA: S. Bliznakov (deposition), S. Boycheva (deposition, surface modification), G. Georgiev (SEM), C. Sippel (deposition, AFM), J.P. Reithmaier (discussion); throughout the world: M.D. Apostolova and N. Milinovik (Bulgarian Academy of Sciences, cell tests), A. Bergmaier (Technical University Munich, ERD), A. Bock (PhysTech GmbH, Moosburg, electrical measurements), J. Bulir (Czech Academy of Sciences, optical measurements), T. Dostalova (Charles University Prague, cell tests); G. Favaro and L. Buforn (CSM Instruments, Switzerland, mechanical testing), D. Grambole (Research Center Dresden-Rossendorf, NRA), C. Hammann and N. Anspach (University of Kassel, RNA and protein interaction), E. Lefterova (Bulgarian Academy of Sciences, electrical measurements), Marco Morra and Clara Casinelli (Nobil Bio Ricerche, Villafranca d'Asti, cell tests), G. Spasov (Bulgarian Academy of Sciences, AES), J. Strnad (Lasak Ltd, Prague, SBF tests), M. Veres (Research Institute for Solid State Physics and Optics, Budapest, IR Raman), V. Vorlicek (Czech Academy of Sciences, Raman), K. Yamamoto, M. Hasegawa and K. Tsugawa (AIST, Tsukuba, TEM, EELS and UV Raman).

## References

[1] K. E. Spear and J. P. Dismukes, editors, *Synthetic Diamond: Emerging CVD Science and Technology*, John Wiley & Sons, New York, 1994.

[2] W. Kulisch, *Deposition of Superhard Diamond–Like Materials*, Springer Tracts on Modern Physics, Heidelberg Berlin, 1999.

[3] P. K. Bachmann and R. Messier, *Chemical & Engineering News* , **25** (1989).

[4] W. A. Yarbrough and R. Messier, *Science* **247**, 688 (1990).

[5] P. Koidl and C.-P. Klages, *Diamond Relat. Mater.* **1**, 1065 (1992).

[6] X. Jiang and C.-P. Klages, *phys. stat. sol. (a)* **154**, 175 (1996).

[7] R. Locher, C. Wild, N. Herres, D. Bohr, and P. Koidl, *Appl. Phys. Lett.* **65**, 34 (1994).

[8] D. M. Gruen, *Mat. Res. Soc. Bull.* **26**, 771 (2001).

[9] J. Butler, *Electrochem. Soc. Interface* **12**, 22 (2003).

[10] D. M. Gruen, *Annu. Rev. Mater. Sci.* **29**, 211 (1999).

[11] W. Kulisch and C. Popov, *Phys. Stat. Sol. (a)* **203**, 203 (2006), Feature Article.

[12] J. E. Field, *The Properties of Natural and Synthetic Diamond*, Academic Press, London, 1992.

[13] W. Kulisch, in *Functional Properties of Nanostructured Materials*, edited by R. Kassing, P. Petkov, W. Kulisch, and C. Popov, page 113, Berlin, 2006, Springer Verlag.

[14] H. Holleck and V. Schier, *Surf. Coat. Technol.* **76–77**, 328 (1995).

[15] J. S. Zabinski and A. A. Voevodin, *J. Vac. Sci. Technol.* **A16**, 1890 (1998).

[16] S. Hogmark, S. Jacobson, and M. Larsson, *Wear* **246**, 20 (2000).

[17] P. B. Mirkarimi, L. Hultman, and S. A. Barnett, *Appl. Phys. Lett.* **57**, 2654 (1990).

[18] P. C. Yashar and W. D. Sproul, *Vacuum* **55**, 179 (1999).

[19] M. Lattemann, S. Ulrich, H. Holleck, M. Stüber, and H. Leiste, *Diamond Relat. Mater.* **11**, 1248 (2002).

[20] A. A. Voevodin and J. S. Żabinski, *J. Mater. Sci.* **33**, 319 (1998).

[21] S. Veprek, *J. Vac. Sci. Technol.* **A 17**, 2401 (1999).

[22] A. A. Voevodin and J. S. Zabinski, *Thin Solid Films* **370**, 223 (2000).

[23] W. Kulisch, C. Popov, V. Vorlicek, P. N. Gibson, and G. Favaro, *Thin Solid Films* **515**, 1005 (2006).

[24] C. Popov, S. Bliznakov, and W. Kulisch, *Diamond Relat. Mater.* **16**, 740 (2007).

[25] S. A. Catlegde and Y. K. Vohra, *J. Appl. Phys.* **86**, 698 (1999).

[26] X. Xiao, J. Wang, C. Liu, J. A. Carlisle, M. Mech, R. Greenberg, D. Guven, R. Freda, M. S. Humayun, J. Weiland, and O. Auciello, *J. Biomed. Mater. Res. B* **77B**, 273 (2006).

[27] P. W. May, *Phil. Trans. R. Soc. Lond.* **A 358**, 473 (2000).

[28] M. Werner and R. Locher, *Rep. Prog. Phys.* **61**, 1665 (1998).

[29] C. Popov and W. Kulisch, in *Proceedings of the CVD XVI and EUROCVD-14 Conference*, page 1079, Pennington, N.J., 2003, The Electrochemical Society.

[30] C. Popov, M. Novotny, M. Jelinek, S. Boycheva, V. Vorlicek, M. Trchova, and W. Kulisch, *Thin Solid Films* **506-507**, 297 (2006).

[31] C. Popov, W. Kulisch, P. N. Gibson, G. Ceccone, and M. Jelinek, *Diamond Relat. Mater.* **13**, 1371 (2004).

[32] T. H. de Keijser, E. J. Mittermeijer, and H. C. Rozendaal, *J. Appl. Cryst.* **16**, 309 (1983).

[33] C. Popov, W. Kulisch, S. Bliznakov, B. Mednikarov, G. Spasov, J. Pirov, M. Jelinek, T. Kocourek, and J. Zemek, *Appl. Phys.* **A 89**, 209 (2007).

[34] I. Montero, L. Galan, A. Laurent, J. Perriere, and J. Spousta, *Thin Solid Films* **228**, 72 (1993).

[35] J. C. Lascovic and S. Scaglione, *Appl. Surf. Sci.* **78**, 17 (1994).

[36] S. T. Jackson and R. G. Nuzzo, *Appl. Surf. Sci.* **95**, 90 (1995).

[37] D. S. Knight and W. B. White, *J. Mater. Res.* **4**, 385 (1989).

[38] J. Wagner, C. Wild, and P. Koidl, *Appl. Phys. Lett.* **59**, 779 (1991).

[39] K. Okada, K. Kanda, S. Komatsu, and S. Matsumoto, *J. Appl. Phys.* **88**, 1674 (2000).

[40] J. R. Nemanich, J. T. Glass, G. Lucovsky, and R. E. Shroder, *J. Vac. Sci. Technol.* **A 6**, 1783 (1988).

[41] S. Prawer, K. W. Nugent, D. N. Jamieson, J. O. Orwa, L. A. Bursill, and J. L. Peng, *Chem. Phys. Lett.* **332**, 93 (2000).

[42] A. C. Ferrari and J. Robertson, *Phys. Rev. B* **63**, 121405 (2001).

[43] R. Pfeiffer, H. Kuzmany, B. Salk, and B. Günther, *Appl. Phys. Lett.* **82**, 4149 (2003).

[44] C. Popov, W. Kulisch, S. Boycheva, K. Yamamoto, G. Ceccone, and Y. Koga, *Diamond Relat. Mater.* **13**, 2071 (2004).

[45] W. Kulisch, C. Popov, S. Bliznakov, G. Ceccone, D. Gilliland, L. Sirghi, and F. Rossi, *Thin Solid Films* **515**, 8407 (2007).

[46] W. Kulisch, C. Popov, T. Sasaki, L. Sirghi, H. Rauscher, D. Grambole, and F. Rossi, Paper submitted to *phys. stat. sol. (a)*, 2008.

[47] W. Kulisch, T. Sasaki, F. Rossi, C. Popov, C. Sippel, and D. Grambole, *phys. stat. sol. (RRL)* **2**, 77 (2008).

[48] K. M. McNamara, K. K. Gleason, and C. J. Robinson, *J. Vac. Sci. Technol.* **A 10**, 3143 (1992).

[49] S. Michaelson, O. Ternyak, A. Hoffman, O. E. Williams, and D. M. Gruen, *Appl. Phys. Lett.* **91**, 103104 (2007).

[50] S. Michaelson, O. Ternyak, R. Akhvlediani, O. E. Williams, D. M. Gruen, and A. Hoffman, *phys. stat. sol. (a)* **204**, 2860 (2007).

[51] C. J. Tang, A. J. Neves, and M. C. Carmo, *Appl. Phys. Lett.* **86**, 223107 (2005).

[52] K. Wu, E. G. Wang, Z. X. Cao, Z. L. Wang, and X. Jiang, *J. Appl. Phys.* **88**, 2867 (2000).

[53] A. Badzian, T. Badzian, and S.-T. Lee, in *Proceedings of the Third International Symposium on Diamond Materials*, edited by J. P. Dismukes, K. V. Ravi, K. E. Spear, B. Lux, and N. Setaka, page 378, Pennington, NJ, 1993, The Electrochemical Society Proc. Vol. 93–17.

[54] D. Zhou, A. R. Krauss, L. C. Qin, T. G. McCauley, D. M. Gruen, T. D. Corrigan, R. P. H. Chang, and H. Gnaser, *J. Appl. Phys.* **82**, 4546 (1997).

[55] K. Wu, E. G. Wang, J. Chen, and N. S. Xu, *J. Vac. Sci. Technol.* **B 17**, 1059 (1999).

[56] S. A. Catlegde and Y. K. Vohra, *J. Appl. Phys.* **84**, 6469 (1998).

[57] S. Bhattacharyya, O. Auciello, J. Birell, J. A. Carlisle, L. A. Curtiss, A. N. Goyette, D. M. Gruen, A. R. Krauss, J. Schlueter, T. Sumant, and P. Zapol, *Appl. Phys. Lett.* **79**, 1441 (2001).

[58] J. Birell, J. A. Carlisle, O. Auciello, D. M. Gruen, and J. M. Gibson, *Appl. Phys. Lett.* **81**, 2235 (2002).

[59] D. Zhou, F. A. Stevie, L. Chow, J. McKinley, H. Gnaeser, and V. H. Desai, *J. Vac. Sci. Technol.* **A 17**, 1135 (1999).

[60] E. Givargizov, A. N. Stepanova, E. S. Mashkova, V. A. Malchonov, F. Shi, P. Hudek, and I. W. Rangelow, *Microlelectronic Engineering* **41/42**, 499 (1998).

[61] W. Kulisch, C. Popov, H. Rauscher, L. Sirghi, T. Sasaki, S. Bliznakov, and F. Rossi, *Diamond Relat. Mater.* , (in press); doi:10.1016/j.diamond.2008.01.077 (2008).

[62] T. P. Ong and R. P. H. Chang, *Appl. Phys. Lett.* **55**, 2063 (1989).

[63] F. C. N. Hong, G. T. Liang, J. J. Wu, D. Chang, and J. C. Hsieh, *Appl. Phys. Lett.* **63**, 3149 (1993).

[64] I. Schmidt, F. Hentschel, and C. Benndorf, *Diamond Relat. Mater.* **5**, 1318 (1996).

[65] X. Xiao, J. Birrell, J. E. Gerbi, O. Auciello, and J. A. Carlisle, *J. Appl. Phys.* **96**, 2232 (2004).

[66] S. A. Catledge, Y. K. Vohra, and P. B. Mirkarimi, *J. Phys. D* **38**, 1410 (2005).

[67] W. Kulisch, C. Popov, S. Boycheva, M. Jelinek, P. N. Gibson, and V. Vorlicek, *Surf. Coat. Technol.* **200**, 4731 (2006).

[68] T. G. McCauley, D. M. Gruen, and A. R. Krauss, *Appl. Phys. Lett.* **73**, 1646 (1998).

[69] S. Takeuchi, S. Oda, and M. Murakawa, *Thin Solid Films* **388/389**, 238 (2001).

[70] C. Wild, P. Koidl, W. Müller-Sebert, H. Walcher, R. Kohl, N. Herres, R. Locher, R. Samlenski, and R. Brenn, *Diamond Relat. Mater.* **2**, 158 (1993).

[71] A. van der Drift, *Philips Res. Rep.* **22**, 267 (1967).

[72] A. Hogmark, P. Hollman, A. Alahelisten, and P. Hedenquist, *Wear* **200**, 225 (1996).

[73] P. K. Bachmann and W. v. Enckevort, *Diamond Relat. Mater.* **1**, 1021 (1992).

[74] W. Kulisch, C. Popov, S. Boycheva, L. Buforn, G. Favaro, and N. Conte, *Diamond Relat. Mater.* **13**, 1997 (2004).

[75] N. Savvides and T. J. Bell, *J. Appl. Phys.* **72**, 2791 (1992).

[76] R. Kuschnereit, P. Hess, D. Albert, and W. Kulisch, *Thin Solid Films* **312**, 66 (1998).

[77] J. Philip, P. Hess, T. Feygelson, J. E. Butler, S. Chattopadhyay, K. H. Chen, and L. C. Chen, *J. Appl. Phys.* **93**, 2164 (2003).

[78] W. B. Yang, X. F. Lü, and Z. X. Cao, *J. Appl. Phys.* **91**, 10068 (2002).

[79] C. Zuiker, A. R. Krauss, D. M. Gruen, X. Pan, J. C. Li, R. Csencsits, A. Erdemir, C. Bindal, and G. Fenske, *Thin Solid Films* **270**, 154 (1995).

[80] C. Popov, W. Kulisch, M. Jelinek, A. Bock, and J. Strnad, *Thin Solid Films* **494**, 92 (2006).

[81] J. Ristein, *Appl. Phys.* **A 82**, 377 (2006).

[82] O. A. Williams, M. D. Whitfield, R. B. Jackman, J. J. Foord, J. E. Butler, and C. E. Nebel, *Appl. Phys. Lett.* **78**, 3460 (2001).

[83] J. J. Mares, P. Hubik, J. Kristofik, J. Ristein, P. Strobel, and L. Ley, *Diamond Relat. Mater.* **17**, doi: 10.1016/j.diamond.2008.01.063 (2008).

[84] M. I. Landstrass and K. V. Ravi, *Appl. Phys. Lett.* **55**, 975 (1989).

[85] A. Härtl, S. Nowy, J. Hernando, J. A. Garrido, J. Hernando, and M. Stutzmann, *Sensors* **5**, 496 (2005).

[86] C. Popov, W. Kulisch, S. Bliznakov, G. Ceccone, D. Gilliland, L. Sirghi, and F. Rossi, *Diamond Relat. Mater.* , (in press); doi:10.1016/j.diamond.2008.01.078 (2008).

[87] M. C. Feliciangeli, G. Conte, M. C. Rossi, L. Giorgi, R. Giorgi, and N. Lisi, *Sensors Actuators B* **126**, 245 (2007).

[88] H. Ye, R. Jackman, and P. Hing, *J. Appl. Phys.* **94**, 7878 (2003).

[89] S. Boycheva, C. Popov, J. Bulir, A. Piegari, and W. Kulisch, *Fullerenes, Nanotubes and Carbon Nanostructures* **13(S1)**, 457 (2005).

[90] T. Sharda, T. Soga, and T. Jimbo, *J. Appl. Phys.* **93**, 101 (2003).

[91] K. H. Chen, D. M. Bhusari, J. R. Yang, S. T. Lin, T. Y. Wang, and L. C. Chen, *Thin Solid Films* **332**, 32 (1998).

[92] L. C. Chen, K. H. Kichambare, J. J. Wu, J. R. Yang, and S. T. Lin, *J. Appl. Phys.* **89**, 753 (2001).

[93] M. J. Papo, S. A. Catlegde, Y. K. Vohra, and C. Machado, *J. Mater. Sci. Mater. Med.* **15**, 773 (2004).

[94] S. A. Catledge, M. D. Fries, Y. K. Vohra, W. R. Lacefield, J. E. Lemons, S. Woodard, and R. Venugopalan, *J. Nanosci. Nanotechnol.* **2**, 1 (2002).

[95] W. Yang, O. Auciello, J. E. Butler, W. Cai, J. A. Carlisle, J. E. Gerbi, D. M. Gruen, T. Knickerbocker, T. L. Lasseter, J. N. Russell Jr, L. M. Smith, and R. J. Hamers, *Nature Mater.* **1**, 294 (2002).

[96] J. Wang, M. A. Firestone, O. Auciello, and J. A. Carlisle, *Langmuir* **20**, 11450 (2004).

[97] A. Härtl, E. Schmich, J. A. Garrido, J. Hernando, S. C. R. Catharino, S. Walter, P. Feulner, A. Kromka, D. Steinmller, and M. Stutzmann, *Nature Mater.* **3**, 736 (2004).

[98] P. Christiaens, V. Vermeeren, S. Wenmackers, M. Daenen, K. Haenen, M. Nesladek, M. vandeVen, M. Ameloot, L. Michiels, and P. Wagner, *Biosens. Bioelectron.* **22**, 170 (2006).

[99] C. Popov, W. Kulisch, J. P. Reithmaier, T. Dostalova, M. Jelinek, N. Anspach, and C. Hammann, *Diamond Relat. Mater.* **16**, 735 (2007).

[100] C. Popov, S. Bliznakov, S. Boycheva, N. Malinovik, M. D. Apostolova, N. Anspach, C. Hammann, W. Nellen, J. P. Reithmaier, and W. Kulisch, *Diamond Relat. Mater.* **17**, 882 (2008).

[101] P. K. Chu, J. Y. Chen, L. P. Wang, and N. Huang, *Mat. Sci. Eng.* **R 36**, 143 (2002).

[102] A. Ohl and K. Schröder, *Surf. Coat. Technol.* **116–119**, 820 (1999).

[103] T. Strother, T. Knickerbocker, J. N. Russell Jr., J. E. Butler, L. M. Smith, and R. J. Hamers, *Langmuir* **18**, 1968 (2002).

[104] T. Knickerbocker, T. Strother, M. P. Schwartz, J. N. Russell Jr, J. E. Butler, L. M. Smith, and R. J. Hamers, *Langmuir* **19**, 1938 (2003).

[105] R. J. Hamers, J. E. Butler, T. Lasseter, B. M. Nichols, J. N. Russell Jr., K. Y. Tse, and W. Yang, *Diamond Relat. Mater.* **14**, 661 (2005).

In: New Research on Nanocomposites
Editors: Luis M. Krause et al, pp. 149-221
ISBN: 978-1-60456-799-1


*Chapter 6*

# Nanostructure, Phase Configuration and Mechanical Behaviors in TiN Based Nanocomposite Superhard Thin Films

***Y.H. Lu***

Department of Materials Physics, University of Science and Technology Beijing, Beijing 100083, P. R. China

## Abstract

Nanostructure and phase configuration are critical to determine mechanical and other properties e.g. thermal stability, oxidation and wear in nanocomposite superhard thin films. This paper presents our results of the nanostructure evolutions and phase configuration transformations in the formation of nanocomposite thin films prepared by unbalanced magnetron sputtering and their corresponding mechanical and other properties. Two typical ternary nanocomposite thin films Ti-B-N, Ti-C-N and a quaternary thin film Ti-B-C-N were investigated. Their nanostructures were determined by X-ray diffraction and high-resolution transmission electron microscopy, while phase configuration by X-ray photoelectron spectroscopy and Raman spectroscopy, and the hardness was examined by microindentation method. The content includes: (1) introduction; (2) deposition and analyse methods of TiN based nanocomposite thin films; phase configuration and nanostructure in (3) Ti-B-N thin films prepared at room and elevated temperatures, (4) Ti-C-N and (5) Ti-B-C-N thin films deposited at elevated temperature. The section 6 in this chapter refers their mechanical and other properties including hardness, residual compressive stress, thermal stability, oxidation and wear. At the end, a brief summary is outlined.

**Keywords**: Nanocomposite, TiN based, thin films, nanostructure, phase configuration, mechanical properties.

# 1. Introduction

## 1.1. Background

Surface engineering is today an increasingly important discipline to industry where performance and life-time of both tools and products are sought to be improved and extended. One important aspect of this discipline is the application of hard thin films in order to improve tribological and other properties of low performance or cheaper substrate materials. Hard thin films are commonly used in many tribological applications, such as cutting and forming tools, bearing, and machine parts. They also exhibit a number of other properties, making them a class of outstanding materials for decorative application, corrosion protection, superconductive components, or diffusion barriers. The hard thin films can be divided into three categories according to its chemical characterization: metallic, ionic and covalent bonds. The bonds and the corresponding material behaviors are plotted in Figure 1-1,[1] whose details are described as follows:

1) Covalent structural materials have the highest hardness;
2) Ionic structural materials have a better thermal stability;
3) Metallic structural materials have better adhesion and toughness.

The hard thin films are usually comprised of transition metal nitrides, carbides, borides and oxides. Based on Figure 1-1 and the corresponding phase diagram, we can find the most suitable thin film materials and film/substrate adhesion.

In the past decades, transition metal nitrides have been extensively studied and widely used in industry because of their attractive properties. Among the transition metal nitrides, TiN, produced either by physical vapor deposition (PVD, including ion plating, magnetron sputtering, activated reactive evaporation) or chemical vapor deposition (CVD), is favored for its high hardness and gold-like appearance, and becomes the most widely used thin films in industrial applications ranging from machine tools and mechanical parts to decorative items. Since the beginning of the 1980, they have been developed industrially on a large scale. Besides, TiN possesses the excellent electrical conductivity because it is a good diffusion barrier and makes it an idea contact material in the microelectronics industry. However, there are two key drawbacks in TiN, lower hardness and limited oxidation resistance at high temperature that can be reached during the process application, which limits its application.[2]

Nowadays, many efforts focused on the development of new hard thin film materials based on TiN to further improve the mechanical and high-temperature properties. More recently, addition of a third element to form ternary nanocomposite materials of (Ti, X)N, where X=B, C, Al, Si, Cr, etc, have been attractive for advanced hard film materials because of their superior properties, in particular, higher hardness and improved oxidation resistance in order to meet the increasing need in applications such as molds, drills, hobs, mills, cutting tools, press stamps, machine parts for motors, and micro-electro-mechanical systems.[3,4]

The nanostructure of nanocomposite thin films has a general characteristic of spinodal decomposition, where thermodynamically driven phase segregation occurs at a characteristic length scale. This characteristic length is determined by a balance between the decrease in chemical potential due to the segregation, and the elastic strain energy due to the incoherency strain. These thin films have two phases which typically consist of a primary nano-crystalline phase whose grains are encompassed by secondary continuous amorphous matrices. For the best known example, namely, by adding small amounts of silicon into TiN, nanocrystalline (nc-) TiN/amorphous (a-) $Si_3N_4$ composite films were formed.[5] Such a novel design of superhard nanocomposites is stable up to high temperatures of ≥1000°C, which is very important for the industrial application.

Such formation of the nanocomposite results in a self-organization of the nanostructure, which minimizes the density of flaws and makes the nanostructure more stable. In the nano-composite, the small (~ nm) grain size hinders the formation of dislocation thus enhancing hardness. In spite of this and due to grain boundary sliding, some deformation does occur. Superhardness is the result of a well-defined interface of high cohesive strength (along with the small grain size), which prevents crack propagation along the grain boundaries. Only compounds exhibiting a certain affinity with each another, combined with a wide miscible gap answer these criteria and exhibit high hardness and thermal stability. Thus, the mechanical properties, such as high hardness, resistance against crack formation and high elastic recovery, are a simple consequence of that self-organized nanostructure. For example, the threshold percolation is decisive in obtaining the highest hardness. A general understanding is that the amorphous phase (a-$Si_3N_4$) surrounded by nanocrystallites (nc-TiN) plays a barrier effect to block dislocation movement and the plastic deformation is significantly suppressed, thus improved the mechanical properties.[6] Their hardness maximum is usually achieved at the percolation threshold mechanism, where the flexible amorphous phase just wets the surface of TiN nanocrystals.[7]

Nanocomposite thin films can be differentiated by their hardness, *H*, into three main categories: hard with $H < 40$ GPa; superhard with $H > 40$ GPa; and ultra-hard thin films with $H > 80$ GPa.

## 1.2. Microstructure of Thin Films

It is well known that the microstructures of thin films are roughly in accordant to the Thornton's substrate zone models (SZMs),[8,9] which classifies the film microstructures into zones I, T (transition), II and III according to the ratio $T_s/T_m$ and to the total sputtering gas pressure $P_T$, as shown in Figure 1-2. Where, $T_s$ is the deposition temperature of the substrate and $T_m$ the melting point of the film material. Thornton's SZMs have proven very useful in providing an overview of the relationship between the microstructure of magnetron sputtered metal films and the most prominent deposition parameters. Form Figure 1-2 it is found that a fibrous, low dense grained thin film with voided boundaries is formed at low temperature zone I of $T/T_m<0.3$. By comparison a crystalline, high-quality, dense sputtered, featureless microstructure with smooth surface is produced in high temperature zone III of $T/T_m>0.5$. Table 1-1 lists the rough microstructures, properties and mechanisms in different zones of the sputtered thin films. On the other hand, due to limited atomic assembly kinetics during the deposition process especially by using an unheated substrate, defects, supersaturated, and metastable phases can easily be obtained in thin films. TiN thin films usually have high $T_m$ values of ~2950°C,[10] while the deposition temperature $T_s$ often has to be kept relatively low (e.g. below 550°C for high-speed steel substrates), which corresponds to a $T_s/T_m$ ratio of about 0.25, i.e. to zone I or T of the SZMs.

## 1.3. Ti-B-N Thin Films

Titanium-boron-nitride (Ti-B-N) thin films are more superior to TiN due to their metallic character and excellent properties, such as extreme hardness, high toughness, chemical inertness and good thermodynamic stability at high temperatures.[11,12] In a defined composition range, Ti-B-N thin films exhibit at least three-fold better wear resistance behavior than TiN or CrN[13] and higher hardness over 50 GPa.[14,15] For these reasons, they have emerged as a very promising candidate for use in hard wear-resistant thin films and protective applications in machining industry for cutting, forming and stamping.

The phase diagram of Ti-B-N experimentally determined at 1500 °C by Novotny et al.[16] gives the phase composition for a particular stiochiometry at thermodynamic equilibrium. The interesting features of Ti-B-N are composed of several elementary and binary phases depending on their compositions. The phase diagram of Ti-B-N can be subdivided into five different zones which have different combination of co-existing phases including the hard TiN, $TiB_2$ and BN (the cubic BN (c-BN) or the soft hexagonal BN (h-BN) phase), as shown in Figure 1-3. Therefore, their contrasting properties make Ti-B-N an attractive material for scientific investigation.

Up to now, Ti-B-N films have been synthesized by a variety of deposition techniques including cathodic arc evaporation,[17] chemical vapor deposition (CVD)/plasma-enhanced CVD,[18] and reactive dc/rf magnetron sputtering.[19] Characterization was primarily based upon microstructure, chemical-bonding configuration of elements, mechanical behavior, and surface morphology as a function of processing conditions. Previous studies showed a good agreement between the phase composition experimentally obtained and that predicted by the phase diagram.[20,21] However, when the sputtering techniques were used specially at room temperature, the microstructure and phase composition of the grown films were found to depend strongly on the deposition parameters such as substrate temperature, bias voltage, sputtering power, and vacuum conditions.[22,23,24,25] In addition, the microstructure of Ti-B-N was also found to be highly related to the amount of B addition into a growing TiN film during deposition. For example, at lower B contents the nanocrystalline (nc-) TiN and amorphous (a-) $TiB_2$ phases were formed, whereas higher B contents led to a microstructure consisting of a-TiN and nc-$TiB_2$.[26,27] It is apparent that the structural properties of Ti-B-N are highly related to their physical and mechanical properties, a systematic investigation into the phase configuration, nanostructure and mechanical behaviors is necessary.

The present investigation was carried out with the primary objective of obtaining a deeper understanding of the nanostructural characteristics and properties of Ti-B-N thin films in terms of their phase configuration, nanostructure, hardness, residual tress, thermal stability, oxidation and wear behaviors.

## 1.4. Ti-C-N Thin Films

Titanium carbide shows an excellent hardness value (HV=32.4GPa), high melting point (3260°C) and high strength at high temperature.[28] As compared with that of the TiN thin films, the relatively higher affinity of TiC thin films to the cemented carbide substrate is well known, leading to improved adhesion between film and substrate.[29] However, it is too brittle, which can be overcome by formation of Ti-C-N by way of incorporation of N into TiC. Ti-C-N also is hard, high melting compounds, and has found application in carbonitride hard metals, the so-called cermets, where it constitutes the hard phase and bonds with Ni and/or Co to form a rough and wear-resistant hard metal, and as a thin film for hard metal cutting tips. The corresponding phase diagram is given in Figure 1-4. Besides, thin films of Ti-C-N have recently received increasing attention worldwide because they have excellent hardness and abrasive wear resistance compared to conventional TiN, and better chemical stability and toughness than TiC.[30,31,32]

**Table 1.1. Structure, properties and mechanisms of sputtered thin films**

| Zone | Structure | Properties | Mechanism |
|---|---|---|---|
| ZIII | large equiaxed grain boundaries, smoth surface | low dislocation density, soft recrystalized grains | high temperature, surface and bulk diffusion |
| ZII | columnar, tight grain boundaries, facet top | hard, low ductility | surface diffusion (Z1>Z2 transition increases with deposition rate) |
| ZT | similar to Z1, no void or domes, dense grain | high dislocation density, hard, high strength, low ductility | negligible surface diffusion |
| ZI | fibrous, low dense grains, voided boundaries, | hard | substrate induced, low temperature (negligible surface diffusion) |

Some basic properties of Ti-C-N materials for example microstructure, composition, and surface morphology have been studied extensively.[33,34] It is generally accepted that Ti-C-N is a solid solution of TiN and TiC that incorporates the advantages and characteristics of both films.[35] The hardness of the corresponding thin films plotted against the composition followed a monotonic trend according to the rule of mixture[36] or the bonding.[37] On the other hand, some researchers synthesized nanocomposite Ti-C-N thin films of nanocrystalline (nc-) TiC embedded into an amorphous (a-) matrix.[38] It is believed that the nanocomposite structure will break the rule of mixture and the bonding, and hence affect mechanical behaviors. For the hardness-microstructure relationship of nanocomposite thin films, the percolation threshold mechanism,[7,36,39,40,41] which has been verified for nanocomposite Ti-Si-N and Ti-B-N thin films,[40,41,42] implies that the greatest hardness is achieved when the flexible amorphous phase just wets the surface of the TiN nanocrystals. However, it has not been clarified in the Ti-C-N thin films. Thus, a systematic investigation of nanostructure evolution with mechanical behaviors is necessary. On the other hand, similar with other nancomposite thin films such as Ti-Si-N and Ti-B-N, the interaction between the embedded nanocrystallites with amorphous matrices during nanostructure formation is interesting and has a great effect on nanostructure and other properties.[36,40,42] For example, the formation of an amorphous phase due to the addition of carbon is believed to influence the grain growth of nanocrystallites. Such a detailed knowledge is beneficial for understanding the nature of the Ti-C-N nanocomposite structure. However, the published research is limited to date.

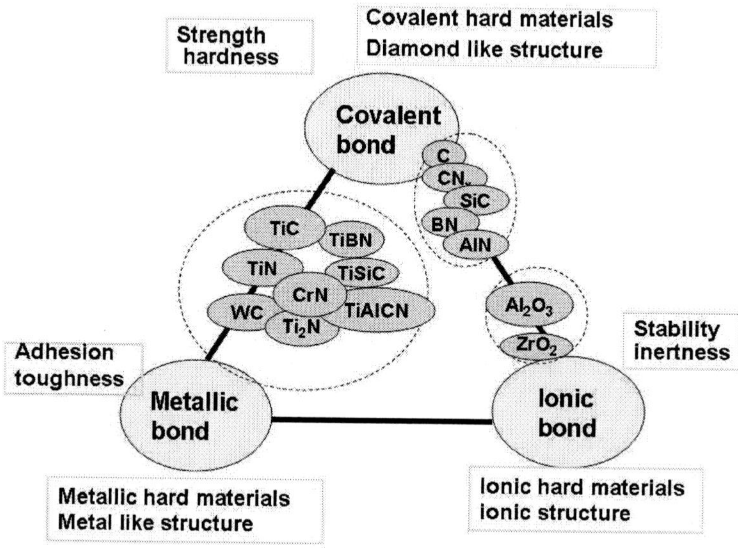


Figure 1.1. The relationship between the chemical bonds and the corresponding material behaviors.

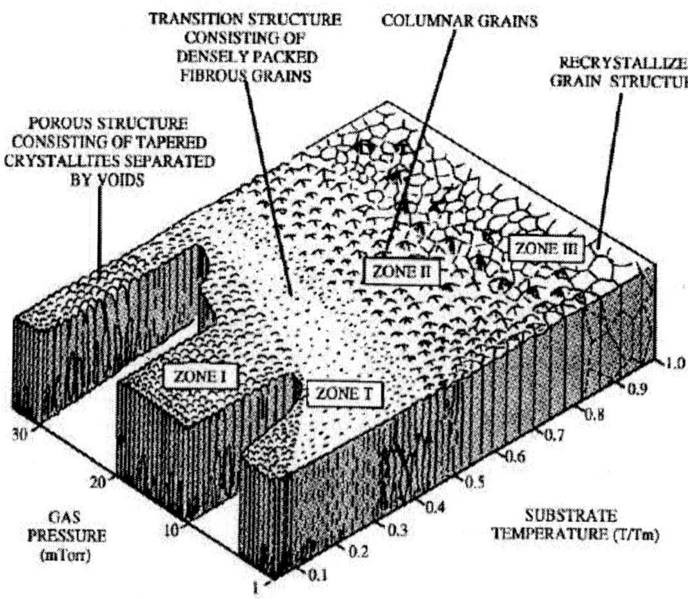


Figure1.2. Thornton's substrate zone models.

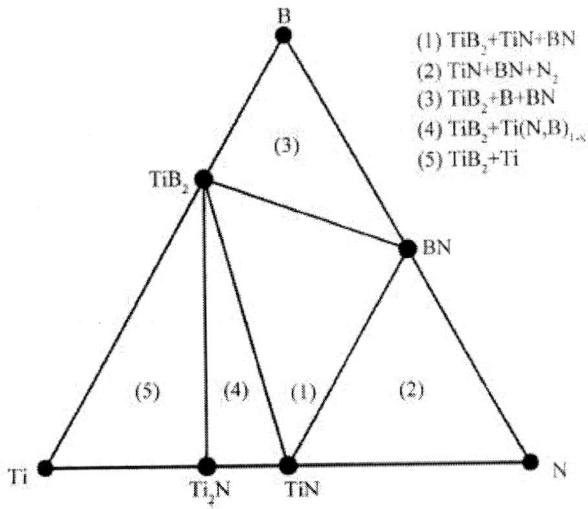


Figure 1.3. Ternary diagram of Ti-B-N.

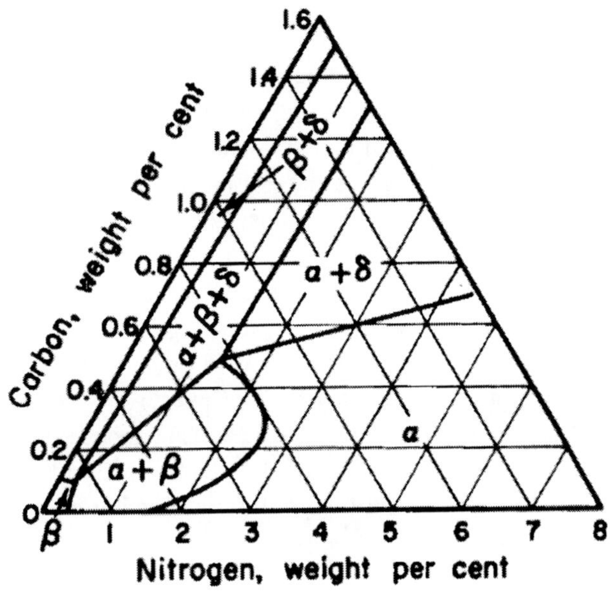


Figure 1.4. Ternary diagram of Ti-C-N.

# 2. Deposition and Analyse Methods of TiN Based Nanocomposite Thin Films

## 2.1. Deposition of Present Studied Film

The closed field unbalanced magnetron sputtering system with four magnetrons to enhance the plasma density was used to fabricate the Ti-B-N, Ti-C-N and Ti-B-C-N thin films. The schematic diagram of the magnetic head is illustrated in Figure 2.1. This method has at least three distinct advantages: (1) Energy and momentum transfer between intense Ar-ion plating and the surface of a growing film at low temperatures induces subplantation processes, which is necessary for the formation of high quality films. (2) This technique has been established in industry because of the easy handling and the possibility of depositing large areas. (3) Tools with rather complex geometries can be deposited.

The ion-cleaning was carried out at low bias voltage to reduce the contamination. The sputtering process was shown in Figure 2.2. Targets at a negative voltage attract argon ions and cause ion bombardment, which makes the film denser in the deposit. Atoms are then knocked out of the target surface and accelerated to the substrate surface.

Sputtering is probably the most attractive one since it is a process of atom-by-atom momentum transfer which could eliminate the possible problems associated with direct heating and preferred directions of sputtered target.

Instantaneously, the space inside the chamber glows, with particularly bright areas in the magnetic traps at the front of each magnetron. This is because the electrons interact with the argon gas, causing it to become excited, and then decay back to its original state, giving off its extra energy as light, producing plasma.

To fabricate Ti-B-N, Ti-C-N and Ti-B-C-N thin films with different compositions, a dc magnetron sputtering system (UDP650/450, Teer Coating Limited) was utilized (Figure 2.3). Their target configurations and deposition temperatures are shown in Table 2.1. The substrates used in present works were commercially available single crystal Si(100) wafers characterized by a root-mean-square (rms) roughness $\sigma \approx 0.2$ nm. The sputtering was carried out in an Ar-$N_2$ gas mixture at a pulsed biased voltage of -60 V (250 kHz) with a substrate rotation speed of 15 rpm and the target-substrate distance of 13 cm. Such a biased voltage is the best beneficial for deposition of TiN thin films. The base pressure in the chamber was $\leq 5.0 \times 10^{-5}$ Pa, and the working pressure, consisting of Ar and $N_2$ with a constant flow rate, was set at 0.26 Pa during all depositions. The typical current targets for Ti-B-N, Ti-C-N and Ti-B-C-N film depositions will be described in separate text. Prior to deposition of thin films, the substrate surface was pre-sputter cleaned with Ar plasma at a negative bias of 500 V with a 250 kHz frequency for 30 min, and a thin Ti buffer layer with a thickness of about 100 nm was subsequently deposited in order to improve adhesion and reduce stress at the interface with the substrate. An optical emission monitor (OEM) system was used to detect and control the flow rate of reactive $N_2$ gas. Table 2.1 also lists all optical emission intensity ratio values for all present film deposition.

## 2.2. Film Characterization

### 2.2.1. Composition and Phase Analyses

#### (1) X-ray Photoelectron Spectroscopy (XPS)

XPS is known as electron spectroscopy for chemical analysis, involves the excitement of photoelectrons from the method by incident x-ray. The emitted electron signal is plotted as a spectrum of binding energies. Different chemical states resulting from compound formation are reflected in the photoelectron peak positions and shapes. Spectral information is collected from a depth of 10-50Å analysis with 10 μm spatial resolution, depending on the studied material. The strength of XPS is its ability to identify different chemical states. Because the electron orbital have individual binding energies, it is possible to identify the binding energy for each electron orbital.

The elemental concentration and binding energy of the films were analyzed by XPS on a PHI 5802 system. The analyzer was operated in the fixed analyzer transmission mode at a constant pass energy of 11.75 eV. The binding energy scale was calibrated against the Cu $3p_{3/2}$ and Cu $2p_{3/2}$, respectively. The instrumental energy resolution, measured using the Cu $3p_{3/2}$ peak, was 0.1 eV. Before examination, the film surface was cleaned using methanol in an ultrasonic bath, then subsequently etched about 30 nm by a 3 keV rastered argon ion beam with a current of 0.2-0.4 μA to remove the surface oxide. The information on the chemical bonding states was obtained by subtracting the background with the Shirley's method[43] and deconvoluting the spectra by a curve-fitting method using a non-linear least squares fitting of a mixed Gaussian-Lorentzian product function.

**(2) Auger Electron Spectroscopy (AES)**

AES is a powerful surface analytical tool to probe surfaces, thin films, and interfaces for its composition radiation. It is a surface specific technique utilizing the emission of low energy electrons in the Auger process and is one of the most commonly employed surface analytical techniques for determining the composition of the surface layers of a sample. This utility arises from the combination of surface specificity (0.5 to 10 nm), good spatial surface resolution (as good as 10 nm), periodic table coverage (except hydrogen and helium), and reasonable sensitivity (100 ppm for most elements).

Auger spectroscopy can be considered as involving three basic steps (see Figure 2.4):

(1) Atomic ionization (by removal of a core electron) (a)
(2) Electron emission (the Auger process) (b)~(c)
(3) Analysis of the emitted Auger electrons

The last stage is simply a technical problem of detecting charged particles with high sensitivity, with the additional requirement that the kinetic energies of the emitted electrons must be determined.

**(3) Fourier Transform Infra Red (FT-IR)**

FT-IR Spectroscopy refers to the IR absorption (or reflection) data being collected and converted from an interference pattern to a spectrum. It is a powerful tool for identifying types of chemical bonds in a molecule by producing an infrared absorption spectrum that is like a molecular "fingerprint".

Molecular bonds vibrate at various frequencies depending on the elements and the type of bonds. For any given bond, there are several specific frequencies at which it can vibrate. According to quantum mechanics, these frequencies correspond to the ground state (lowest frequency) and several excited states (higher frequencies). One way to cause the frequency of a molecular vibration to increase is to excite the bond by having it absorb light energy. For any given transition between two states the light energy (determined by the wavelength) must be exactly equal to the difference in the energy between the two states (usually ground state ($E_0$) and the first excited state ($E_1$)), i.e. difference in Energy States = Energy of Light Absorbed

$$E_1 - E_0 = h\,c\,/\,\lambda$$

where,

h = Planks constant, c = speed of light, λ = wavelength of light.

A Nicolet FT-IR spectroscope was employed to identify the BN phase. After the measurements, data were collected by the spectrometer and sent to the computer as an interferogram. The computer determined the frequency and intensity of each component wave of the interferogram by performing a Fourier Transform of the data. These frequencies and intensities make up the peaks of an IR spectrum.

### 2.2.2. Structure Analyses

**(1) X-ray Diffraction (XRD)**

XRD is a powerful non-destructive technique for investigation of structural properties of crystalline materials. A diffraction pattern is produced when a crystalline material is irradiated with a collimated beam of x-ray. The diffraction pattern and the intensity of each diffracted x-ray beam as a function of the diffraction angle can provide information such as crystal structures, phase purity, grain size, lattice constants etc.

X-rays have wavelengths on the order of a few angstroms, the same as typical interatomic distances in crystalline solids, so x-rays can be diffracted from minerals which, by definition, are crystalline and have regularly repeating atomic structures. When certain geometric requirements are met, x-rays scattered from a crystalline solid can constructively interfere, producing a diffracted beam. In 1912, W. L. Bragg recognized a predictable relationship among several factors:

$$n\lambda = 2d\sin\theta,$$

where θ is the Bragg angle and d is the spacing between atomic planes in the crystalline phase and n is an integer. The intensity of the diffracted x-ray is measured as a function of the diffraction angle 2θ and the specimen's orientation.

For determining the degree of preferred orientation, the TC developed by Harris was introduced. The TC values of diffraction peaks were calculated from the XRD patterns by the following Equation:[44]

$$TC_{hkl} = \frac{\dfrac{I_{hkl}}{I^0_{hkl}}}{\dfrac{1}{n}\sum \dfrac{I_{hkl}}{I^0_{hkl}}}$$

Where $I^0_{hkl}$ is the relative intensity of $hkl$ plane of power diffraction obtained from JCPDS card, $I_{hkl}$ is the integrated intensity of $hkl$ plane diffraction peak of sample.

The value of mean grain size (GS) of nanocrystalline TiN was obtained by averaging the grain size values calculated from (111) and (200) diffraction peaks using the corresponding full width at half maximum (FWHM) value and a Scherrer relation.[45] The Scherrer equation can be expressed as:

$$GS = \frac{0.9\lambda}{\cos\theta \bullet FWHM}$$

Crystallographic analysis in this work were conducted at room temperature by X-ray diffraction (XRD) using a Bragg-Brentano diffractometer (Simens D500) in θ-2θ configuration. The diffractometer used Cu $K_\alpha$ line at 0.15418 nm as the source for diffraction. The typical scanning range was from 20 ° to 60 ° (2θ) with a speed of 0.03° per second in the continuous mode.

**(2) Scanning Electron Microscopy (SEM)**

SEM is perhaps the most widely employed thin-film characterization instrument. In SEM, electrons thermionically emitted from the cathode filament are drawn to an anode, focused into a fine probe (~5nm) and scanned over a rectangular area of specimen surface. As the beam interacts with the samples, it creates various signals, such as secondary electrons, internal currents, photon emission, etc. By using this signal to modulate the brightness of a cathode ray tube, a highly magnified image can be formed on the screen. JEOL SEMS (JSM-820, field emission JSM-6335F) were used to examine the cross-sectional microstructure and columnar features as well as the wear surface in these works.

**(3) High-resolution transmission electron microscopy (HRTEM)**

HRTEM and associated energy dispersive X-ray spectroscopy (EDS) and electron energy-loss spectroscopy (EELS) are powerful tools for structure characterization of nanomaterials because it provides not only atomic-resolution lattice images, but also chemical information at a spatial resolution of 1 nm or better, allowing direct identification the chemistry of a single nanocrystal. As the imaging resolution approaches 0.1 nm, HRTEM is fully capable of revealing a real space image on the atom distribution at atomic-scale structures in nanomaterials.[46] With a finely focused electron probe, the structural characteristics of a single nanoparticle can be fully characterized. Phase contrast, which is produced by the phase modulation of the incident electron wave when transmits through a crystal potential, is sensitive to the atom distribution in the specimen and is the basis of high-resolution TEM.

The nanostructure was analyzed by cross-sectional and plan-view high resolution transmission electron microscopy (HRTEM, field emission JEOL 2010F) operated at 200 kV. For plan-view TEM observations, the films were first mechanically thinned from Si side to a thickness of about 10 μm. Further thinning to a thickness of electron transparency was carried out by a Gatan Model 600 ion-polishing system using one gun with 5 k eV Ar ions at a beam current of 5.0 mA and an incident angle of 15°. During ion-milling the film surface was protected. Cross-sectional TEM specimens were prepared by gluing two samples face-to-face with epoxy resin, mechanically thinned from both transverse sides to ~10 μm, and then subsequently ion milled.

### 2.2.3. Surface Analyses

**(1) Atomic force microscope (AFM)**

AFM was invented in 1986 by Binnig, Quate and Gerber. AFM can image the surface with atomic resolution. It utilizes a sharp probe moving over the surface of a sample in a raster scan. The probe is a tip on the end of a cantilever which bends in response to the force between the tip and the sample. AFM provides extraordinary topographic contrast direct height measurements and unobscured views of surface features without requiring further coating.

Several surface roughness parameters can help to quantify the surface texture of a material. In present study, $R_{rms}$, which is the root mean square deviation of the assessed profile, helps to quantify the surface texture of nanocomposite TiN thin films. $R_{rms}$ is determined from:

$$R_{rms} = \sqrt{\frac{\sum_{n=1}^{N}\left(z_n - \bar{z}\right)^2}{N-1}}$$

Where $\bar{z}$ mean z height.

In present study, the surface topography of the films was acquired by scanning the samples in air with AFM (Auto-Probe CP, Park Scientific Instruments) operated in a contact mode. The radius of the tip is about 10 nm, and the side angle is about 35°. This instrument is capable of atomic resolution in the vertical and horizontal directions. The scan areas were 5 x 5 μm$^2$ and 1 x 1 μm$^2$ with a resolution of 256 × 256 pixels. The rms roughness was measured and calculated by the above equation.

### 2.2.4. Mechanical Behavior Test

**(1) Microhardness**

Hardness of a material is a measure of its ability to resist deformation upon load. The most universal measure of the hardness is the energy of the plastic deformation which is calculated from the ratio of the applied load to the area of that deformation.

During the hardness measurement a diamond indenter is pressed into the material with a given load L and, after unloading, the area of the remaining plastic deformation A is measured. The hardness in GPa or kg/mm$^2$ is then calculated from the equation H = aL/A, where the constant a depends on the type of the indenter (Vickers, Knoop and others) which differs by the angles between the opposite faces of the diamond pyramid.[47]

Nowadays, the measurements are performed by automatic load-displacement sensing indentometers which determine the hardness and elastic modulus automatically (see Figs.2.5 (a) and (b) and 2.6). Microhardness measurements with a load of 20 to 200 mN were commonly used for thin films. For even thinner films of less than ~100nm nanoindentation with loads less than a few mN is preferred.

The microhardness of thin films was measured by a Fischer microhardness instrument. A four-sided diamond pyramid tip (Berkovich tip) was used with a single loading-unloading cycle. The indenter tip shape (the area function) was calibrated using the method described in

Ref..[48] The calibration of the hardness value was frequently checked by measuring on a fused silica sample. Testing was done using the constant-load-rate mode until a load of 20 mN was reached. The corresponding indentation depth was 0.15~0.25 μm, less than 20 % of thin film thickness in order to avoid the substrate effect. Ten separated measurements were taken for each sample in order to get a mean value.

**(2) Residual stress**

The residual stress (σ) is composed of two components: thermal stress ($\sigma_{th}$) and intrinsic stress ($\sigma_{in}$). That is,

$$\sigma = \sigma_{th} + \sigma_{in}$$

The thermal stress $\sigma_{th}$ occurs due to the difference in thermal expansion coefficients of the thin films $\alpha_c$ and the substrate $\alpha_s$, which can be calculated from the following formula:[49]

$$\sigma_{th} = [E_c/(1-\nu_c)](\alpha_s - \alpha_c)(T - T_s)$$

Where $E_C/(1-\nu_C)$ is the biaxial elastic modulus of the thin films. $T_S$ and T are the substrate temperature during film deposition and the temperature at which the σ is measured, respectively. The intrinsic stress $\sigma_i$ occurs as a sequence of the accumulation of crystallographic defects that are built into the film during its deposition. Usually, $\sigma_{th}$ is much less than $\sigma_i$.

The Newton ring method is usually used to measure the average intrinsic stress of the films by way of measuring the curvature of the silicon wafer on an interferometer. For the case where the film thickness, $t_f$, is much smaller than the substrate thickness, $t_s$, the average in-plane stress, σ, can be calculated from the curvature by the well-known Stoney's equation,[50]

$$\sigma = \frac{E_s}{6(1-\nu_s)}\frac{t_s^2}{t_f}\left[\frac{1}{R} - \frac{1}{R_o}\right]$$

where $Es/(1-v_s)$ is the biaxial modulus of the substrate (for Si (100) substrate, $Es/(1-v_s)$=180.5 GPa), $R_o$ and R demote the curvature radii before and after film deposition, respectively. The radii of curvature are determined from

$$R = \frac{1}{4\lambda}\frac{\Delta d_m^2}{\Delta_m}$$

where d is the diameter of the dark Newton's ring, m is the ring order, and λ is the wavelength of the monochromatic light source (600nm).

**(2) Wear test (Pin-on-disc Test)**

A computer controlled pin-on-disc tester can provide a quantitative measure of the wear properties of a thin film by creating wear tracks. Testing can be carried out under a wide range of loads and speeds, with the use of different wear track diameters allowing a number of tests to be performed on one sample. The load can be varied by adjusting the amount of dead weight hung at the end of the loading beam as shown in Figure 2.7.

In present study, the wear pin is a 5 mm diameter ball made from AISI440C. Before a test the wear pin is started a clean ball surface with no debris is essential - this is ensured by moving the wear pin (ball) in its holder until a clean surface can be seen under a microscope. The ball holder is then tightened to ensure no movement of the ball during testing. The wear pin creates a circular wear track of the required diameter by offsetting the pin relative to the sample's centre of rotation. The sliding motion of the sample under the wear pin provides a frictional force. Frictional force is detected, recorded and displayed by the computer both during and after the test. The test is automatically stopped if the maximum friction force or time is reached. From the resulting plot of friction versus time the friction coefficient and lifetime of the thin film can be determined.

Specific Wear Rate can be expressed as follows:

$$\text{Specific Wear Rate}\left(\frac{m^3}{Nm}\right)=\frac{Volume}{Load\ \times\ Sliding\ \ dis\tan ce}$$

where, Volume $=\frac{wt}{2}\times\ \pi\ \ d$,

w = track width

t = track depth

$\pi$ d = circumference of wear track.

Sliding distance = Circumference × number of revolutions=$\pi$ d ×number of revolutions

Number of revolutions = rpm×testing times in minutes

Finally,

$$\text{Specific Wear Rate}\left(\frac{m^3}{Nm}\right)=\frac{\frac{wt}{2}\ \times\ (\pi d)}{Load\ \times\ (\pi d)\ \times\ no.\ \ of\ \ revolutions}$$

Therefore, the specific wear rate of the film after sliding can be determined by the above equation.

**Table 2.1. Deposition parameters for deposition of Ti-B-N, Ti-C-N and Ti-B-C-N thin films.**

| Deposition parameters | Ti-B-N | Ti-C-N | Ti-B-C-N |
|---|---|---|---|
| Deposition temperature | Room temperature, 500°C | 500°C | 500°C |
| Target configuration | Ti-$TiB_2$ | Ti-graphite | Ti-$TiB_2$-graphite |
| OEM | 55%, variable | 50%, variable | variable |

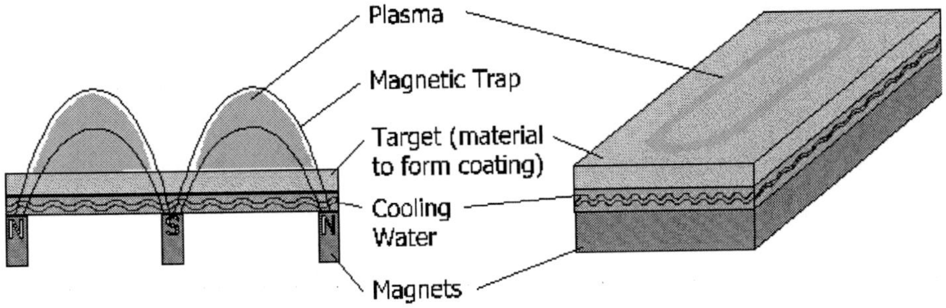


Figure 2.1. Schematic diagram of magnetron.

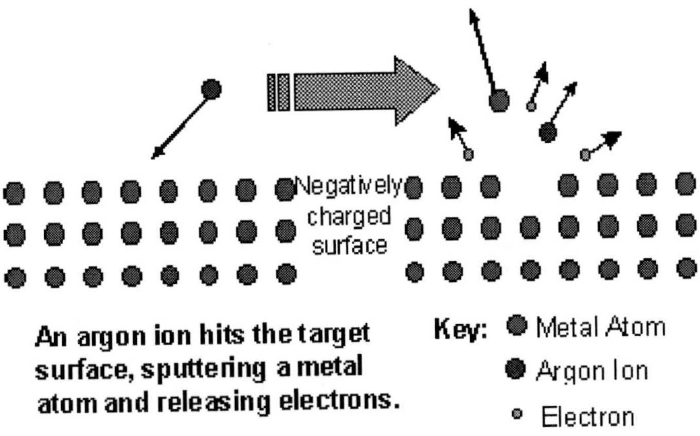


Figure 2.2. Schematic diagram of the sputtering process.

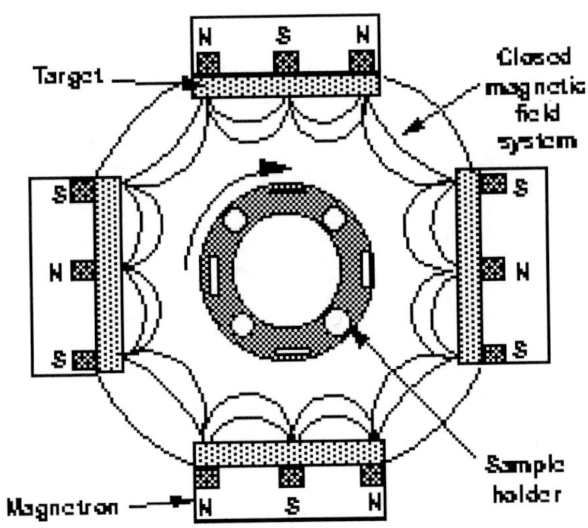


Figure 2.3. Schematic diagram of unbalance magnetron sputtering.

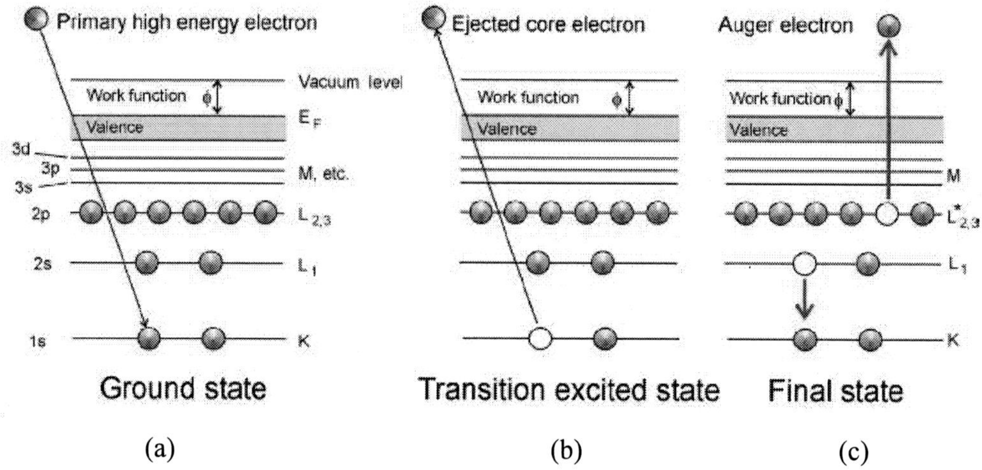


Figure 2.4. Schematic diagram of Auger process.

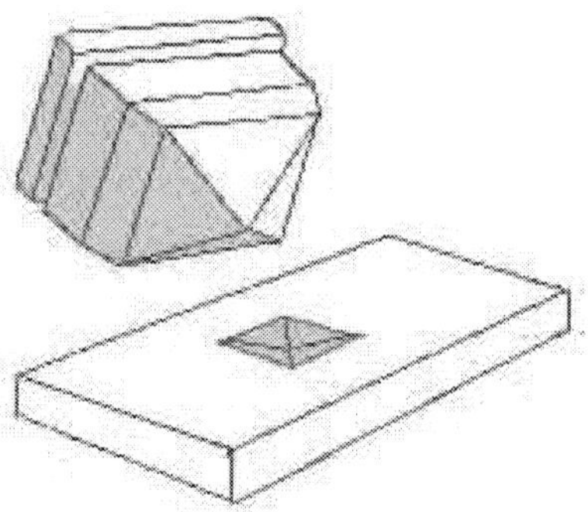

(a)

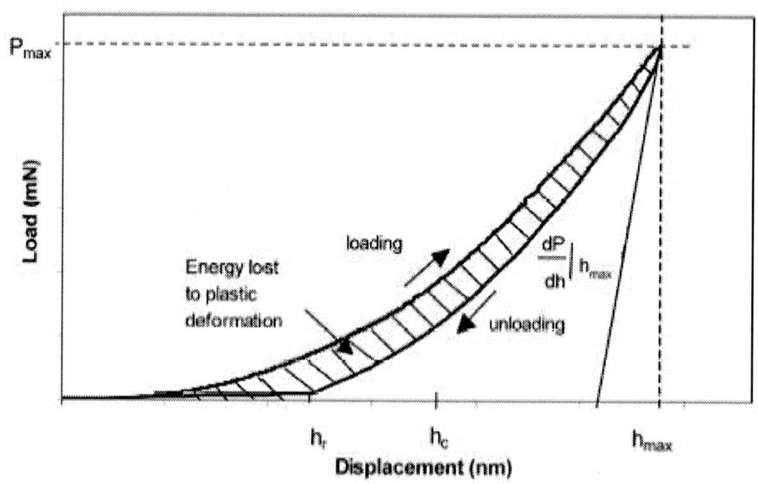


(b)

Figure 2.5. (a) A diamond indenter, and (b) indentation test procedure. A loading sequence is followed by an unloading sequence. $h_r$ is the depth of the residual impression, $h_c$ is the intercept of the tangent to the initial unloading curve, $h_{max}$ is the maximum penetration beneath the specimen surface, and $P_{max}$ is the maximum load applied to the indenter.

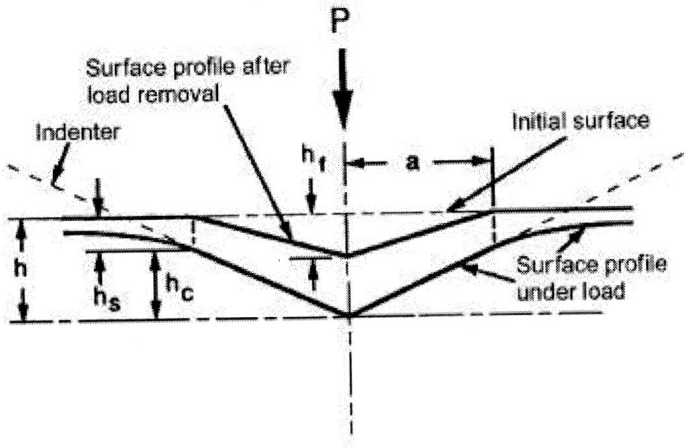


Figure 2.6. The deformation pattern of an elastic-plastic sample during and after indentation is H = $P_{max}$/A, where $P_{max}$ is the maximum load and A is the projected area. The Young's modulus can be calculated by following equations:

$$s = \frac{dP}{dh} = \frac{2}{\sqrt{\pi}}.E_r.\sqrt{A},$$

$$\frac{1}{E_r} = \frac{(1-v^2)}{E} + \frac{(1-v_i^2)}{E_i},$$

where S is stiffness of system;

$E_r$ is reduced Young's modulus;

$E_i$ is Young's modulus of indenter;

E is Young's modulus of film;

and $v_i$ and v are Poission's ratio of indenter and film, respectively.

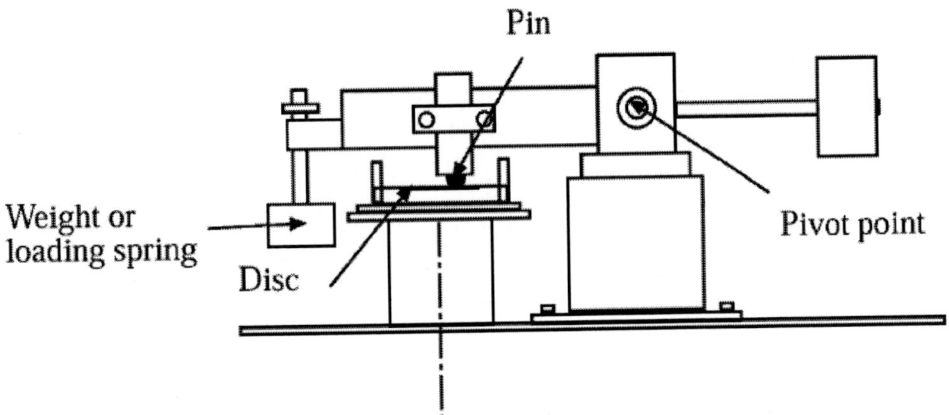


Figure 2.7. Schematic diagram of pin on disc apparatus viewed at the side.

# 3. Phase Configuration and Nanostructure in Ti-B-N Thin Films

## 3.1. Ti-B-N Thin Films with B Contents Deposited at Room Temperature

From view of microstructure, there is a high quenching rate during physical vapor deposition (PVD) process, the microstructure, phase composition of thin films prepared by PVD are considered to be nonequilibrium, and found to strongly depend on the deposition parameters such as substrate temperature, bias voltage, sputtering power and vacuum conditions.[51,52,53] Especially, when we deposited TiN based nanocomposite thin films at low temperature, limited atomic assembly kinetics leads to material conditions far from equilibrium. Thus, a supersaturated NaCl-structural $Ti(B_xN_y)$ solid solution where B substitutes N thus might be formed.[54,55]

The present $Ti\text{-}B_x\text{-}N_y$ thin films with different B contents were deposited at room temperature in an unbalanced reactive dc-magnetron sputtering system (UDP650, Teer Coating Limited) using three Ti and one $TiB_2$ targets. These thin films and the corresponding target configuration and current are listed in Table 3.1. The elemental content of $Ti\text{-}B_x\text{-}N_y$ films is governed to a large extent by the reactions occurring on the target surface. In reactive unbalanced magnetron sputtering, the plasma extends over both the target and the substrate, so that the Ar ions provide both the sputtering flux to both Ti and $TiB_2$ targets and the ion plating flux on the growing film. Under the current deposition conditions, the Ti, B and N contents of the films are governed to a large extent by the reactions occurring on the Ti and $TiB_2$ target surfaces. The Ti, B, and N contents of the present films, analyzed by XPS, are shown in Table 3.1. These values were obtained based on repeated measurements involving many freshly deposited samples. The good agreement between the elemental content data lends confidence to the concentration values and the reliability of the experimental data. By XPS about 2-5 at.% oxygen was found through all the films (not shown here). Close examination of the deposition processing revealed that the most likely source of oxygen contamination was due to high Ti and $TiB_2$ reactivities with residual water in vacuum chamber during deposition.

The chemical bonding configuration of the present $Ti\text{-}B_x\text{-}N_y$ films was characterized by XPS. The results reveal that depending on B contents two or three B-related bonds are observed in all samples. By a deconvolution curve-fitting method using a mixed Gaussian-Lorentzian product function, small amounts of TiB, $TiB_2$ and BN can be identified in the films. For lower B contents, a main TiN bonding (not shown) accompanying with a small amount of TiB and BN bonds was formed [see Figure 3.1(a)]. With increasing B content, TiB bonding gradually transformed to $TiB_2$, indicating that the fractions of $TiB_2$ bonding increased at cost of TiN bonds with B content. When the B content reached about 27.4 at.%, the $TiB_2$ bonding dominated among B-related bonds (see Figure 3.1(a)). Further increase of B content made BN bonding increase faster, which might form both $TiB_2$ and BN phases. The observation that an amorphous h-BN phase was formed at higher B contents was also evidenced by a FT-IR analysis (Figure 3.1(b)). The above observations indicated that phase segregation took place at room temperature during deposition, which are in agreement with the results reported by other research groups.[20,21]

In order to further clarify the nanostructure of $Ti\text{-}B_x\text{-}N_y$ thin films, an HRTEM analysis was carried out. Our HRTEM investigation reveals that the pure TiN film comprises well crystallized nanocrystallites (i.e. A~D), and no amorphous phase is found in grain boundaries

(Figure 3.2(a)). The corresponding selected area diffraction (SAD) pattern, including Debye-Scherrer rings of (111), (200) and (220) Bragg reflections of TiN, clearly reveals the fcc-like B1-NaCl structure (Figure 3.2(b)). This observation indicates that the nitrogen activity and activation energy provided by the room-temperature deposition were enough for the formation of fully-crystallized TiN thin films. However, addition of B made film nanostructure change. Once Ti and B from targets and N from reactive gases arrived the substrate, a thermodynamically driven spinodal segregation occurred

Figs.3.3(a) and (b) show low magnification TEM images of $TiB_{0.36}N_{0.85}$ and $TiB_{1.86}N_{1.51}$, which indicate formation of nanocomposite structure of nanocrystallites embedded into amorphous matrices. On the other hand, during deposition B atoms linked to Ti and/or N, leading to the formation of amorphous $TiB_2$ and/or BN matrices, which depended on composition of thin films. With increasing B content the volume fraction of nanocrystallite decreases, which results in decrease of the distribution density of nanocrystallites (see Figs. 3.3(a) and (b)).

When B content was relatively low (for example 16 at.% for $TiB_{0.36}N_{0.85}$), less amount of amorphous phase was used to stop grain growth of nanocrystallites, resulting in the formation of nanocomposite thin films with relatively large nanocrystallites, as shown in Figure 3.4(a). Obviously, growth and coalescence processes by way of long- range diffusions of atoms occurred. Meanwhile some new nucleation processes took place, resulting in formation of small-sized nanocrystallites (circled in Figure 3.4(a)) and some short-range-order (SRO) areas (squared in Figure 3.4(a)). Therefore, there are two-sized nanocrystallites of ~9 or ~3 nm respectively in $TiB_{0.36}N_{0.85}$. In the large nanocrystallites, the {200} lattice fringes with a lattice spacing of about 2.1 Å are dominant; while in the small-sized nanocrystallites B~G with their grain sizes in the range of 2~4 nm, the single {111} lattice fringes with a lattice spacing of 2.45 Å are observed. They are consistent with {111} and {200} rings of the corresponding fast Fourier transform (FFT) spectrum (see Figure 3.4(b)). The fluctuant spots along the {111} and {200} rings indicate that there is a slight difference in lattice spacing for different nanocrystallites, as a result from the formation of nc-Ti(N,B) due to incorporation of B into TiN lattice. It is noted that dislocations and twins are seldom found inside the some grains. The SAD ring pattern shown in Figure 3.4 (b) can be indexed in terms of a face center cubic (f.c.c) structure, which is the same as that of pure TiN.

The formation of nanocomposite thin films is connected with segregation of amorphous phases to strong grain boundaries by way of thermodynamically driven spinodal segregation. This effect is responsible for stopping grain growth of nanocrystallites. With increasing B ($TiB_2$) content more thermodynamically driven segregation reactions occurred on the substrate surface, nucleation gradually became the major process compared to the subsequent growth and coalescence. As a result, more uniform and finer grains of ~5 mm were formed, as shown in HRTEM images of $TiB_{0.79}N_{1.09}$ and $TiB_{1.86}N_{1.51}$ (see Figs. 3.5 (a) and (b)). The decrease of grain size with B content was further verified by XRD (not shown here). In Figs. 3.5 (a), {200} lattice fringes are usually observed in most nanocrystallites, such as A, B, C and D, while cross {111} lattice fringes are found in only a few nanocrystallites such as E, which is in agreement with its FFT spectrum. Due to the same effects the FFT spectrum in Figure 3.5(a) also shows fluctuant spots along {200} ring. In addition, the {200} Bragg reflection of the SAD pattern (Figure 3.5 (a)) becomes stronger, which reveals formation of more [200]-oriented nc-Ti(N, B) nanocrystallites.

It is reported that the nanocrystals of a few nanometers are presumed to be free of any defects because of a relatively high, destabilizing contribution of defect to the total free energy in nanocrystals. This effect is combined with a short diffusion length to grain boundaries to produce nearly defect-free nanocrystalline grains.[56] However, our HRTEM study revealed that a small amount of B still dissolved into nanocrystalline lattice of TiN to form Ti(N,B) nanocrystallites when the grain size was decreased to ~5 nm in diameter. The detailed issue of incorporation of a small amount of B into TiN is still not well understood.

Another interesting phenomenon is presence of a large strain in the nanocrystallite which is due to twin deformations as labeled in Figs. 3.4 (a) and 3.5 (a)~(b). It has been reported that small grain size is favorable for occurrence of twinning deformation because the critical stress required to activate a twining dislocation becomes less than that for an ordinary one when the grain size is less than a critical value.[57] As a result, the decrease of grain size of nc-Ti(N, B) because of addition of B is favorable for occurrence of twinning deformations in nanocrystallites, as shown in Figure 3.5. However, in this study, the grain size is not unique factor controlling the twinning deformations. The dissolution of B into TiN is believed to be another factor favorable for activation of twinning deformations because such deformation is seldom found in B-free TiN. By comparing Figure 3.5(a) with Figure 3.5(b), $TiB_{1.86}N_{1.51}$ shows more severe twinning deformations than $TiB_{0.79}N_{1.09}$ although they have a similar grain size value. Dissolve of B into TiN possibly resulted in the lattice distortion and twinning deformations of nanocrystallites, and hence changed the internal stress pattern. This is similar to a study carried out by Gray,[58] in which the deformation twins were observed in Al-4.8 wt% Mg subjected to shock loading at -180 °C while they were not seen in pure Al under the same deformation condition. The acceleration effect of solid solution on twin deformation was also found in bcc lattice alloys,[59] which is attributed to solid solution rather than to intrinsic SFE (stacking fault energy) effects.[60] It is believed that the twinning deformations in nanocrystallites are favorable for relaxation of residual compressive stress produced during the film deposition process.

In summary, the resulting effects due to the incorporation of B are summarized as follows: (1) formation of solid solution nc-Ti(N,B) by way of a small amount of B dissolving into nc-TiN lattice; (2) formation of a nanocomposite structure of nc-Ti(T,B) grains embedded into amorphous matrices; (3) decrease in the grain size of nanocrystallites; (4) decrease in volume fraction of nanocrystallites; (5) formation of more [200]-oriented nanocrystallites; and (6) activation of twin deformations in the nanocrystallites. In addition, two different-sized nanocrystallites are formed in the thin films with low B content; while uniform nanocrystallites usually appear in the thin films with high B content.

## 3.2. Ti-B-N Thin Films with N Contents Deposited at Room Temperature

Incorporation of different amounts of nitrogen into amorphous $TiB_{0.65}$ led to formation of Ti-B-N thin films with different N contents (see Table 3.2). These films were deposited at room temperature in a reactive unbalanced dc magnetron sputtering system (UDP450, Teer Coating Limited) using one $TiB_2$ and three Ti targets at different OEM values (see Table 3.2). During deposition the $TiB_2$ target current was fixed at 4.0 A, while the opposite Ti target current was 6.0 A. The electric currents of the other two Ti targets were arranged at 1.0 A. The fractions of the main three phases, $TiB_2$, TiN and BN, were calculated based on the following relation supposing that metallic titanium and oxide phases are discarded:[20,21]

$$a[TiB_2] + b[TiN] + c[BN] = TiB_xN_y. \quad (1)$$

Solvable expressions for a, b, c can be derived using simultaneous equations with various x and y components. From Table 3.2 it could be concluded that the fraction of BN increased, while that of $TiB_2$ decreased with increasing N content.

XPS measurements showed that both Ti and $TiB_2$ bonds were detected in $TiB_{0.65}$, revealing that B and Ti coming from $TiB_2$ target re-bonded to $TiB_2$ on the substrate surface (see Figure 3.6). By XRD (see Figure 3.7) no diffraction peak representing either crystalline Ti or $TiB_2$ was found in $TiB_{0.65}$, indicating that an amorphous thin film consisting of Ti and $TiB_2$ was formed. XRD analysis of the film structure is in accord with the SEM and HRTEM observations (see Figs. 3.8 (a) and (b)).

Incorporation of nitrogen into amorphous compound $TiB_{0.65}$ obviously promoted phase segregation, forming nanocomposite structure characterized by nc-TiN particles embedded into amorphous ($TiB_2$, BN) matrices. After incorporation of about 28 at.% nitrogen, the most N linked Ti to form TiN, and small amount of N reacted with $TiB_2$ to produce TiN and TiB, therefore dominant $TiB_2$ and TiN bonds accompanying with a small amount of TiB bonding were formed (see Figs. 3.6 (a) and (b)). Increase of nitrogen content promoted formation of TiN at cost of $TiB_2$. Further increase of N content obviously promoted formation of BN bonding at cost of $TiB_2$. The BN bonding gradually becomes dominant in high N contented films such as $TiB_{0.50}N_{1.12}$ and $TiB_{0.66}N_{1.38}$.

With incorporation of increasing amount of nitrogen, the fracture surface of TiBN became rough. TiBN thin films incorporating 28 at.% nitrogen shows a little rough, however still keeps amorphous-like in structure. Addition of more nitrogen (42.7 at.% N) gradually produced dense and fine column morphology, as shown in Figure 3.9(a). According to HRTEM observations (Figure 3.9(b)), our present film exhibits nanocomposite, which is characterized by fcc B1-NaCl structural nanocrystallites embedded into amorphous matrices. The average grain size of crystallites is ~8 nm.

In summary, nitrogen had a great role in phase segregation and nanostructure. $TiB_{0.65}$ thin films exhibited to be an amorphous compound consisting of Ti and $TiB_2$. Addition of nitrogen promoted phase segregation by way of preferentially linking with Ti, which resulted in formation of two-phase nanocomposite nc-TiN/a-$TiB_2$ thin films. More nitrogen incorporation accelerated formation of BN bonding at cost of $TiB_2$, and produced nanocomposite nc-TiN/a-($TiB_2$, BN) thin films with densified fine column morphology. The grain size slightly increased with increasing N content.

### 3.3. $TiB_xN_y$ Thin Films with B Content Deposited at 500°C

The Ti-$B_x$-$N_y$ thin films, deposited at 500°C, are shown in Table 3.3. The corresponding target configuration and B contents for every thin films are also shown in Table 3.3. Their compositions in the Ti-B-N phase diagram were plotted in Figure 3.10. It is found that the present films containing less than 12 at.% B are located in the two-phase zone of TiN and

$TiB_2$, more B incorporation shifts the film composition to the three-phase zone of TiN, $TiB_2$ and BN, which agrees with their XPS spectra (shown in Figure 3.11).

XRD and the corresponding texture coefficient analyses reveal that the present deposited $TiN_{0.81}$ shows a highly (111) preferred oriented ploycrystallite (see Figure 3.12), which indicates that the present deposition parameters such as plasma energy and deposition temperature were enough for formation of crystallized TiN. It is apparent that the preferred orientation (texture coefficient value) and grain size remained when a small amount of B (less than ~9 at.%) was incorporated; while a swift decrease in preferred orientation and grain size was followed when B content was slightly increased (up to 15.6 at.%). One possible explanation is that the present $TiN_{0.81}$ is substiochiometric and shows some vacancy concentrations on the nitrogen sublattice. However its ordering can be remained.[61] Thus, a small amount of incorporated B (up to ~9 at.%) was expected to dissolve into the vacancies on the nitrogen sublattice of $TiN_{0.81}$, forming a Ti(N,B) solid solution. The similar phenomenon is also found by other researchers.[62,63]

When B content was further increased to more than ~20 at.%, a nanocomposite structure was formed, as evidenced by HRTEM shown in Figure 3.13. Obviously, the swift change in preferred orientation and TiN column size with B content was obviously contributed to formation of nanocomposite structure (see Figs.3.12 and 3.13). The possible explanations list as follows: (1) the formation of nc-/a- structure obviously decreased energies of grain boundary and structure system and hence promoted growth of low energy plane (200);[64] (2) the amorphous matrices surrounded nanocrystallites strongly hindered growth and coalescence of the nano-columns, hence strongly decreasing the column size. Obviously, such effects were enhanced with increase of B content.

In summary, phase configurations and nanostructure of Ti-B-N thin films with different B contents deposited at 500°C strongly depended on their composition (B content here). When the film compositions were located in two phase zone of TiN and $TiB_2$ (less than ~12 at.% B content in this study), the films usually consisted of highly (111) orientated TiN, accompanying with a small amount of TiB. As the film compositions shifted to three phase zone of TiN, $TiB_2$ and BN (above ~12 at.% B content), TiB gradually transformed to $TiB_2$, simultaneously the preferred orientation of nanocrystallites swiftly changed from (111) to (200), accompanying with an obvious decrease of nanocolumn size.

## 3.4. $TiB_xN_y$ Thin Films with N Content Deposited at 500°C

As listed in Table 3.4, Ti-B-N thin films with different N contents were deposited at 500° C by way of incorporation of different amounts of N into $TiB_{0.65}$ using two $TiB_2$ and two Ti targets. During deposition the current of both $TiB_2$ targets was fixed at 3.0 A while that of both Ti targets at 6.0 A. It is apparent that different nitrogen flows resulted in formation of different microstructural thin films. By XRD (not shown here) no any peak representing either crystalline Ti or $TiB_2$ was found in nitrogen-free $TiB_{0.65}$, which reveals formation of amorphous $TiB_{0.65}$. By XPS (see Figure 3.14) both Ti and $TiB_2$ bonds were detected in $TiB_{0.65}$, which indicates that B and Ti coming from $TiB_2$ target re-bonded to $TiB_2$, which resulted in formation of amorphous compound comprising Ti and $TiB_2$ (Ti-$TiB_2$ compound). This result is the same as that of the films deposited in the room temperature. However, incorporation of 10 at.% nitrogen ($TiB_{0.62}N_{0.18}$) accelerated crystallization of thin films and phase segregation, which caused formation of nanocomposite nc-Ti(N)/a-$TiB_2$ thin films.

When nitrogen flow was increased to a definite level, during deposition nitrogen activity was sufficiently high for segregation into separating nitride phases. For example, in this study, the film structures of $TiB_{0.53}N_{0.54}$ and $TiB_{0.54}N_{0.63}$ were comprised of a compound of TiN and $TiB_2$, which have been verified by XPS (Figure 3.14). Further increase of nitrogen accelerated formation of BN phase, which resulted in formation of small amount of BN besides TiN and $TiB_2$ in $TiB_{0.51}N_{0.82}$. This effect was enhanced with further increase of N content in $TiB_{0.61}N_{1.04}$ and $TiB_{0.61}N_{1.26}$, which has been verified by present XPS (Figure 3.14). Thus, nanocomposite nc-TiN/a-$TiB_2$/a-BN thin films were formed. These findings are consistent with the Ti-B-N phase diagram.[65]

Formation of amorphous nitrogen-free $TiB_{0.65}$ was further demonstrated by its HRTRM image shown in Figure 3.15. With introduction of N into a growing $TiB_{0.65}$ film, two- or three-phase nanocomposite films with microstructures comprising of nanocrystalline grains embedded in an amorphous matrix have also further been evidenced (Figure 3.16).

In summary, depending on the amount of introduced N, three are three nanocomposite features: (1) with addition of small amount of nitrogen (10 at.% in this study) a nanocomposite structure of solid solution Ti(N) nanocrystallites embedded into amorphous $TiB_2$ matrix was formed; (2) with increasing nitrogen content, the films were predominantly built up of a two-phase structure of fine TiN particles together with remaining a-$TiB_2$ phase; (3) with further increasing N content, 2~5 nm TiN nanocrystallites were separated by a thin layer amorphous phase consisting of $TiB_2$ and BN. On the other hand, nitrogen incorporation was a prerequisite of spinodal segregation and hence formation of the nanocomposite structure. From thermodynamic consideration,[66] it is believed that the decrease of free energy because of formation of Ti(N), TiN, $TiB_2$ and BN drove spinodal phase segregation.

**Table 3.1. Deposition conditions and properties of Ti-$B_x$-$N_y$ films used in this study**

| Sample | B content (at.%) | Target configuration (target current) | B/Ti ratio | Roughness rms (nm) |
|---|---|---|---|---|
| TiN | 0 | 3Ti (6.0) - 0$TiB_2$ | 0 | 16.6 |
| $TiB_{0.06}N_{0.82}$ | 3.2 | 3Ti (6.0) - 1$TiB_2$ (2.0) | 0.06 | 7.9 |
| $TiB_{0.18}N_{0.80}$ | 9.1 | 3Ti (6.0) - 1$TiB_2$ (4.0) | 0.18 | 7.1 |
| $TiB_{0.36}N_{0.85}$ | 16.3 | 3Ti (6.0) - 1$TiB_2$ (6.0) | 0.36 | 6.6 |
| $TiB_{0.41}N_{0.76}$ | 18.9 | 3Ti (6.0) - 1$TiB_2$ (8.0) | 0.41 | 5.8 |
| $TiB_{0.79}N_{1.09}$ | 27.4 | 1Ti (6.0) - 1$TiB_2$ (6.0) | 0.79 | 4.2 |
| $TiB_{1.86}N_{1.51}$* | 42.6 | 0Ti - 1$TiB_2$ (6.0) | 1.86 | 1.5 |

**Table 3.2. Deposited films, OEM ratio and corresponding N contents as well as estimated mole fractions of various phases.**

| Film | OEM ratio (%) | N content (at.%) | Fraction (mol%) $TiB_2$ | TiN | BN |
|---|---|---|---|---|---|
| $TiB_{0.65}$ | 100 | 0 | 32.5 | 0 | 0 |
| $TiB_{0.74}N_{0.67}$ | 95 | 27.8 | 38.1 | 58.7 | 3.2 |
| $TiB_{0.46}N_{0.86}$ | 90 | 37.1 | 18.9 | 75.5 | 5.6 |
| $TiB_{0.45}N_{0.89}$ | 80 | 38.0 | 17.3 | 75.2 | 7.5 |
| $TiB_{0.50}N_{1.12}$ | 60 | 42.7 | 10.1 | 70.0 | 19.9 |
| $TiB_{0.56}N_{1.20}$ | 50 | 43.5 | 9.1 | 67.0 | 23.9 |
| $TiB_{0.66}N_{1.38}$ | 40 | 45.5 | 6.3 | 61.5 | 32.2 |

For $TiB_{0.65}$, the rest is metallic Ti.

**Table 3.3. Deposition conditions and properties of $TiB_xN_y$ films used in this study.**

| Sample | B content (at.%) | Target configuration (target current) | Sample | B content (at.%) | *Target configuration (target current)* |
|---|---|---|---|---|---|
| $TiN_{0.81}$ | 0 | 2Ti (6.0) - 0$TiB_2$ | $TiB_{0.49}N_{1.05}$ | 19.3 | 2Ti (6.0) - 2$TiB_2$ (3.0) |
| $TiB_{0.03}N_{0.83}$ | 1.6 | 2Ti (6.0) - 2$TiB_2$ (0.5) | $TiB_{0.53}N_{0.97}$ | 21.2 | 2Ti (6.0) - 2$TiB_2$ (3.5) |
| $TiB_{0.13}N_{0.78}$ | 6.8 | 2Ti (6.0) - 2$TiB_2$ (1.0) | $TiB_{0.63}N_{1.08}$ | 23.2 | 2Ti (6.0) - 2$TiB_2$ (4.0) |
| $TiB_{0.19}N_{0.92}$ | 9.0 | 2Ti (6.0) - 2$TiB_2$ (1.5) | $TiB_{0.96}N_{1.26}$ | 29.8 | 2Ti(6.0,3.0)-2$TiB_2$(4.0) |
| $TiB_{0.25}N_{0.88}$ | 11.7 | 2Ti (6.0) - 2$TiB_2$ (2.0) | $TiB_{1.15}N_{1.23}$ | 34.0 | 2Ti (6.0, 0)-2$TiB_2$ (4.0) |
| $TiB_{0.37}N_{1.0}$ | 15.6 | 2Ti (6.0) - 2$TiB_2$ (2.5) | | | |

**Table 3.4. The studied thin films and the corresponding N contents.**

| **Samples** | **$TiB_{0.65}$** | **$TiB_{0.62}N_{0.18}$** | **$TiB_{0.40}N_{0.30}$** | **$TiB_{0.53}N_{0.54}$** | **$TiB_{0.54}N_{0.63}$** | **$TiB_{0.51}N_{0.82}$** |
|---|---|---|---|---|---|---|
| N content | 0 | 10.0 | 17.7 | 25.4 | 29.0 | 35.2 |
| Samples | $TiB_{0.61}N_{1.04}$ | $TiB_{0.61}N_{1.26}$ | | | | |
| N content | 39.2 | 43.9 | | | | |

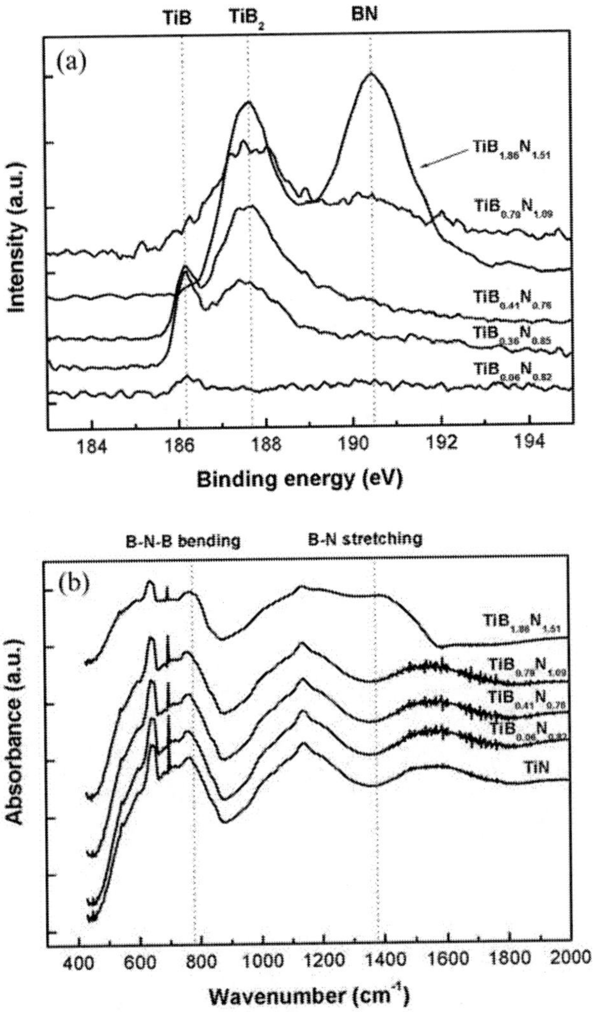


Figure 3.1. Typical (a) XPS B 1s and (b) FT-IR spectra of several Ti-$B_x$-$N_y$ thin films with different B contents.

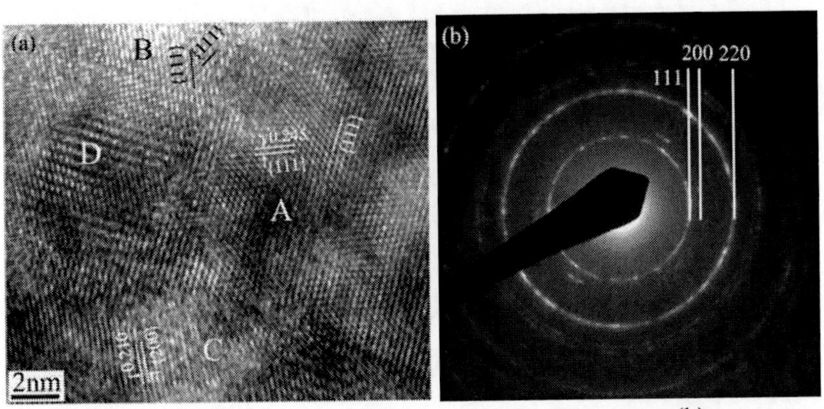


(a) (b)

Figure 3.2. (a) Plan-view HRTEM morphology and (b) SAD pattern of a pure TiN thin film deposited at room temperature.

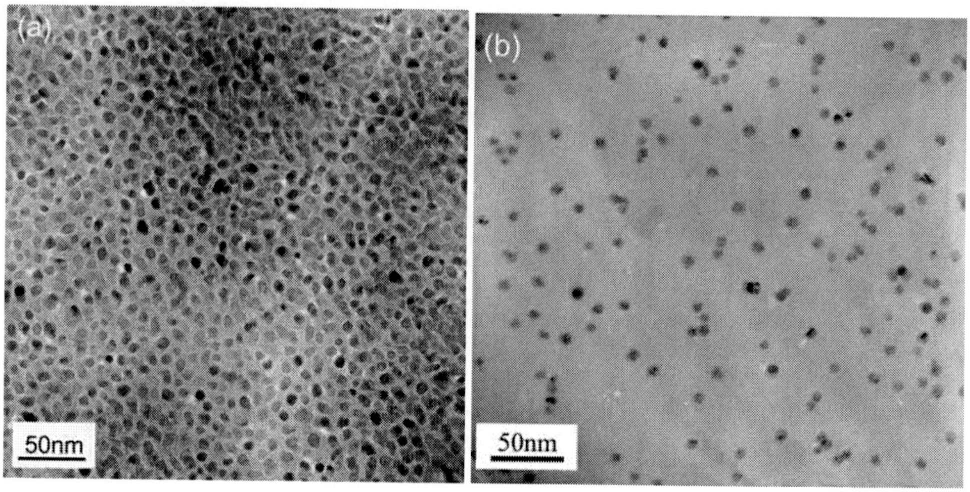


Figure 3.3. Low magnification TEM images of (a) $TiB_{0.36}N_{0.85}$ and (b) $TiB_{1.86}N_{1.51}$.

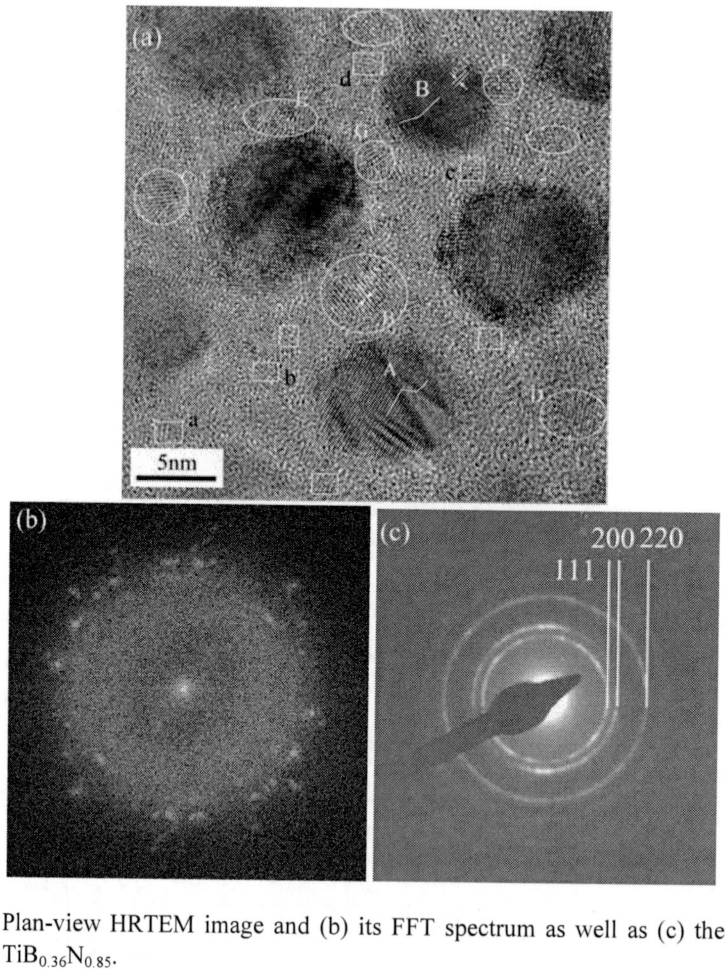


Figure 3.4. (a) Plan-view HRTEM image and (b) its FFT spectrum as well as (c) the corresponding SAD pattern of $TiB_{0.36}N_{0.85}$.

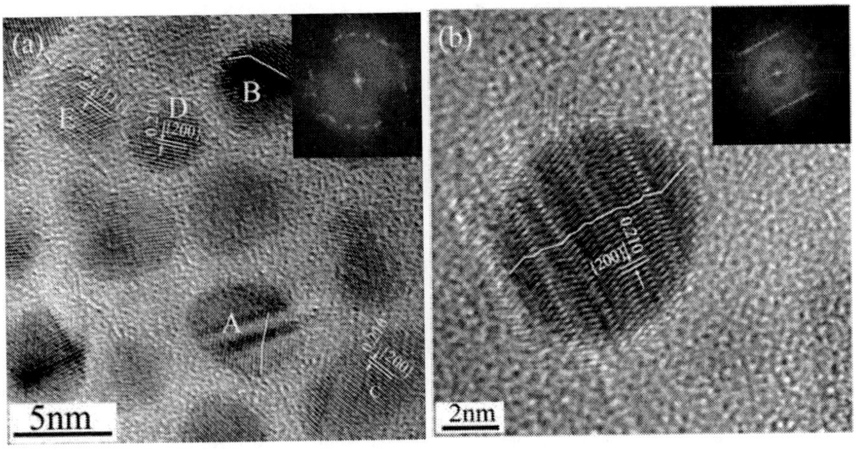


Figure 3.5. Plan-view HRTEM images of (a) $TiB_{0.79}N_{1.09}$ and (b) $TiB_{1.86}N_{1.51}$. Their FFT spectra are inserted.

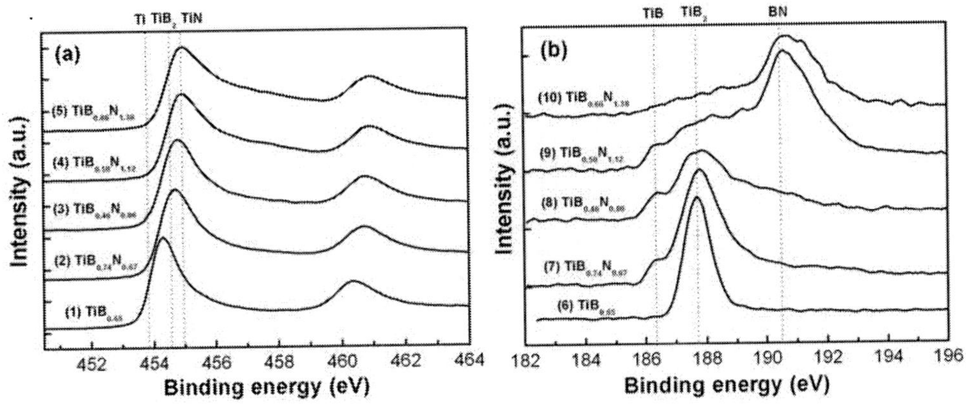


Figure 3.6. Typical XPS (a) Ti 2p and (b) B 1s spectra of Ti-B-N thin films with different N contents.

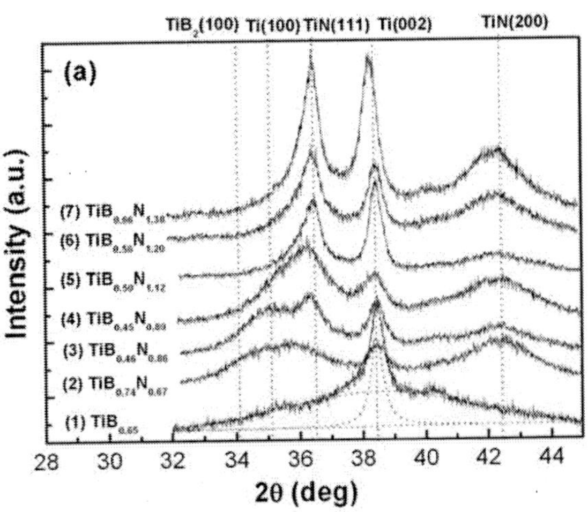


Figure 3.7. XRD spectra of Ti-B-N thin films with different N contents.

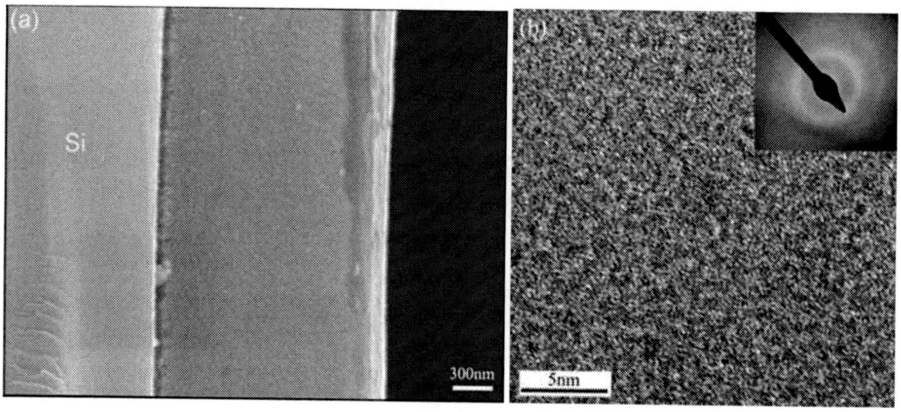


Figure 3.8. (a) Cross-sectional SEM morphology and (b) plan-view HRTEM image of $TiB_{0.65}$. SAD pattern is inserted.

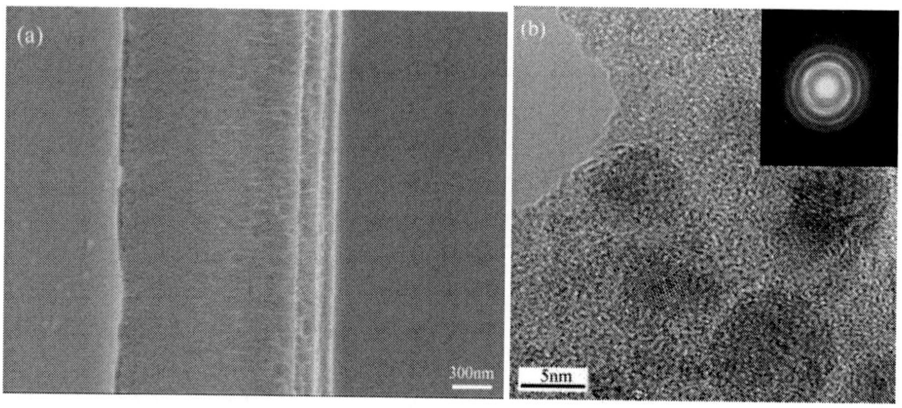


Figure 3.9. Cross-sectional (a) SEM morphology and (b) HRTEM image of $TiB_{0.50}N_{1.12}$. The corresponding SAD pattern is inserted.

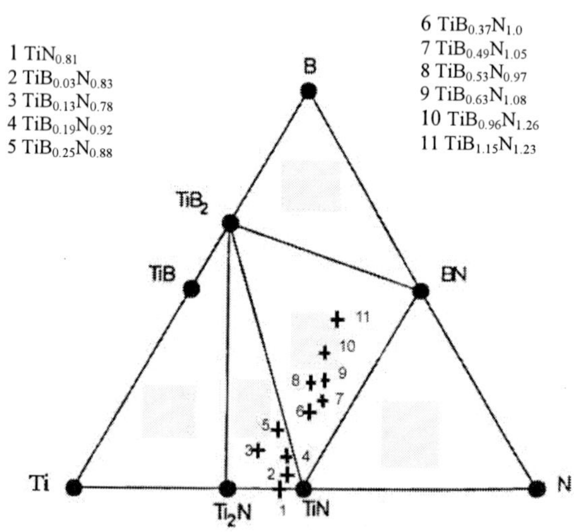


Figure 3.10. The composition positions of the present studied thin films in Ti-B-N phase diagram.

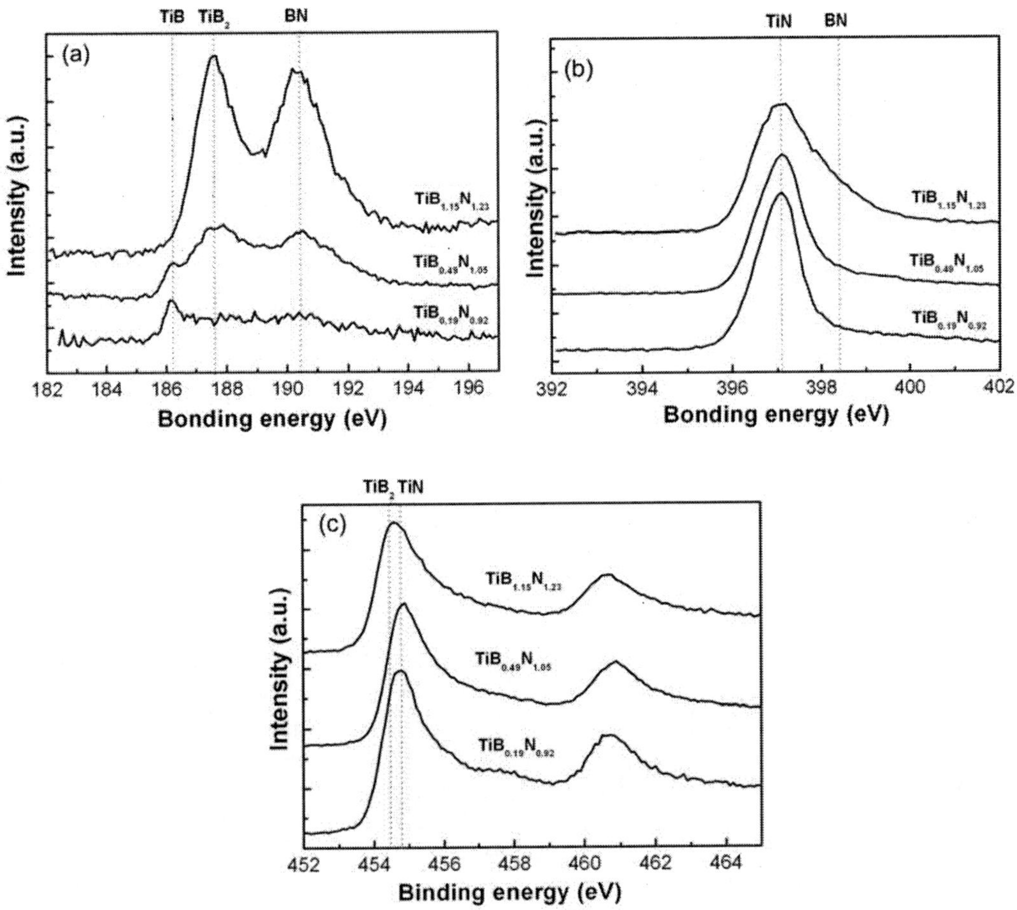


Figure 3.11. XPS (a) B 1s, (b) N1s and (c) Ti 2p spectra of $TiB_xN_y$ thin films with different B contents.

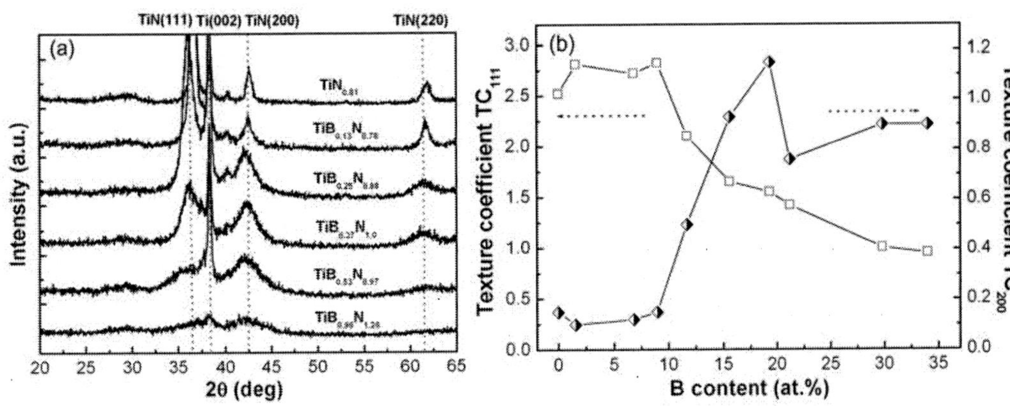


Figure 3.12. Continued on next page.

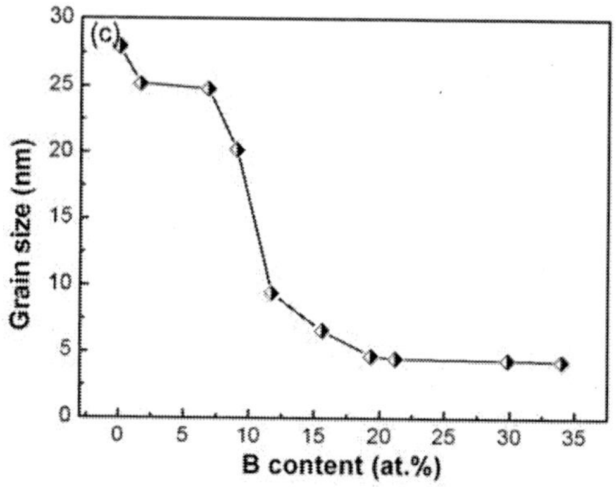


Figure 3.12. (a) XRD spectra and (b) the corresponding (111) and (200) texture coefficients as well as (c) nanocrystallite sizes in $TiB_xN_y$ thin films deposited at 500°C.

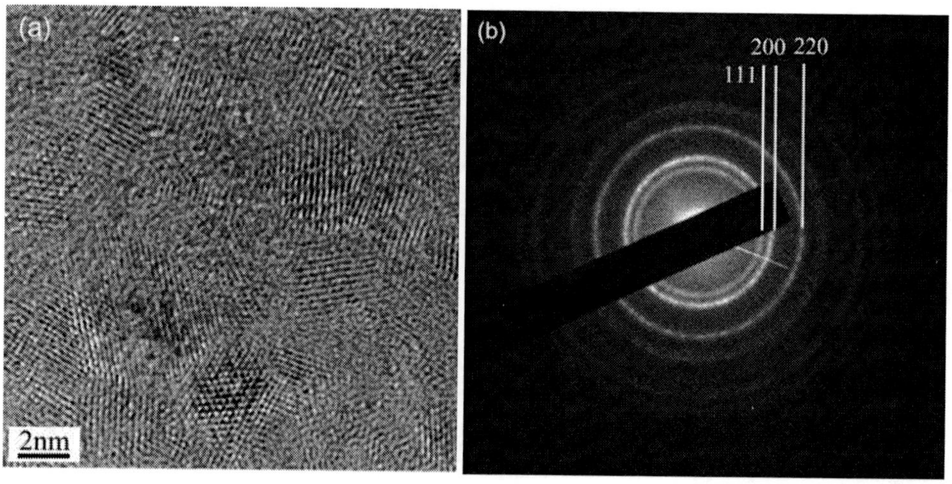


Figure 3.13. (a) An HRTEM image showing nanocomposite structure and (b) the SAD pattern of $TiB_{0.49}N_{1.05}$.

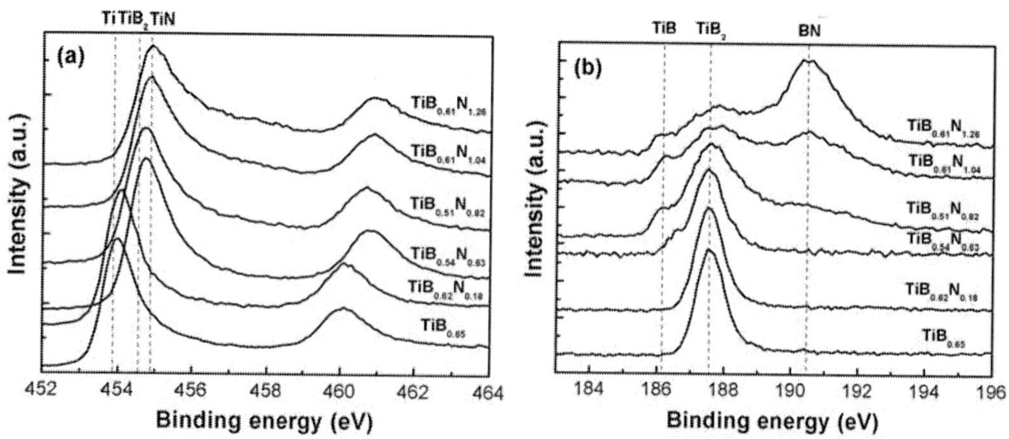


Figure 3.14. XPS (a) Ti 2p and (b) B 1s spectra of Ti-B-N thin films with different N contents.

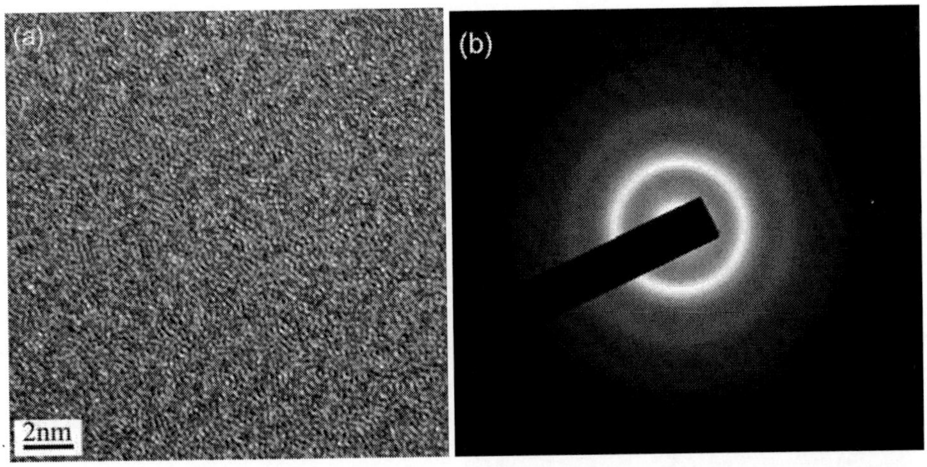


Figure 3.15. (a) A HRTEM morphology and (b) the corresponding SAD pattern of $TiB_{0.65}$.

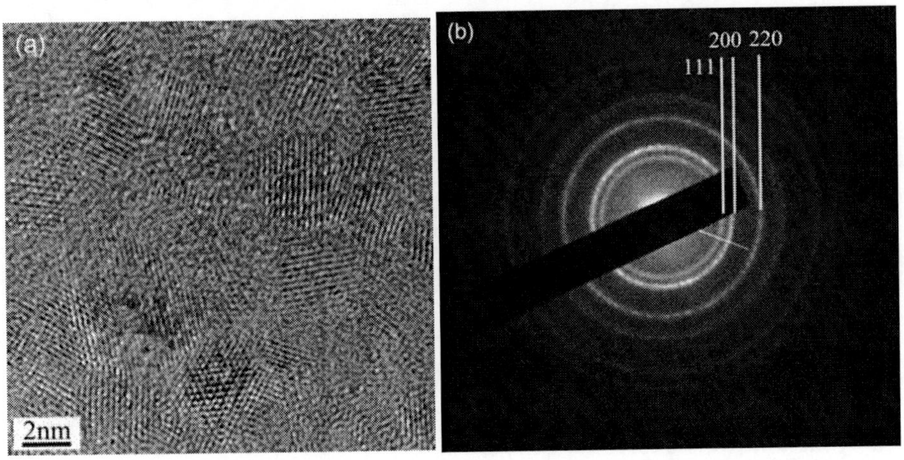


Figure 3.16. (a) A HRTEM morphology and (b) the corresponding SAD pattern of $TiB_{0.61}N_{1.04}$.

# 4. Phase Configuration and Nanostructure in Ti-C-N Thin Films

## 4.1. Ti-C-N Thin Films with C Content Deposited at 500°C

Table 4.1 lists thin films of Ti-$C_x$-$N_y$ with different C contents deposited on polished Si(100) substrates at 500 °C using two graphite and two Ti targets by way of adjusting electric current of graphite targets under fixed Ti target currents of 6.0 A. Under the present investigation, the C-free $TiN_{0.87}$ is largely non-stiochiometric and can show large vacancy concentrations on the nitrogen sublattice such that ordering might be expected.[61] Nitrogen-deficit $TiN_{0.87}$ thus shows the same structure as pure TiN, which is evidenced by its XRD spectrum and HRTEM image (Figure 4.1). Incorporation of a small amount of carbon filled nitrogen sublattice of $TiN_{0.87}$ to comprise Ti-$C_x$-$N_y$ thin films. When the relative atomic ratio X of (C+N)/Ti was less than one unit, a Ti(C, N) solid solution was formed till carbon filled the all nitrogen sublattice vacancies (see Figs. 4.2 and 4.3). Correspondingly, there are only TiN and TiC bonds in Ti(C, N) solid solution thin films, and neither C-C nor C-N bond was detected by XPS (see Figure 4.2). When the atomic ratio of (C+N)/Ti was more than one unit, a distinguished different phase configuration was yielded. As shown in Figure 4.2, both C-C and C-N bonds are observed besides Ti-N and Ti-C bonds, although only one crystalline phase is found by XRD (see Figure 4.1). This indicates that after the carbon occupied all sublattice vacancies of nitrogen, excessive carbon linked each other or to N, forming C-C (disordered C and graphite) or C-N bonding, as shown in Figure 4.2. Thus, two-phase nanocomposite nc-Ti(C,N)/a-(C,$CN_x$) thin films with microstructures comprising of nanocrystalline phase in an amorphous matrix were formed (see Figure 4.2). It should be pointed out that the C-C bonding predominated in the amorphous matrices according to XPS (Figure 4.2).

Raman method is a powerful technique to examine the C-reference matrix phase. Figure 4.3 reveals that adding a small amount of carbon into a growing TiN film did not make Raman vibration active in the range of 1000~2000 $cm^{-1}$. It indicates that there is no free carbon in the films because the doped carbon all entered into nitrogen sub-lattice of TiN. The Raman vibration was active when the excessive carbon formed C-C bonds in $TiC_{0.24}N_{0.71}$, $TiC_{0.39}N_{0.66}$, $TiC_{0.81}N_{0.63}$ and $TiC_{1.61}N_{0.73}$ (Figure 4.3). The G band, coming from the zone center $E_{2g}$ mode, is sensitive to the degree of the two-dimensional ordering, while the disordered-induced (D) band, corresponding to phonons near the K and M points of the Brillouin zone, is usually associated with nonplanar microstructural distortions.[67] More C incorporation not only gave rise to downshift of the G peak position, but also broadened the G peak, as shown in Figure 4.3, which implies that C incorporation promoted the formation of bond-angle disorder and therefore softened the vibrational modes.[68]

A quantitative analysis of the Raman spectra was based on fitting of detected Raman spectra. It is reported that the Raman spectra with G (graphitic) and D (disordered) peaks were fitted with a Breit-Wigner-Fano (BWF) G peak, a Lorentzian D peak, and a linear background using a nonlinear least-squares fitter (NLSF) program. The BWF line shape is described by the following expression:[69]

$$I(\omega) = \frac{I_0[1+2(\omega-\omega_0)/q\Gamma]^2}{1+[2+(\omega-\omega_0)/\Gamma]^2}$$

where $I(\omega)$ is the intensity as a function of frequency, $I_0$ is the peak intensity, $\omega_0$ and $\Gamma$ are the peak position and FWHM (full width at half maximum) value, respectively. $q^{-1}$ is the BWF coupling coefficient. It is reported that $I_D/I_G$ is generally a measure of the density of defects and the degree of disorder due to zone edges or border phonons of the carbon cluster.[70,71,72] Therefore, it is concluded that an increase of $I_D/I_G$ values [Fig. 4-4(b)] is rationally related to more fractional disordered C or higher density defects in amorphous matrices with more carbon incorporation, which is consistent with the behaviors of the G peak [Fig. 4-4(a)], and was proved by our HRTEM observation.

The nanostructure evolution in nanocomposite films with different amounts of C incorporation was unambiguously verified by HRTEM observations (see Figs.4.6 and 4.6). The HRTEM image in Figure 4.5 (a) reveals that the C-free $TiN_{0.87}$ shows well-crystallized polycrystallite. There is no amorphous phase observed in grain boundaries. This indicates that the present plasma conditions such as nitrogen activity and activation energy were enough for the fully crystallization of TiN. However, C incorporation made some conditions change. Once Ti and C from targets and N from reactive gases arrived the substrate, on one hand, the fcc-like solid solution Ti(C, N) nanocrystallites were formed (Figure 4.5). In Figure 4.5(a), only one high-angle grain boundary and two adjacent grains (A and B) as well as one Morié pattern zone C with a large fringe spacing are observed, indicating that $TiN_{0.87}$ is a well crystallized polycrystallite comprising of grains in various sizes, shapes and orientations. Incorporation of a small amounts of C (up to 12.4 at.%, X=0.95) fully dissolved sublattice vacancies of nitrogen in $TiN_{0.87}$, forming Ti(C,N) solid solution polycrystallites. Several grains A~D in Figure 4.5 (b) and A~C in Figure 4.5(c) are observed. There is no amorphous phase along the grain boundaries. On the other hand, if the amount of deposited carbon exceeded the needed for formation of Ti(C, N) (X≥1), excessive carbon was extruded to grain boundary zone. After all nanocrystallites for example A~G were fully wet, a nanocomposite structure of nc-Ti(C,N) embedded into amorphous matrices was formed (see Figure 4.6 (a)). Thicker amorphous matrix walls and finer nanocrystallites for example A~E were followed when more C was extruded and more CN was formed with C incorporation (see Figure 4.6 (b)). When the C content was increased to a high level, for example ~48 at.% in this study (X=2.34), the nucleation and growth of nc-Ti(C,N) were heavily hindered and disturbed due to blocking of carbon on diffusion of Ti and N atoms, some small sized grains for example A~D and various types of defects were therefore produced (see Figure 4.6 (c)).

Obviously, carbon incorporation greatly decreased the grain size (Figure 4.1 (b)) by way of the formation of solid solution or two-phase nanocomposite structure. The carbon atoms incorporated in the films to form solid solution at low carbon content and nanocomposite structure at high carbon content can prevent TiN crystalline growth in several ways: (1) carbon can act as a roadblock to the diffusing Ti and N or can trap diffusing Ti and N, or can serve as a nucleation site for new grains, thus leading to the formation of small sized grains; (2) the amorphous carbon extruded from Ti(C,N) and CN formed during reactive deposition hindered grain growth, and hence decreasing the grain size. These effects are further enhanced with more C incorporation. On the other hand, carbon incorporation promoted the formation of more (200) oriented films, which were proved by XRD and SAD patterns (see Figure 4.1 and Figure 4.6). From Figs.4.5 and 4.6 it is found that the {200} Bragg reflection in the SAD pattern looks stronger compared with those of $TiC_{0.24}N_{0.71}$ and $TiC_{0.39}N_{0.66}$, which was activated by enhanced preferred [200] orientation. This process is presumed to relate to the following two ways: direct deposition of carbon on {111} plane of TiN and formation of nanocomposite structure. Direct deposition of carbon on {111} prohibited atom growth on {111} plane, as a result depressing formation of [111] orientation and promoting growth of [200] orientation. This thought is supported by Barna's research[73] in which the impurity oxygen was prone to precipitate along {111} plane of Al thin films, therefore decreased the inclination of formation of preferred [111] orientation. On the other hand, the formation of nc-/a- nanostructure obviously decreased energies of grain boundary and nanostructure system, and hence promoted growth of low energy (002) plane, [74] which contributed to the change in nanocrystalline orientation [see Figs. 4.1(a) and 4.5].

## 4.2. Ti-C-N Thin Films with N Contents Deposited at 500°C

In this section, we report the effect of incorporation of different amounts of nitrogen into near-stiochiometric TiC ($TiC_{1.02}$) on phase configuration and nanostructure. XPS measurements showed that TiC bonding was detected in $TiC_{1.02}$, revealing that C and Ti coming from Ti and graphite targets respectively bonded to TiC on the substrate surface (Figure 4.7). By XRD (Figure 4.8) only TiC characterized diffraction peak representing crystalline TiC was found in $TiC_{1.02}$, indicating that a polycrystalline TiC thin film was formed. XRD and XPS structural analyses of $TiC_{1.02}$ are in accord with that of HRTEM observation (see Figure 4.9(a)). In Figure 4.9, nano-grains A-H with different orientations comprised nano-polycryalline $TiC_{1.02}$. Incorporated N obviously dissolved into the lattice of $TiC_{1.02}$ and substituted part of C. Some C was substituted, and linked each other to form C-C, or linked to N to form C-N, as shown in Figure 4.7. Thus, nanocomposite thin films comprising nc-$TiC_xN_y$ and a-(C, $CN_x$) were formed (see Figs. 4.9 (b) and (c)). In Figure 4.9(b) nanocrystallites A~E with an average size of ~3 nm are found to be embedded into amorphous matrices. More N incorporation enhanced this effect, which increased the amorphous matrix wall thickness (see Figure 4.9 (c)).

**Table 1. Depositions and properties of Ti-$C_x$-$N_y$ films deposited at 500 °C.**

| Sample | C content (at.%) | Target configuration (target current) | (C+N)/Ti ratio |
|---|---|---|---|
| $TiN_{0.87}$ | 0 | 2Ti (6.0) – 0C | 0.87 |
| $TiC_{0.13}N_{0.77}$ | 6.7 | 2Ti (6.0) – 2C (0.5) | 0.90 |
| $TiC_{0.14}N_{0.74}$ | 7.3 | 2Ti (6.0)–2C (1.0) | 0.88 |
| $TiC_{0.24}N_{0.71}$ | 12.4 | 2Ti (6.0)–2C (1.5) | 0.95 |
| $TiC_{0.39}N_{0.66}$ | 19.1 | 2Ti (6.0) – 2C (2.0) | 1.05 |
| $TiC_{0.81}N_{0.63}$ | 33.2 | 2Ti (6.0) – 2C (3.5) | 1.44 |
| $TiC_{1.61}N_{0.73}$ | 48.2 | 2Ti (3.0, 6.0) –2C (4.0) | 2.34 |

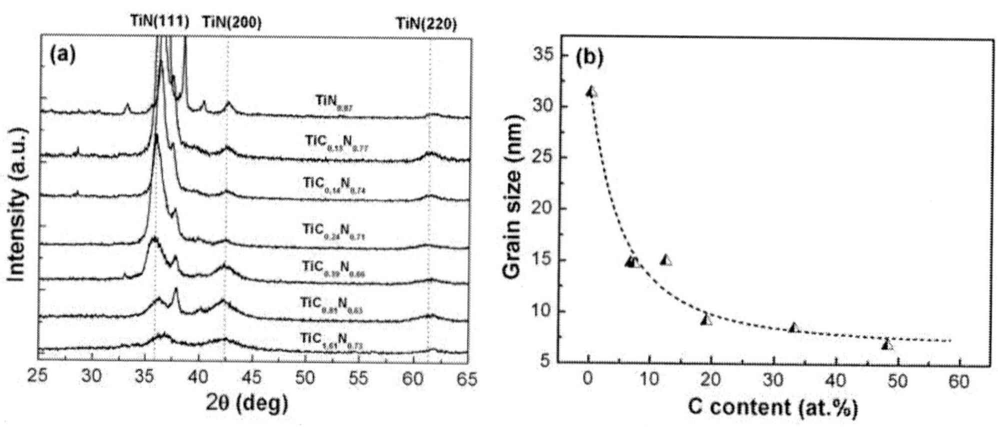


Figure 4.1. (a) XRD θ-2θ spectra of Ti-$C_x$-$N_y$ thin films with different C content and (b) the dependence of grain size on C content.

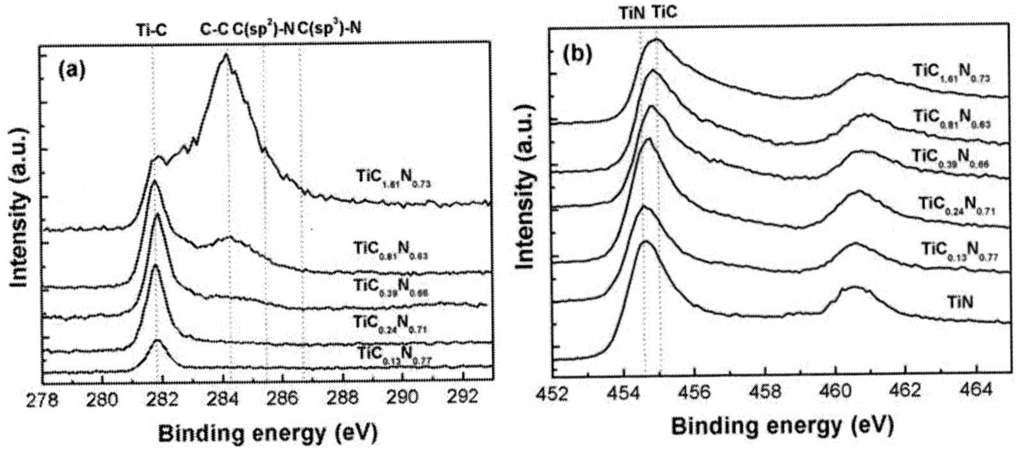


Figure 4.2. XPS (a) C 1s and (b) Ti 2p core-level spectra of Ti-$C_x$-$N_y$ thin films with different C contents.

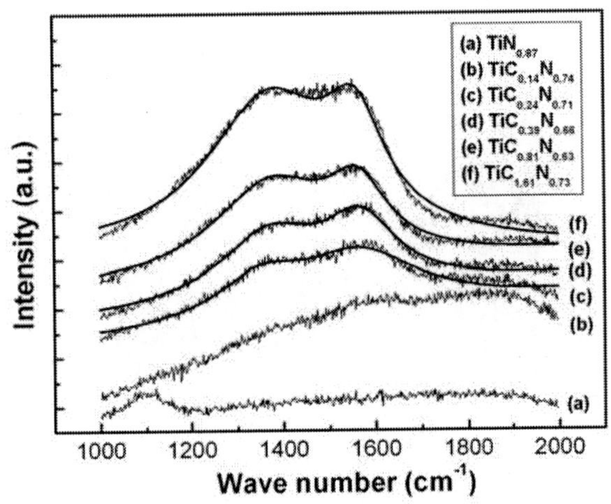


Figure 4.3. Raman spectra of Ti-$C_x$-$N_y$ thin films with different C contents. The spectra with D and G peaks have been fitted with one Lorentzian line shape for the D peak and one BWF line shape for the G peak using the NLSF program.

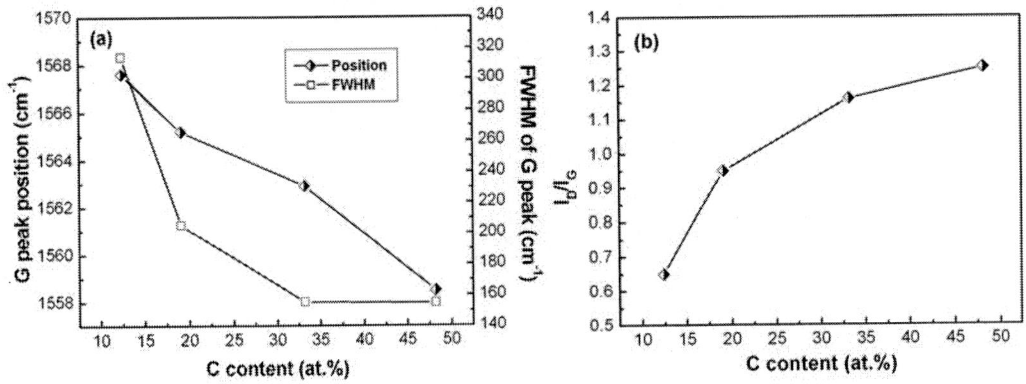


Figure 4.4. (a) The positions and FWHM of G peaks as well as (b) $I_D/I_G$ ratio as a function of C content.

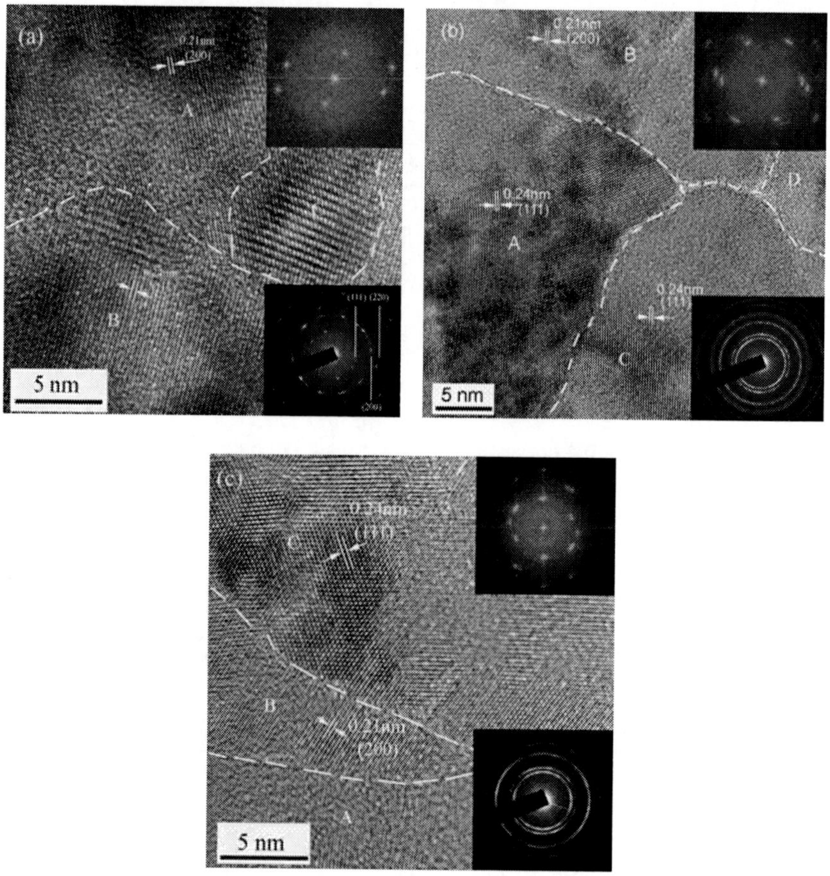


Figure 4.5. Plan-view HRTEM images showing SS nanostructures with several different X values. (a) X=0.87 ($TiN_{0.87}$), (b) X=0.88 (7.3 at.% C, $TiC_{0.14}N_{0.74}$), and (c) X=0.95 (12.4 at.% C, $TiC_{0.24}N_{0.71}$). Their FFT spectra and the corresponding SAD patterns are inserted respectively.

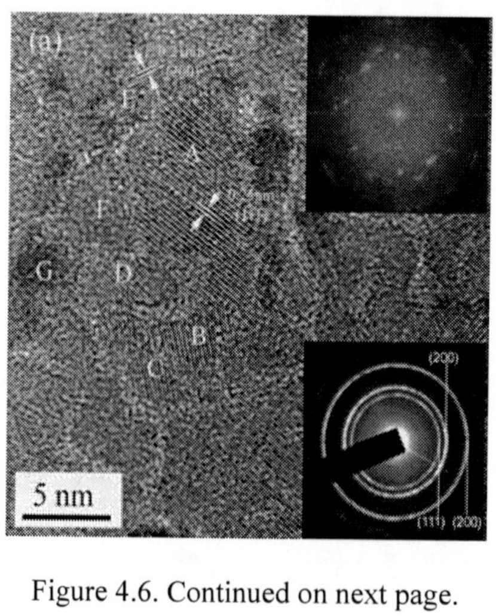


Figure 4.6. Continued on next page.

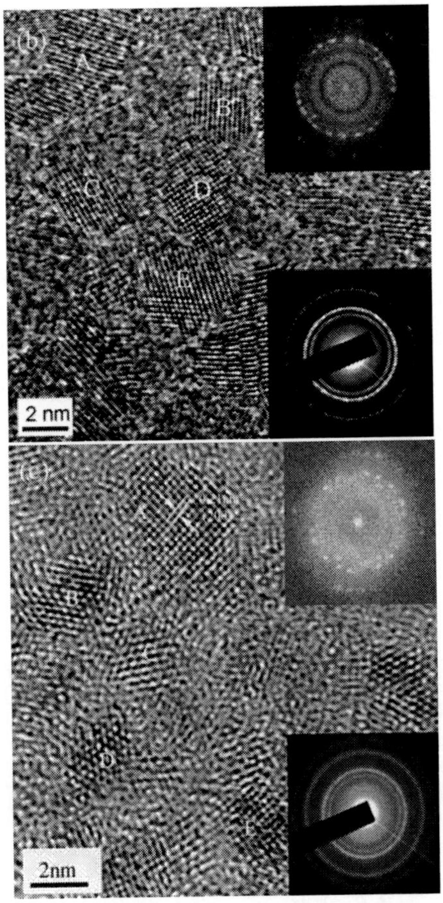


Figure 4.6. Plan-view HRTEM images showing two-phase nanocomposite structures with several different X values. (a) X=1.05 (19.1 at.% C, $TiC_{0.39}N_{0.66}$), (b) X=1.44 (33.2 at.% C, $TiC_{0.81}N_{0.63}$) and (c) X=2.34 (48.2 at.% C, $TiC_{1.61}N_{0.73}$). Their FFT spectra and the corresponding SAD patterns are inserted respectively.

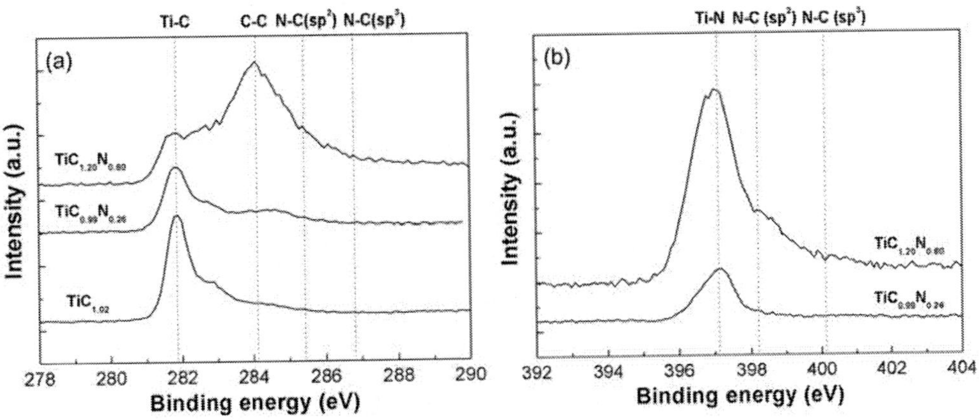


Figure 4.7. XPS (a) C 1s and (b) N 1s core-level spectra of Ti-$C_x$-$N_y$ thin films with different N contents.

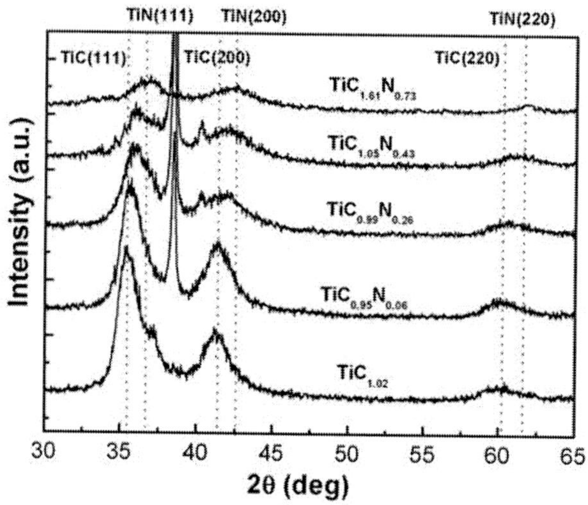


Figure 4.8. XRD spectra of Ti-B-N thin films with different N contents.

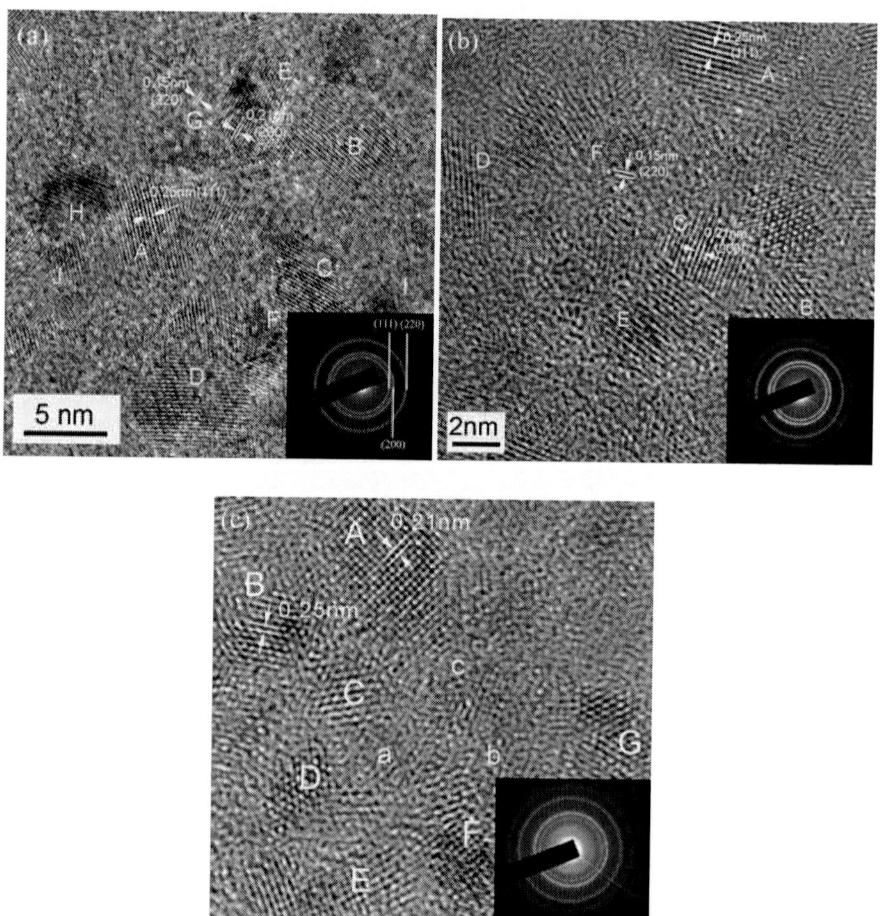


Figure 4.9. Plan-view HRTEM images of (a) $TiC_{1.02}$, (b) $TiC_{0.99}N_{0.26}$ and (c) $TiC_{1.61}N_{0.73}$.

# 5. Phase Configuration and Nanostructure in Ti-B-C-N Thin Films

Recently, quaternary system of nanocomposite hard thin films are receiving many attentions because it may combine good properties of different ternary systems through nanostructure design.[75] Ti-C-N films have been extensively used as industry film materials especially for various tools because of its better mechanical properties and wear characterizations.[76,77,78] While Ti-B-N thin films showed superhard hardness value and high thermal stability.[79,80] Thus, good structural and mechanical properties were expected to be obtained in Ti-B-C-N thin films.[81,82]

In this section, we report some research results of Ti-B-C-N thin films. Present thin films of Ti-B-C-N with different N contents by way of incorporation of different amounts of N into Ti-B-C were deposited on polished Si(100) substrates at 500 °C using two opposed C, one Ti and one $TiB_2$ targets. Our investigation based on XRD and HRTEM analyses have revealed that the present pure Ti-B-C exhibits a quasi-amorphous structure comprising TiC and $TiB_2$ (see Figs.5.1 and 5.2). This result is similar with that of Mitterer's study.[83] Just like $TiN_x$, $TiC_x$ is non-stiochiometric and can show some vacancy concentrations on the carbon sublattice, however keeps its fcc ordering. Incorporation of a small amounts of N is believed to dissolve into $TiC_x$, which results in formation of solid solution $TiC_xN_y$ ($x+y<1$). As a result, $TiC_xN_y$ fraction kept almost unchanged although TiN bonding was successively increased (Figure 5.3(c)), while TiC kept almost unchanged (Figure 5.3). The above phenomena remained till N fully filled the all vacancies in $TiC_xN_y$ (at that time, $x+y=1$). When N content was further increased, it is believed that the following two reactions occurred during deposition:

$$TiB_2+N_2=TiN+2BN \quad \Delta G=-424.8 \text{ KJ/mol} \tag{1}$$

$$TiC+N_2=TiN+CN \quad \Delta G=-172.4 \text{ KJ/mol} \tag{2}$$

Reaction (1) has the bigger negative Gibbs free energy according to the existing thermodynamic parameters,[84] and took place automatically, while reaction (2) was restrained till $TiB_2$ was used out. As a result, further increase of N promoted formation of BN and TiN at cost of $TiB_2$ (see Figure 5.3 and Table 5.1). On the other hand, CN was restrained to be formed in our compositional range, as shown in Figure 5.3. According to the nanostructure and phase analyses three main phases, Ti(C,N), $TiB_2$ and/or BN, of Ti-B-C-N thin films with different N contents were quantitatively estimated. For $TiB_xC_yN_z$ samples with different stoichiometry as determined by XPS, the three phases were calculated based on the following relation:

(1) for the BN-free thin films (N≤11.2 at%),

$$a[TiC_mN_n] + b[TiB_2] = TiB_xC_yN_z$$

(2) for the thin films (N≥15.3 at.%) with BN phase

$$a[TiC_mN_{1-m}]+b[TiB_2]+c[BN]=TiB_xC_yN_z$$

Solvable expressions for a, b, c, m and n can be derived using simultaneous equations with various x, y and z components. The calculation results are listed in Table 5.1. Based on the calculation results, it is found that increase of N content results in the following effects:

(1) $TiC_xN_y$ fraction keeps almost unchanged at N content less than 18.7 at.%, however rapidly increases with further increase of N content;

(2) $TiB_2$ fraction slightly decreases at N content less than 18.7 at.% ($TiB_{0.60}C_{0.36}N_{0.45}$), a swift decrease in $TiB_2$ fraction is followed with further increase of N content (after BN formation);

(3) BN fraction greatly increases after it forms.

Their dependences on N content are plotted in Figure 5.4.

Thus, the resulting effects due to incorporation of N are summarized as follows. (1) In the case of N content less than 11.2 at.%, incorporated N successively dissolved into $TiC_x$, which formed solid solution $Ti(C_x,N_y)$. Thus, nanocomposite thin films of nc-$Ti(C_x,N_y)$/a-$TiB_2$ was formed. (2) In the case of N content more than 15.3 at.%, additional BN was formed. Nanocomposite thin films of nc-$Ti(C_xN_{1-x})$ embedded into amorphous ($TiB_2$, BN) were hence produced. On the other hand, incorporation of N obviously promoted crystallization of nanocrystallites. However, a decrease in nanocrystalline sizes was followed after formation of BN. From view of nanostructure, with incorporation of N into a growing Ti-B-C film, nitrogen activity obviously promoted phase segregation and hence accelerated the nano-grain growth in Ti-B-C-N thin films. In this study, the film $TiB_{0.65}C_{0.36}$ was a quasi-amorphous compound comprising TiC and $TiB_2$; while incorporation of 11.2 at.% nitrogen ($TiB_{0.60}C_{0.38}N_{0.25}$), on one hand, made the interface between nanocrystallits and amorphous matrices clear (phase segregation); on the other hand, promoted the grain growth to ~2 nm (see Figure 5.5). The above phenomena indicate that existence of nitrogen was a prerequisite of spinodal segregation and hence formation of the nanocomposite structure. From thermodynamic consideration, it is believed that the decrease of free energy because of formation of Ti(C,N), $TiB_2$ and BN drove spinodal segregation. On the other hand, when the nitrogen content was further increased, the film structure became more chaos, which was believed to be due to formation of BN phase (see Figure 5.6). The similar effect was found in the Ti-B-N thin films,[42] where formation of BN phase made the short range ordering (SRO) vanish and promoted matrix structure more chaos.

In summary, the pure Ti-B-C was two-phased quasi-amorphous thin films comprising $TiC_x$ and $TiB_2$. Incorporation of a small amount of N not only dissolved into $TiC_x$ but also promoted growth of $TiC_x$ nano-grains. As a result, nanocomposite thin films of nanocrystalline (nc-) $TiC_x(N_y)$ ($x+y<1$) embedded into amorphous (a-) $TiB_2$ were observed till nitrogen fully filled all carbon vacancy lattice (at that time $x+y=1$). Further increase of N content promoted formation of a-BN at cost of $TiB_2$, which produced nanocomposite thin films of nc-$Ti(C_x,N_{1-x})$ embedded into a-($TiB_2$, BN). Formation of BN also decreased nanocrystalline size.

**Table 5.1. The deposited Ti-B-C-N thin films, the corresponding N content and (B+C+N)/Ti ratio**

| | $TiB_{0.65}C_{0.36}$ | $TiB_{0.63}C_{0.40}N_{0.07}$ | $TiB_{0.53}C_{0.34}N_{0.17}$ | $TiB_{0.60}C_{0.38}N_{0.25}$ |
|---|---|---|---|---|
| N content | 0 | 3.3 | 8.3 | 11.2 |
| Nanostructure | $0.67TiC_{0.53}$, $0.33TiB_2$ | $0.68TiC_{0.59}(N_{0.10})$, $0.32TiB_2$ | $0.73TiC_{0.47}(N_{0.23})$, $0.27TiB_2$ | $0.70TiC_{0.54}N_{0.36}$, $0.30TiB_2$ |
| Thin films | $TiB_{0.63}C_{0.36}N_{0.36}$ | $TiB_{0.60}C_{0.36}N_{0.45}$ | $TiB_{0.58}C_{0.40}N_{0.63}$ | $TiB_{0.60}C_{0.40}N_{0.79}$ |
| N content | 15.3 | 18.7 | 24.1 | 28.3 |
| Nanostructure | $0.70TiC_{0.51}N_{0.49}$ $0.30TiB_2+0.03BN$ | $0.70TiC_{0.51}N_{0.49}$ $0.30TiB_2+0.15BN$ | $0.82TiC_{0.49}N_{0.51}$, $0.18TiB_2+0.22BN$ | $0.86TiC_{0.47}N_{0.53}$, $0.14TiB_2+0.32BN$ |

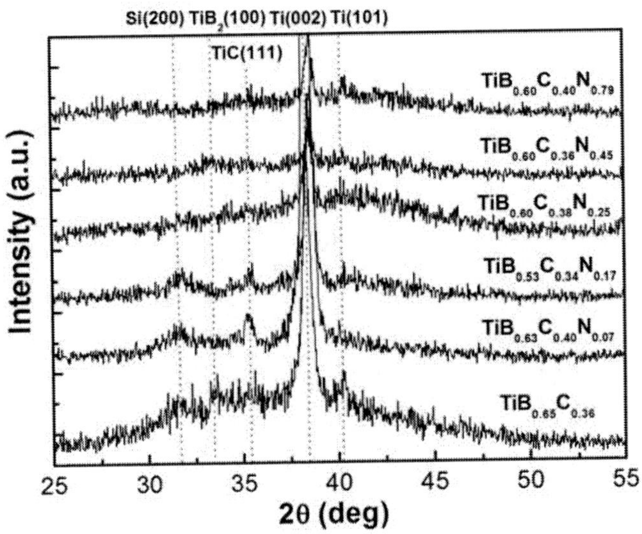


Figure 5.1. XRD θ~2θ spectra of present deposited Ti-B-C and Ti-B-C-N thin films with different N contents.

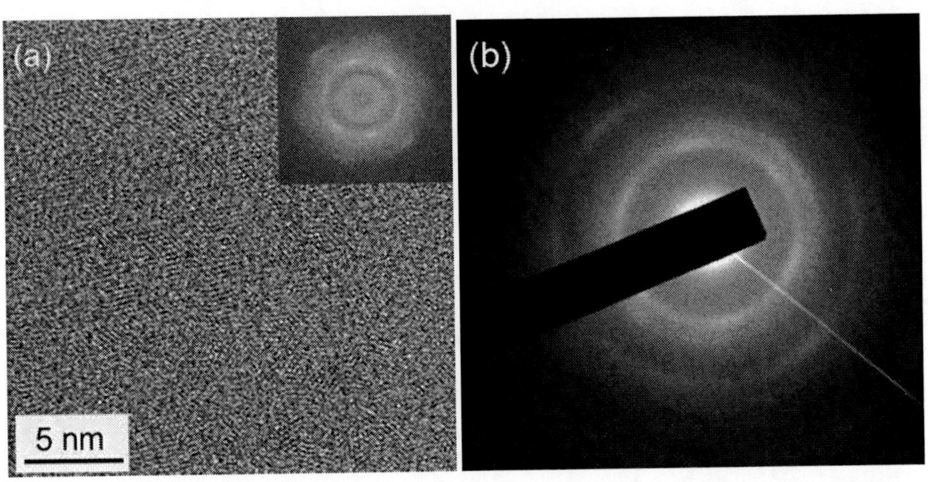


Figure 5.2. (a) A HRTEM image and the corresponding FFT spectrum in the inset, and (b) SAD pattern of $TiB_{0.65}C_{0.35}$.

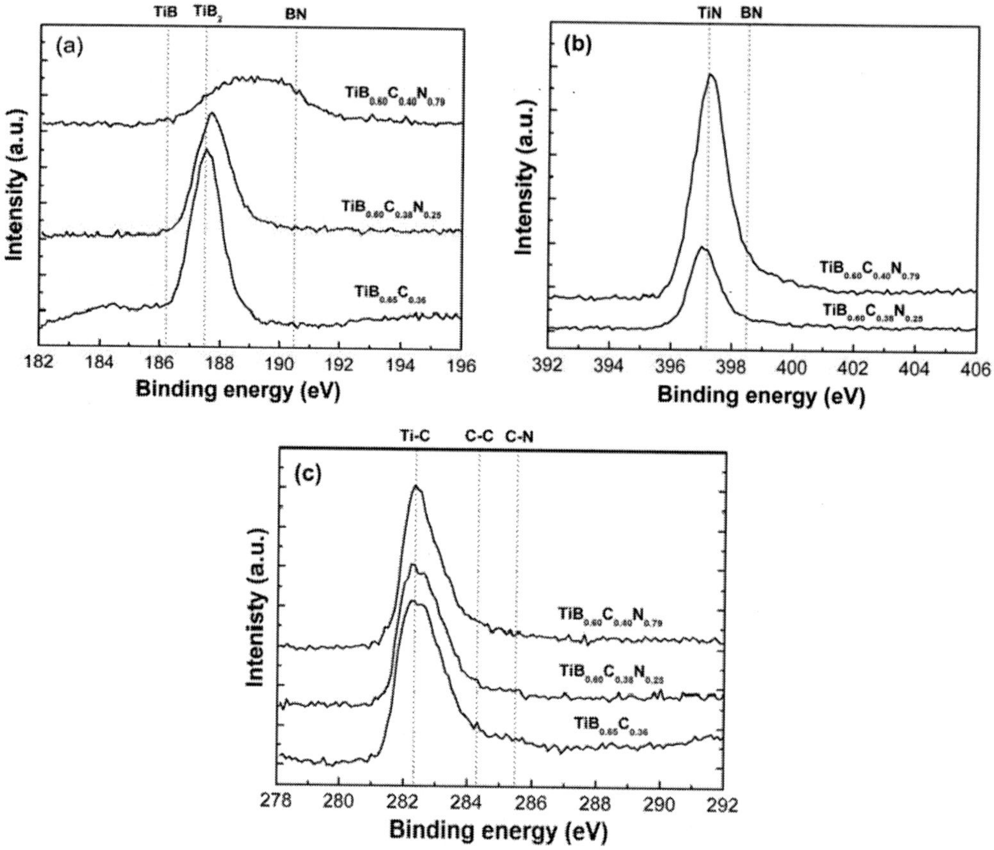


Figure 5.3. Typical XPS (a) B 1s, (b) N 1s and (c) C 1s spectra in Ti-B-C-N thin films with different N contents.

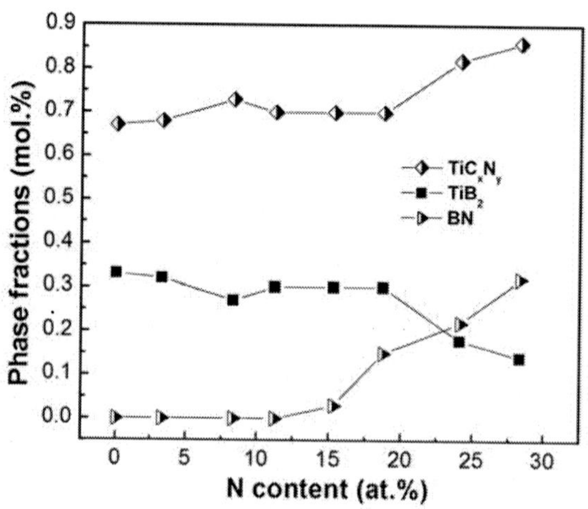


Figure 5.4. The mole fractions of $TiC_xN_y$, $TiB_2$ and BN in Ti-B-N thin films.

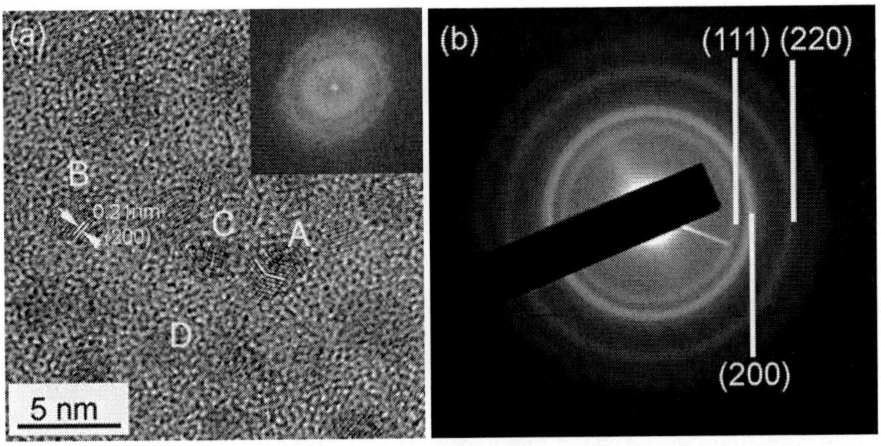


Figure 5.5. (a) A HRTEM image and the corresponding FFT spectrum in the inset, (b) SAD pattern of $TiB_{0.60}C_{0.38}N_{0.25}$.

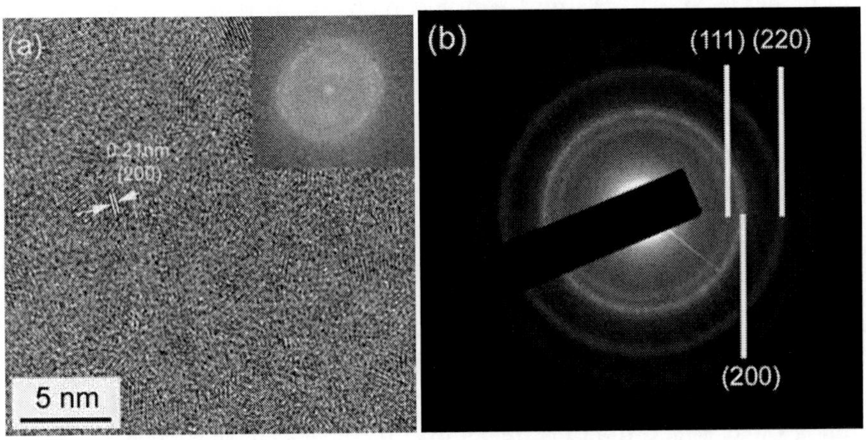


Figure 5.6. (a) A HRTEM image and the corresponding FFT spectrum in the inset, (b) SAD pattern of $TiB_{0.60}C_{0.40}N_{0.79}$.

# 6. Properties of TiN Based Nanocomposite Thin Films

## 6.1. Hardness and Residual Compressive Stress

Up to now, according to the published results no consensus has yet been reached on the exact amount of B addition into TiN for obtaining maximum hardness in Ti-$B_x$-$N_y$ films. Maximum hardness between 30-55 GPa was achieved by using PECVD and reactive magnetron sputtering.[85,86,87,88] In this work the hardness of the Ti-$B_x$-$N_y$ films showed a similar nonlinear trend with pervious work by other researchers using different deposition methods. The hardness of the film was found to increase with small amounts of B addition in thin films deposited at both room and elevated temperatures. In the Ti-B-N thin films deposited at room temperature, the maximum hardness was measured to be ~44 GPa at B content of 18.9 at.% ($TiB_{0.41}N_{0.76}$), as shown in Figure 6.1.

It is noticed that the enhancement of hardness is not consistent to, and less contributed to the residual compressive stress by comparison between Figs.6.1 (a) and (b) as well as Figs.6.2 (a) and (b). Thus, our understanding is that the greatly enhanced hardness of Ti-$B_x$-$N_y$ films is rationally considered to be due to the development of fine grain size and phase configuration. On one hand, improved hardness coincides with a reduced grain size, which implies a high density of grain boundaries. On the other hand, from Figure 6.3 it can be seen that the hardness maximum corresponds to the microstructure possessing a mol fraction of 73 mol% TiN, which fits to percolation threshold mechanism. This result is similar to the report in nc-TiN/a-BN thin films by Andrievski in $TiB_2$/TiN thin films,[89] where the highest hardness was found for a composition of 75 mol% $TiB_2$ and 25 mol% TiN. The work on nc-TiN/a-$Si_3N_4$ also revealed the maximum of hardness was correlated with $Si_3N_4$ content of 17-23 mol.%, which was attributed to the percolation threshold mechanism when the structurally flexible $Si_3N_4$ wets the surface of the TiN nanocrystals.[7,90]

However, there is a special example in the Ti-B-N thin films with different B contents deposited at 500°C, which did not fit to percolation threshold mechanism. Instead, improved mechanical properties of $TiB_xN_y$ films with incorporation of B into TiN were attributed to highly preferred orientation and solid solution hardening. As shown in Figure 6.2, the microhardness value is greatly increased with incorporation of B, and gets to hardness maximum of ~49 GPa when B content gets to ~9 at.%. In this case, all B dissolved into $TiN_{0.81}$ to form Ti($B_xN_y$) solid solution with a highly preferred (111) orientation (see section 3.3). A decrease of hardness value is followed due to formation of BN when B content is further increased.

Another typical example, which did not fit to percolation threshold mechanism, is about solid solution hardening in $TiC_xN_y$ thin films. For example there are large amounts of vacancies existed in carbon sublattice vacancies for the substiochiometry $TiN_{0.45}$ polycrystallites. With incorporation of nitrogen into $TiC_{0.45}$, incorporated nitrogen successive dissolved into carbon sublattice vacancies, which resulted in continuous increase of microhardness value, as shown in Figure 6.4 (a). A similar phenomenon is $TiC_xN_y$ thin films with incorporation of C into nitrogen-deficial $TiN_{0.81}$. As shown in Figure 6.5, the microhardness value was increased with incorporation of C into $TiN_{0.81}$ till all nitrogen sublattice vacancies were filled and formed stiochiometry Ti($C_x$,$N_{1-x}$). This trend was followed by a decrease after the formation of nanocomposite thin films.

Residual stress in thin films is an important science issue for engineering thin films from the viewpoint of performance and integrity because it may affect a variety of phenomena, including physical properties, adhesion, generation of crystallographic defects, and formation of surface growth such as hillock and whiskers. Excessive tensile stress may produce cracking of the film and/or substrate, and the compressive stress can cause bucking. The compressive stress observed for a TiN film at a low working-gas pressure was attributed to the bombardment of the growing film surface by energetic particles such as sputtered atoms and working-gas atoms or ions through the atomic peening process.[91] On one hand, the crystal lattice is deformed and excess atoms are forced into interstitial sites of TiN lattices to expand the crystal lattice, resulting in compressive stress in the films. With incorporation of N into amorphous $TiB_{0.65}$, the mole fraction of TiN crystallites was increased, resulting in increase of residual compressive stress, as shown in Figure 6.2 (b). A similar effect occurs in solid solution thin films. With successive dissolution of solute into solid solution, the residual compressive stress was continuously increased. For example, a continuous increase in residual compressive stress is found with successive dissolution of N into $TiC_{0.45}$, as shown in Figure 6.4(b). A similar phenomenon occurs in $TiC_xN_y$ thin films with incorporation of C into $TiN_{0.81}$ till C filled all nitrogen sublattice vacancies (see Figure 6.5). On the other hand, formation of nanocomposite structure had different effects on the residual compressive stress. For the $TiB_xN_y$ nanocomposite thin films, increasing compressive stress due to incorporation of B is related to formation of amorphous phase and interfacial defect between nano-grain and amorphous phase. It is rationally presumed that the a-$TiB_2$ formed followed by nc-TiN during sputtering process of $TiB_xN_y$. The sequent formation of a-$TiB_2$ followed by TiN nano-grains gave rise to high compressive stress due to its volume expansion during formation. This effect was enhanced at higher B contents because more amorphous phase was produced, as shown in Figs.6.1(b) and 6.2(b). On the other hand, as shown in Figure 6.5, when C content got to 12.7 at.%, the dissolution of C into $TiN_{0.81}$ reached saturation, more carbon incorporation formed nanocomposite thin films of nc-Ti(C, N) embedded into amorphous matrices comprising of disordered C, graphite and $CN_x$ (detailed structure has been described in the section 2.3.1). Amorphous phase surrounded nanocrystallites relaxed part of the residual stress, which resulted in great decrease of the residual compressive stress.

In summary, the hardness of nanocomposite thin films depended on not only phase configuration, nanostructure but also matrix type. Percolation threshold mechanism took place under immiscible hard amorphous matrices, which was sometime broken by solid solution hardening with incorporation of solutes into nitrogen-deficial $TiN_x$. Miscible solute e.g. C dissolved into nitrogen lattice vacancies of $TiN_x$, which increased the film hardness by solid solution hardening and highly preferred orientation. On the other hand, solid solution due to incorporation of solutes into $TiN_x$ increased the residual compressive stress, while formation of nanocomposite structure had different effects on the residual compressive stress.

## 6.2. Thermal Stability

High thermal stability is one of many important properties required for industrial applications of hard and superhard thin films on cuttering tool. Unfortunately, not all superhard nanocomposite thin films meet this condition. It is necessary to find the origin of the high thermal stability of the superhard nanocomposite thin films. Our results revealed that the thermal stability of TiN based nanocomposite thin films depended on that of nc-TiN, amorphous matrices and grain boundaries between them. All TiN based nanocomposite thin films have the same nanocrystallites of nc-TiN, therefore the amorphous matrices and the corresponding grain boundary played an important role in thermal stability. In this section, two types of amorphous matrices, ($TiB_2$, BN) and C, were discussed.

### 6.2.1. Ti-B-N Thin Films with a-($TiB_2$,BN) Matrices

As studied above, Ti-B-N thin films exhibited nanocomposite of nc-TiN embedded into a-($TiB_2$, BN) matrices. It is apparent that a-($TiB_2$, BN) matrices usually contained a lot of defects, its formation during deposition, recovery (structure relaxation) at low temperature and recrystallization at high temperature during annealing had special effects on the thermal stability of nanostructure and hardness.

As shown in Figs. 6.6, 6.7 and 6.8, it is apparent that annealing below 600°C did not obviously change microstructure and binding, but only relaxed part of the residual compressive stress, which belonged to recovery procedure. These changes were linked to structure relaxation (i.e. point defect migration, recombination and annihilation) in the nanocrystallites and matrices, which was not contributed to XPS and FT-IR examinations (see Figure 6.9).

After recovery, the crystallization of amorphous matrices was followed by grain growth of nc-TiN occurred with increasing annealing temperature. In Ti-B-N thin films, TiB phase first precipitated at 800°C because of its lower crystallization temperature according to the Ti-B phase diagram, as determined by their XRD and XPS spectra (see Figs.6.6 and 6.9(a)). Concurrently, B rejected from TiB particles and enriched in remaining amorphous phase, which made the amorphous matrix phase more stable. The similar effect had been observed in some other bulk mixed structures consisting of nanocrystalline particles embedded in amorphous phase matrix.[92] Then $TiB_2$ crystalline phase precipitated during subsequent 1000°C-annealing, as shown in Figure 6.6. It is noted that Ti-B-N thin films with different B contents showed similar results. The general trend was that the TiB phase was dominant in low B-contented samples, whereas $TiB_2$ phase took over in high B-contented samples. For intermediate B content both TiB and $TiB_2$ phases were found to co-exist. Obviously, crystallization of amorphous phase contracted volume of the films, which made the residual stress further decrease and its direction transform from compression to tension at 800° C, as shown in Figure 6.4. Under effect of such a tension stress, some microcracks are found on the post-annealing film surfaces (not shown here). On the other hand, crystallization of TiB and $TiB_2$ needed atom diffusion. During this process some N was excluded, and concentrated. They linked to B, forming BN bonding. Thus BN bonding increased in the condition of high annealing temperature and high B content, as shown in Figure 6.8.

The grain boundary had a great importance in thermal stability of the films. As shown in Figure 6.10, the grain size of pure TiN was successively increased with increasing annealing temperature, which indicates that nc-TiN/nc-TiN simple grain boundary had lost its stability at 400°C. However, addition of enough B made grain boundary stable, hence the grain size remained unchanged up to a high temperature. The stability degree was dependent on B content, as shown in Figure 6.10. It is believed that addition of enough B changed structure of grain boundaries, resulting in increase of recrystallization temperature.[93] On one hand, the incorporation of B into TiN produced nanocomposite structure of nc-TiN particles embedded into amorphous matrix. The amorphous phase stopped the grain growth of nc-TiN before it crystallized. After crystallization the grain boundary became nc-/nc- simple type, the grain growth mode was hence similar to that of pure TiN (see Figure 6.10). On the other hand, the grain growth was believed to be related to segregation of boron along grain boundary just after nanocrystalline TiN was formed and grow. This is because the B was continually excluded during nucleation and growth of TiN, and segregated along the grain boundary. Karvankova also suggested the similar distribution of B atoms on the surfaces of the TiN nanocrystals less than approximately 10-20 nm in size.[94] As a barrier, the B atoms segregated at grain boundary hindered grain growth by way of decreasing boundary energy, leading to increase of recrystallization temperature. Increasing B content enhanced this effect. As shown in Figure 6.11 (a), B-free TiN showed less microstructure stability, an obvious recrystallization was found in 800 °C-annealed sample. In contrast, incorporation of 27 at.%

B made the microstructure more stable, as shown in Figure 6.11 (b). Even after 1000°C-annealing the microstructure morphology still remained the column structure.

According to the hardness-grain size relationship described by Veprek et al ,[95] it is apparent that the decrease of hardness for example in the pure TiN films with annealing temperature is attributed to the increase of grain size with annealing temperature. Addition of enough B strengthened the grain boundary, and hence hindered grain growth, as a result, the hardness decreased slowly, as shown in Figs. 6.12. When B content was increased to about 19 at.%, corresponding to $TiB_{0.41}N_{0.76}$ with the maximum hardness value, the mole fraction of amorphous $TiB_2$ was about 21 mol% which just wet the TiN nanocrystallite, fitting to the percolation threshold mechanism.[36] The film had a strong grain boundary free of void and defect, hence restrained crystallization of amorphous matrices and hindered grain growth to a great extent. As a result, this film showed the highest hardness stability (Figure 6.12(a)). However, when B content was further increased, the amorphous phase along grain boundary became much more, defect recovery and structure relaxation were presumed to make hardness decrease despite the grain size of nc-TiN remained unchanged upon annealing (see Figure 6.12(d)). Thus, both lack of and excessive amorphous phase resulted in decrease of hardness stability. It is concluded that the stability of hardness depended on not only the stability of microstructure but also recovery of amorphous matrix phase.

Incorporation of N into amorphous $TiB_{0.65}$ promoted phase segregation and formation of nanocomposite thin films, which increased thermal stability of the thin films. That is, phase configuration and nanostructure formed because of incorporation of nitrogen into $TiB_{0.65}$ controlled the thermal stability of thin films. For clarity, we divided the present thin films into three groups. For the first group films, $TiB_{0.65}$ shows amorphous microstructure, while $TiB_{0.65}N_{0.18}$ exhibits a nanocomposite structure of solid solution Ti(N) embedded into high mol fraction amorphous matrix $TiB_2$. Their microstructures are believed to be typical of incomplete phase segregation due to lack of enough nitrogen incorporation. The self-hardening in this group (Figs. 6.13 (a) and (b)) is related to some changes of this type of microstructure at low temperature (600°C) and sequent phase segregation above 800°C. This effect is similar with a description by Cahn.[96] For the second group films $TiB_{0.53}N_{0.52}$ and $TiB_{0.53}N_{0.63}$, their compositions are located in the two-phase zone of TiN and $TiB_2$ in the Ti-B-N phase diagram, corresponding to the relatively high hardness values of about 44 ($TiB_{0.53}N_{0.52}$) and 52 GPa (the hardness maximum, $TiB_{0.53}N_{0.63}$) respectively. Their microstructures are very stable and do not show obvious change after annealing at 600 and 800°C. Hence, a high thermal stability occurred, as shown in Figs. 6.13 (c) and (d). Thus, a high thermal stability is expected to be related with the optimum composition, which is corresponding to a high hardness value. For the third group films $TiB_{0.61}N_{1.04}$ and $TiB_{0.61}N_{1.26}$, their compositions are located in the three-phase zone of TiN, $TiB_2$ and BN in the Ti-B-N phase diagram, corresponding to low hardness values. The mole fraction of amorphous matrices phase increased, and showed in excess of that needed for the optimum composition, as a result the grain boundary became unstable. With increasing temperature the nanostructral relaxation of amorphous matrix at low temperature, crystallization of amorphous matrix and recrystallization of nanocrystallites at high temperature took place. As a result, the microhardness values continuously decreased with annealing temperature, as shown Figs. 6.13 (e) and (f).

### 6.2.2. Ti-C-N Thin Films with C Matrices

Four thin films of Ti-$C_x$-$N_y$ with different C contents and nanostructures were deposited on polished Si (100) substrates at 500 °C in an unbalanced reactive dc magnetron sputtering system (UDP450, Teer Coating Limited) using two graphite and two Ti targets. The film compositions and nanostructures have been investigated in detailed and listed in Table 1. Their phase configurations and nanostructures were investigated at section 2.2.1.

The recovery procedure of Ti-$C_x$-$N_y$ thin films at low temperature was obviously related to the activated defect annihilation in nanocrystallites and matrices, which resulted in the following effects: (1) decrease of FWHM value and right shift of XRD peak position due to reduced inhomogeneous strain (see Figure 6.14); (2) upshift of the G peak position and decrease of G peak width in Raman spectra (see Figs.6.15 and 6.16 (a) and (b)). The latter implies that the annealing decreased bond-angle disorder and density of defects in amorphous matrices and therefore hardened the vibrational modes,[97] which was further verified by the decrease of $I_D/I_G$ value (see Figure 6.16(b)). $I_D/I_G$ is generally a measure of the density of defects and the degree of disorder due to zone edges or border phonons of the carbon cluster.[98,99] Figure 6.16 (b) also indicates that an obvious decrease in bond-angle disorder and density of defects took place after 800 °C annealing, which is consistent with structural change detected from XRD measurement. It indicates that there is consistent in nanostructure recovery between nnocrystallites and amorphous matrices during annealing process.

It is apparent that for SS thin films ($TiC_{0.24}N_{0.71}$) annealing did not precipitate C or form C-N, and the annealed films kept the same structure as the deposited ones till 1000°C (by comparison with Figs.6.17(a) and (b)) except for the contented defect, residual stress and nanocrystalline size. However for nanocomposite thin films, the condition was different. When annealing temperature got to 1000 °C, a recrystallization process took place, which resulted in formation of a distinguished nanostructure (see Figure 6.18). As an example, Figs.6.18 (a) and (b) show nanostructures of as-deposited and 1000°C-annealed $TiC_{1.61}N_{0.73}$, respectively. From Figure 6.18 (a), it is seen that $TiC_{1.61}N_{0.73}$ exhibits a lower nc-grain distribution density and a smaller nanocrystalline size of ~2 nm. Besides, some circled short range ordering (SRO) areas are found in the amorphous matrices, and a lot of lattice distortions or defects appear in the nanocrystallites and amorphous matrices. After 1000 °C-annealing, as shown in Figure 6.18 (b), $TiC_{1.61}N_{0.73}$ becomes to be composed of different oriented nanocrystallites A~E with an average size of ~5 nm and amorphous grain F. These nanocrystallites are separated by either simple grain boundary or amorphous phase with a thickness of less than 1 nm. Besides, SRO and lattice defects disappeared. This annealing effect is further verified by the obvious changes in XRD and Raman spectra at that temperature (see Figure 6.14 and 6.15).

The change in grain size with temperature is believed to be related with the grain boundary in TiN and SS Ti(N,C) as well as amorphous matrices in the case of nanocomposite thin films. As shown in Figure 6.19, the grain size of pure TiN is continuously increased with temperature, which indicates that a simple nc-TiN/nc-TiN grain boundary did not hinder the grain growth during annealing. For $TiC_{0.24}N_{0.71}$, 12.4 at.% C incorporation was believed to change the grain boundary structure although no amorphous matrices was precipitated, hence restraining the grain growth. As a result a slow grain growth rate is observed in $TiC_{0.24}N_{0.71}$ (see Figure 6.19). Incorporation of a third element strengthening grain boundary increased recrystallization temperature has been reported in Ti-B-N thin films.[100] Such restraining effect was obviously further enhanced after nanocomposite structure was formed in $TiC_{0.68}N_{0.67}$ and $TiC_{1.61}N_{0.73}$ (see Figure 6.19), which was because that the surrounded amorphous matrices strongly hindered the grain growth of nc-Ti(C,N).

According to hardness-grain size relationship increase of grain size decreases the films hardness. Thus, when grain size of $TiN_{0.87}$ is increased from ~28 nm of as-deposited sample to ~62 nm of 1000°C-annealed one, the corresponding hardness value is decreased from 23.7 to 15.3 GPa (see Figure 6.20). By comparison, a great decrease in hardness from 22.6 to 11.8 GPa takes place in $TiC_{1.61}N_{0.73}$, however its corresponding grain size is only slightly increased from ~6.5 nm of as-deposited sample to ~9.5 nm of 1000 °C-annealed one (see Figure 6.20). This indicates that the grain size is not unique factor controlling film hardness. Defect annihilation with annealing temperature is believed to be another factor for decrease in both hardness and residual compressive stress. Defect annihilation with annealing temperature has been proved by XRD, Raman and HRTEM analyses (see Figs.14, 15, 17-18), which is believed to result in decrease in both defect density and lattice distortion in the films. As a result, both residual compressive stress and hardness values are decreased.

In summary, from above two examples the following results can be drawn:

(1) The thermal stability of nanocomposite thin films strongly depended on not only phase configuration and nanostructure, but also the matrix type.

(2) Nanocomposite structures had a positive effect on microstructure stability, which resulted in an elevated recrystallization temperature of thin films.

(3) Recrystallization of amorphous matrices in nanocomposite thin films greatly decreased the residual stress, even transformed from compression to tension.

(4) The hardness stability did not show consistency with that of microstructure, and it also depended on phase configuration and composition.

## 6.3. Oxidation Resistance

The oxidation of metals or alloys at high temperature can be treated as a special case of metallic corrosion, which considers the material destruction due to chemical causes. Four thin films with different compositions and nanostructures were deposited at 500 °C using two $TiB_2$ and two Ti targets. Table 6.3 shows the as-deposited thin films, the corresponding atomic concentrations, and nanostructures. It is apparent that different nitrogen flows resulted in the formation of different microstructural thin films, which has been investigated in detail in section 3.3.

Figure 6.21 shows mass gain per unit surface area as a function of annealing temperature of TiN, $TiB_{0.65}$, $TiB_{0.62}N_{0.18}$ and $TiB_{0.61}N_{1.04}$. It was found that the oxidation rates of $TiB_{0.65}$, $TiB_{0.62}N_{0.18}$ and TiN increased rapidly above 600 °C, while that of $TiB_{0.61}N_{1.04}$ showed a large retardation. It is interesting to note that a two-stage oxidation process was observed between 600 and 1000 °C in $TiB_{0.61}N_{1.04}$. Below 800 °C, a slow linear growth law was determined, while above 800 °C the oxidation proceeded quickly. On the other hand, at each oxidation temperature $TiB_{0.65}$ showed the highest oxidation rate; while the reverse trend was seen for the $TiB_{0.61}N_{1.04}$ thin film.

The nanocomposite $TiB_{0.61}N_{1.04}$ thin film showed a higher oxidation resistance than the other three thin films of $TiB_{0.65}$, $TiB_{0.62}N_{0.18}$ and TiN (Figure 6.21). A two-step oxidation process governed the oxidation behaviour in the regime between 600 and 1000 °C in $TiB_{0.61}N_{1.04}$. The first process, being dominant at low temperatures up to about 800 °C, was characterized by a slow increase of mass gain and formation of uniform and fine Ti-oxide (by comparison with Figs.6.22 (a) and (b)). The second process, being dominant at high temperature above 800 °C, was characterized by a high oxidation rate and formation of coarse Ti-oxide (see Figs.6.22 (c) and (d)). In contrast to that of $TiB_{0.61}N_{1.04}$, the oxidation of pure TiN, amorphous $TiB_{0.65}$ compound and nanocomposite $TiB_{0.62}N_{0.18}$ were found in all XRD spectra.

The microstructure of film oxide is different with oxidation temperature. Figure 6.23 shows the XRD θ-2θ scans for a typical $TiB_{0.61}N_{1.04}$ film oxidized at 600, 800 and 1000 °C respectively. After annealing to 600 °C, the major peaks observed at 2θ angles are ascribed to various types of polycrystalline Ti-oxides, such as $TiO_2$, $Ti_2O_3$ and $TiN_xO_y$ ($TiN_{0.74}O_{0.34}$), which indicates that the surface of $TiB_{0.61}N_{1.04}$ was oxidized after 600 °C annealing although the oxidation kinetics curve did not show any mass gain (see Figure 6.21). The possible reason for this is because the mass gain was too small to be detected. When oxidation temperature increased to 800 °C, all reflection peaks observed at 2θ angles became narrow and their intensities were increased significantly, indicating that the grain size of both Ti-oxides and TiN increased after annealing (see Figure 6.24). After annealing to 1000 °C, the XRD peaks were further narrowed, and simultaneously the peaks contributed to TiN vanished, which reveals that grain size of well-ordered Ti-oxides further increased, accompanying with full oxidation of almost the whole film. It is noted that no reflection peaks representing B-oxides were found in all XRD spectra.

Oxidation gave rise to change in elemental concentration on the film surface (see Figure 6.25). It is believed that, during oxidation in $TiB_{0.61}N_{1.04}$ thin films, the following reactions

$$TiN + O_2 \rightarrow TiO_2 + N_2$$

$$TiB_2 + O_2 \rightarrow TiO_2 + B$$

$$BN + O_2 \rightarrow B_2O_3 + N_2$$

are also thermodynamically favorable at room temperature because $TiO_2$ and $B_2O_3$ are more stable.[84] Likewise, the other suboxides of Ti such as TiO, $Ti_2O_3$ and $TiN_xO_y$ are also more stable than TiN, $TiB_2$ and BN, resulting in formation of Ti oxides of various types, as shown in Figure 6.26. It is believed that the reaction production $N_2$ gave off during oxidation, which resulted in low N content in the oxide, as shown in Figure 6.25. Likewise, B is believed to be prone to form $B_2O_3$ because $B_2O_3$ is also more thermodynamically stable at room temperature. $B_2O_3$ is reported to have a low melting temperature of about 450 °C and presumed to be easy to evaporate at high temperature. After formation of $B_2O_3$ it also gave off above 450 °C, as a result nearly no B was detected in the oxide formed above 600 °C by XPS (see Figure 6.25), which is similar to the N behavior.

It is believed that two factors, oxidation resistance of nanocrystallites and thermal stability of amorphous matrices, had great effects on oxidation behaviors, which resulted in their difference in oxidation resistance. For nanocrystallite Ti and TiN the following oxidation reactions

$$Ti+O_2 \rightarrow TiO_2 \ \Delta G = -888.8 \text{ kJ/mol } TiO_2 \quad (1)$$

$$TiN+O_2 \rightarrow TiO_2+N_2 \ \Delta G = -645 \text{ kJ/mol } TiO_2 \quad (2)$$

are thermodynamically favorable at room temperature. The changes in Gibbs free energies ($\Delta G$) during reactions were calculated according to the existing thermodynamic parameters.[84] From thermodynamic consideration, reaction (1) took place more easily than reaction (2). As a result, thin films of $TiB_{0.65}$ compound and nanocomposite $TiB_{0.62}N_{0.18}$ showed a higher oxidation rate than $TiB_{0.61}N_{1.04}$ thin films. On the other hand, amorphous $TiB_2$ or/and BN matrix phase which separated the nanocrystalline TiN or Ti(N) acted as an efficient diffusion against oxygen diffusion, which protected nc-TiN or Ti(N) nanocrystallites from oxidation. From this view, pure TiN without amorphous matrix phase protection exhibited weak oxidation resistance. Obviously the thermal stability and crystallization of amorphous matrix phase had crucial effects on oxidation process, which was similar to that of Al or Si incorporation into TiN.[101,102,103]

The fractography and morphology of oxides are different with oxidation temperature. When the temperature was below than 800 °C, the occurrence of oxidation was only restricted on the surface of samples (Figure 6.27 (b)). For a comparison, the fractography of the corresponding as-deposited sample is given in Figure 6.27 (a). Correspondingly, the surface roughness value increased slowly (see Figure 6.28). Above 800 °C, the crystallization of amorphous $TiB_2$ phase took place, and the amorphous barrier did not exist. Thus, the Ti-oxides of various types were formed, accompanying with a great increase of surface roughness value (see Figure 6.28) and further acceleration of oxidation because the oxygen diffusion easily took place by way of the intergranular grain boundaries. As a result, the oxidation took place inwards (Figs. 6.27 (c) and (d)), accompanying with fast growth of oxides (Figs. 6.24, 6.27 (c) and (d)). It is found from Figure 6.29 that the oxide in $TiB_{0.61}N_{1.04}$ has the smallest thickness and grain size compared with the other three ones when the all four thin films were oxidized at 800°C, which further proved that the nanocomposite $TiB_{0.61}N_{1.04}$ exhibited a superior oxidation resistance.

In order to clearly illustrate effect of crystallization of amorphous matrix phase on oxidation process, an annealing experiment on $TiB_{0.61}N_{1.04}$ nanocomposite thin films at 600, 800 and 1000 °C was carried out. The results indicated that the microstructure after annealing at 600 °C kept stable, that is, the 600 °C-annealed XRD spectrum (not shown here) was just the same as that of as-deposited sample. However, when annealing temperature increased to 800 °C, crystallization of amorphous matrix phase occurred, and several TiN, $Ti_2N$ and $TiB_2$ crystallites were found (see Figure 6.30). This indicates that crystallization of $TiB_2$ matrix phase just began to occur after vacuum-annealing at 800°C. Below 800 °C amorphous $TiB_2$ and BN kept stable, which acted as an efficient diffusion barrier against oxygen diffusion, protecting TiN grains from oxidation, resulting in oxidation retardment below 800 °C.

In summary, the following important conclusions can be drawn:

(1) Nanocomposite $TiB_{0.61}N_{1.04}$ thin films exhibited much higher high-temperature oxidation resistance than TiN, $TiB_{0.65}$ and $TiB_{0.62}N_{0.18}$ thin films. The nanocomposite nc-TiN/a-$TiB_2$/a-BN thin films showed a superior oxidation resistance than other three films, which was characterized by a two-stage oxidation process. Below 800 °C, a low oxidation rate accompanying with formation of fine Ti oxides occurred; while above 800 °C a rapid oxidation process accompanying with formation of large-grained Ti oxides took place.

(2) It is presumed that the two-stage oxidation process of nanocomposite nc-TiN/a-$TiB_2$/a-BN thin films was related to oxidation resistance of nanocrystallites and thermal stability of amorphous matrix phase. The diffusion barrier of amorphous phase below 800 °C was responsible for low oxidation rate, while above 800 °C crystallization of amorphous $TiB_2$ phase took place, and a rapid diffusion of oxygen along grain boundary resulted in high oxidation rate.

(3) Only Ti oxides of various types, TiO, $TiO_2$, $Ti_2O_3$ and $TiN_xO_y$, were found in oxide formed above 600 °C. No boron concentration was found, which was due to a low melting temperature of $B_2O_3$.

## 6.4. Wear Behaviours

Wear is defined as the progressive loss of a substance from the operating surface of a body occurring as a resulting of the relative motion between surfaces. In general, the friction and wear processes occur jointly when two surfaces undergo sliding or rolling under load. On the other hand, it has been reported that incorporation of B into a growing TiN film resulted in the formation of two- and/or three-phase Ti-B-N nanocomposites with microstructures comprising of nanocrystalline phase in an amorphous matrices.[104,105] Such Ti-B-N nanocomposite films improved dry sliding wear resistance behavior as well as showed a lower friction coefficient compared to TiN.[106,107,108,109] However up to now wear mechanism change due to incorporation of B into TiN is seldom reported although the reference wear mechanism of TiN thin films has been well studied.[110]

In the present study, several nanocomposite thin films of Ti-B-N with different B contents, prepared on Si(100) at room temperature, were used to study the wear mechanism. The detail microstructure of these films has been investigated in detail, [111] which is described in the section 4.1. The results indicated that the all samples exhibited nanocrystalline (nc-) TiN/ amorphous (a-) $TiB_2$ nanocomposites except sample $TiB_{1.86}N_{1.51}$, which contained additional h-BN phase, showed an nc-TiN/a-$TiB_2$/a-BN nanocomposite. The maximum hardness occurred at $TiB_{0.41}N_{0.76}$ which consists of nc-TiN and a-$TiB_2$ phases.

Our results revealed that the wear behaviors strongly depended on the phase configuration rather than hardness. The wear behaviors of TiN films presented in this work show analogous features to the former studies of TiN thin films, where polishing wear modes are observed at low temperatures, sliding speeds and contact loads.[112] Its characteristic is micro-abrasion (micro-cutting), i.e. a continued formation of asperities on the contact body generates grooves in the wear track (see Figure 6.31 (a)). Therefore, TiN thin films are easily removed by the abrasive wear, as a result, yielding a high wear rate (see Figure 6.32).

After incorporation of a small amount of B, the nc-TiN/a-$TiB_2$ thin films possessed higher hardness and residual stress.[111] As a result, a harder nc-TiN/a-$TiB_2$ thin film exhibits higher resistance to plastic deformation, and is more effective to resist such a penetration, leading to a lower wear rate (see Figure 6.32). On the other hand, a higher contacting stresses at the asperities will generate the defect region beneath the surface of thin films by way of generation and pile-up of dislocations and crack nucleation.[113] These defected regions are easily fractured into very fine particles or sometimes power-like debris. They can be attached on the pin tip and the wear track, which is possibly related to a-$TiB_2$. It is believed that the amorphous debris easily agglomerates due to its large surface area, as shown in Figure 6.31. Obviously, incorporation of B changes wear mechanism. With increasing B content, the amount of a-$TiB_2$ increases, and more powder-like wear particles generate (see Figure 6.31). When the B content reaches 42 at.%, the $sp^2$ type h-BN is formed ,[111, 114] which accelerates further formation of wear debris due to its graphitic structure,[115] as shown in Figure 6.31 (f).

AES examination on wear debris of $TiB_{0.39}N_{0.84}$ indicates that oxygen of $H_2O$ or $H_2O$ itself in air participated in reaction with the coupled films during sliding. However, no oxygen was detected on worn surface. In order to detect the oxygen species in debris and worn surfaces accurately, AES comparison analyses on debris, worn surfaces and film surfaces were carried out respectively. Detected regions 1, 2 and 3 shown in Figure 6.33 (a) represent wear debris, worn and film surfaces, respectively. It is found that a large oxygen concentration more than 40 at. % exists within the debris, while no obvious oxygen concentration was detected on worn surface (Figure 6.33 (b)). Besides, a small amount of Fe was found in debris, which came from wear ball.

XPS examination reveals the presence of various types of Ti-oxide, including TiO, $Ti_2O_3$ and $TiO_2$, in wear debris, as shown in Figure 6.34(a). Only a small amount of TiN and $TiB_2$ bonds is found in the debris. This indicates that most of Ti-B-N thin films has been oxidized, forming different types of oxides. Likewise, oxidization of B and Fe took place to form $B_2O_3$ and FeO respectively (see Figs.6.34(b) and (c)). This means that the Ti oxide dominates in debris although almost all metallic elements have been oxidized to various degrees.

Obviously, the debris particles trapped can produce further damage to both contacting surfaces through causing abrasive wear.[116] Simultaneously, an adhesive wear mechanism takes place by way of transformation of tribo-film from the ball onto wear track. Therefore, both wear mechanisms coexist during sliding. After formation of a transfer layer on the wear track, wear behavior gradually changes from metal-coating contact to coating-coating contact, which is contributed to the changes in friction coefficient and wear rate. Thus, a steady reduction of friction coefficient and wear rate with B content is concluded to be due to the following two factors. First, the tribo-film acts as protective layer and only an occasional metal to coating contact occurs. Secondly, the tribo-film contains a lot of oxygen, and is mainly comprised of various Ti oxides. These oxides usually have a lower strength and hardness, which decreases the friction coefficient. This friction coefficient-reducing effect of oxides has been reported early.[117,118] Obviously, the more the B content, the more tribo-films exist, therefore lower friction coefficient and wear rate are obtained. It was observed that the shearing and tearing because of adhesive wear resulted in plastic deformation in tribo-film, that direction was accordant with the sliding direction.

In summary, the following results can be drawn:

(1) Both friction coefficient and specific wear rate decreased with incorporation of B into TiN or increasing B content, accompanying with transformation of corresponding wear modes from abrasive wear (ploughing) to powder formation, subsequently to plastic deformation. When the B content reached a critical value (~42 at. %), an inverse trend in the wear rate occurred due to formation of soft h-BN phase in Ti-B-N thin films.

(2) Oxygen of $H_2O$ in air participated in and reacted with films, and formed mainly Ti oxide consisting of TiO, $TiO_2$ and $Ti_2O_3$, accompanying with minor amount of FeO in debris. The formation of oxide in debris was believed to have an effect on increasing wear resistance.

**Table 6.1. Present four studied thin films, the corresponding C contents, (C+N)/Ti ratios and structures.**

| Samples | C content (at.%) | (C+N)/Ti | Structure |
|---|---|---|---|
| $TiN_{0.87}$ | 0 | 0.87 | f.c.c nc-TiN |
| $TiC_{0.24}N_{0.71}$ | 12.4 | 0.95 | f.c.c nc-Ti(N,C) |
| $TiC_{0.68}N_{0.67}$ | 28.9 | 1.35 | nc-Ti(N,C)/a-(C,$CN_x$) |
| $TiC_{1.61}N_{0.73}$ | 48.2 | 2.34 | nc-Ti(N,C)/a-(C,$CN_x$) |

**Table 6.2. Studied films, phase configuration, and corresponding element concentrations measured by XPS.**

| Samples | Phase configuration | Ti (at.%) | B (at.%) | N (at.%) | O (at.%) |
|---|---|---|---|---|---|
| TiN | Pure TiN | 52.3 | 0 | 43.4 | 4.3 |
| $TiB_{0.65}$ | Ti+$TiB_2$ compound | 58.2 | 37.8 | 0 | 4.0 |
| $TiB_{0.62}N_{0.18}$ | nc-Ti(N)/a-$TiB_2$ | 54.8 | 34.0 | 10.0 | 1.4 |
| $TiB_{0.61}N_{1.04}$ | nc-TiN/a-$TiB_2$/a-BN | 36.5 | 22.4 | 38.0 | 3.2 |

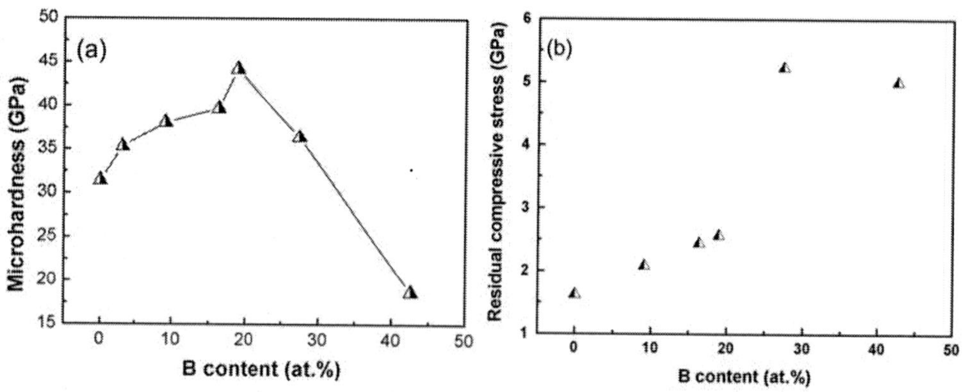


Figure 6.1. The dependence of (a) microhardness and (b) residual compressive stress on B content in Ti-Bx-Ny thin films deposited at room temperature.

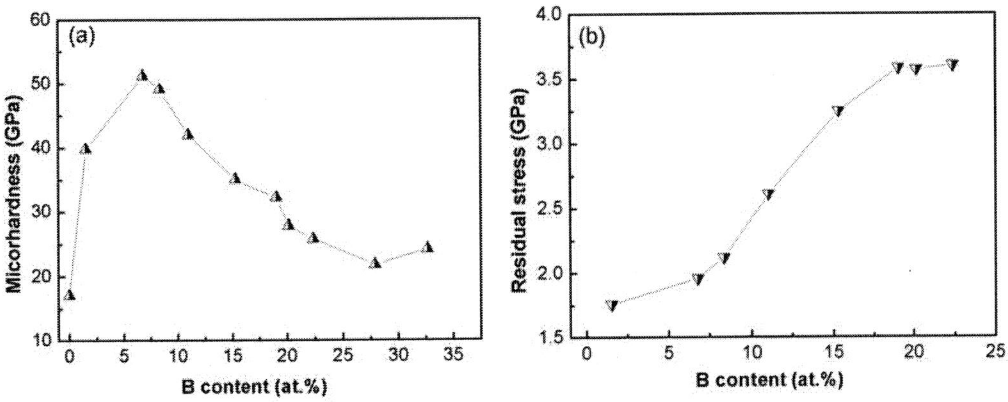


Figure 6.2. The dependences of (a) microhardness and (b) residual compressive stress on B content in Ti-B-N thin films deposited at 500°C.

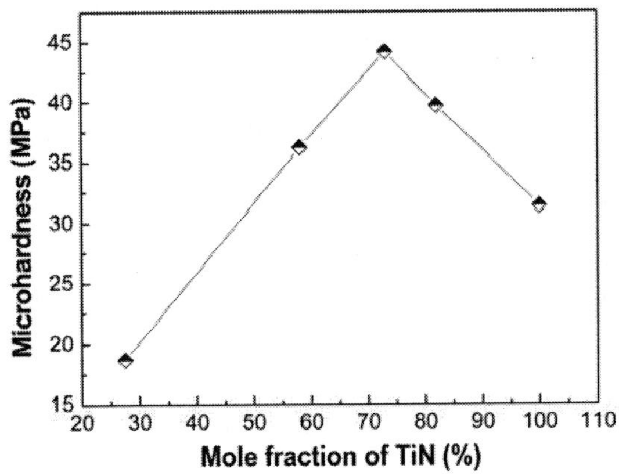


Figure 6.3. Microhardness as a function of mol fraction of nc-TiN in Ti-B-N thin films deposited at room temperature.

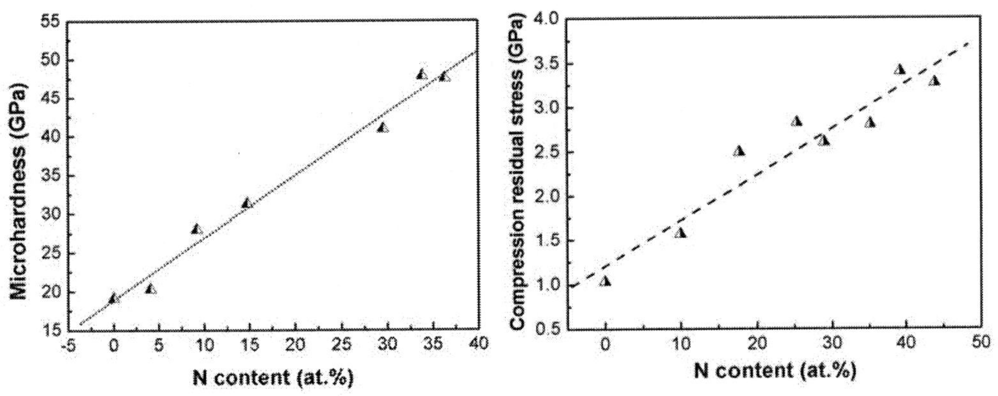


Figure 6.4. (a) Microhardness and (b) residual compressive stress in $TiC_xN_y$ thin films with incorporation of different amounts of nitrogen into $TiC_{0.45}$.

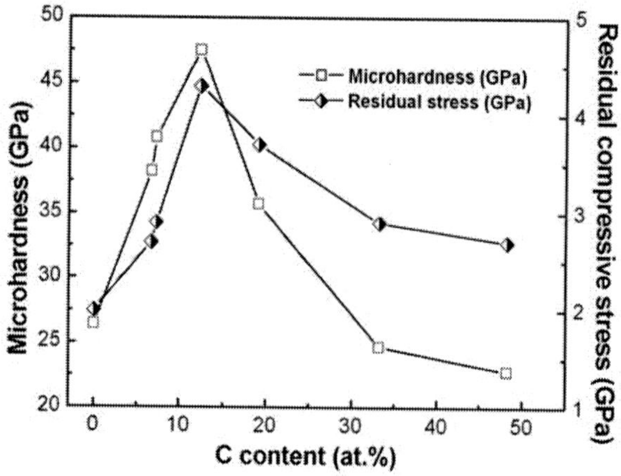


Figure 6.5. Microhardness and residual compressive stress values as a function of C content.

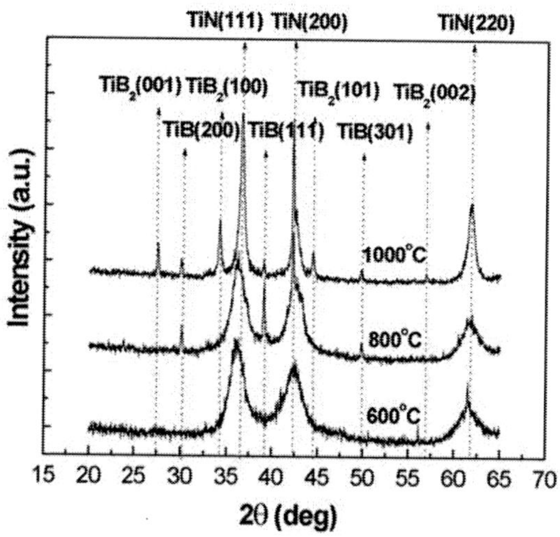


Figure 6.6. XRD spectra of $TiB_{0.35}N_{0.85}$ annealed at different temperatures.

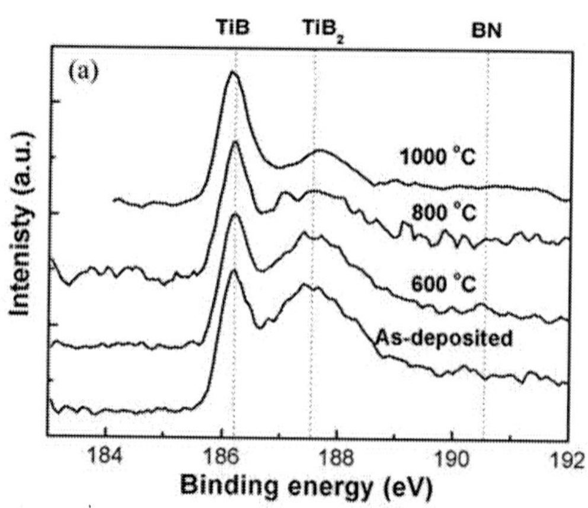


Figure 6.7. Continued on next page.

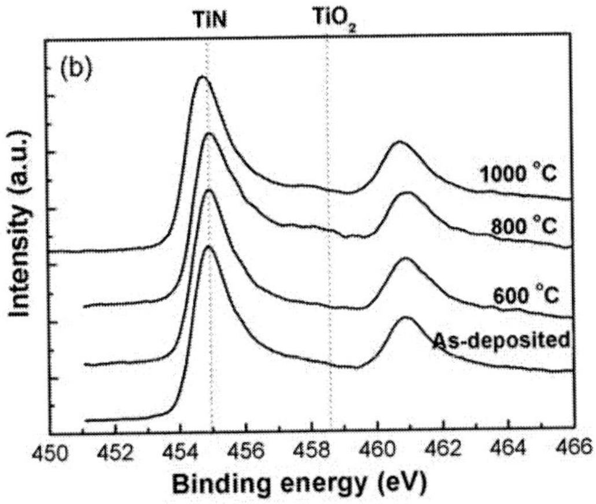


Figure 6.7. XPS core-electron level (a) B1s and (b) Ti 2p spectra of $TiB_{0.36}N_{0.85}$ after annealing at different temperatures.

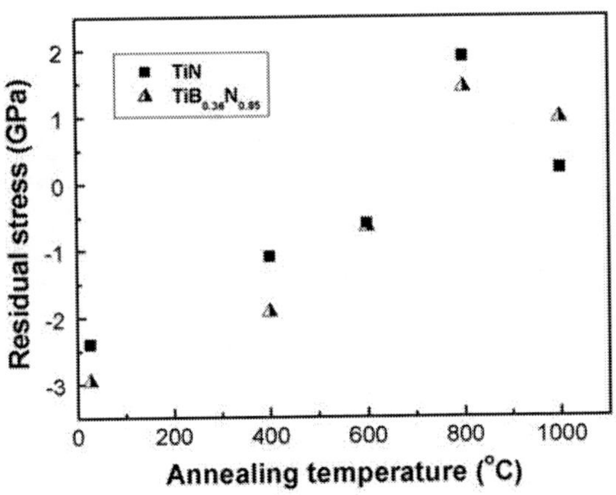


Figure 6.8. Residual stress as a function of annealing temperature in $TiB_{0.36}N_{0.85}$ and TiN.

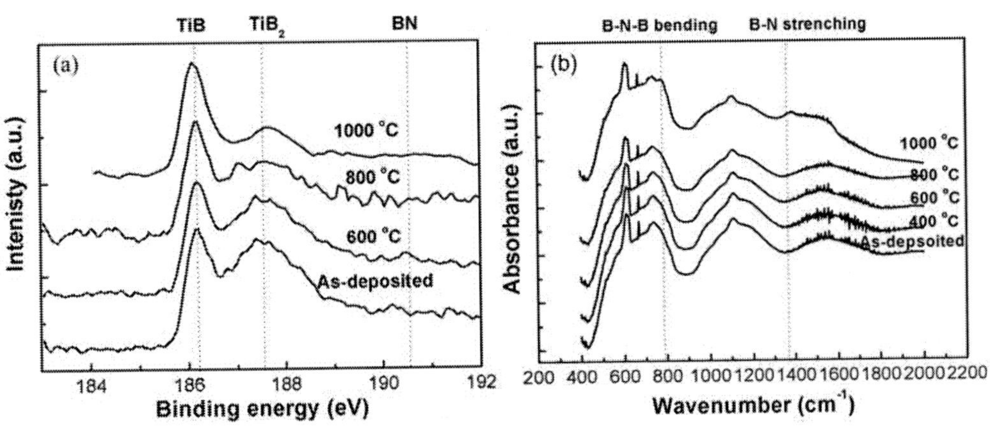


Figure 6.9. (a) XPS B1s spectra of $TiB_{0.36}N_{0.85}$ and (b) FT-IR spectra of $TiB_{0.79}N_{1.09}$ after annealing at different temperatures.

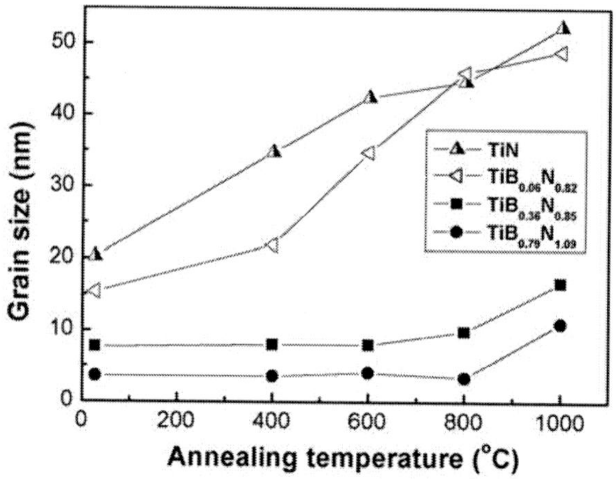


Figure 6.10. Grain size of nc-TiN in Ti-B-N thin films with different B contents as a function of annealing temperature.

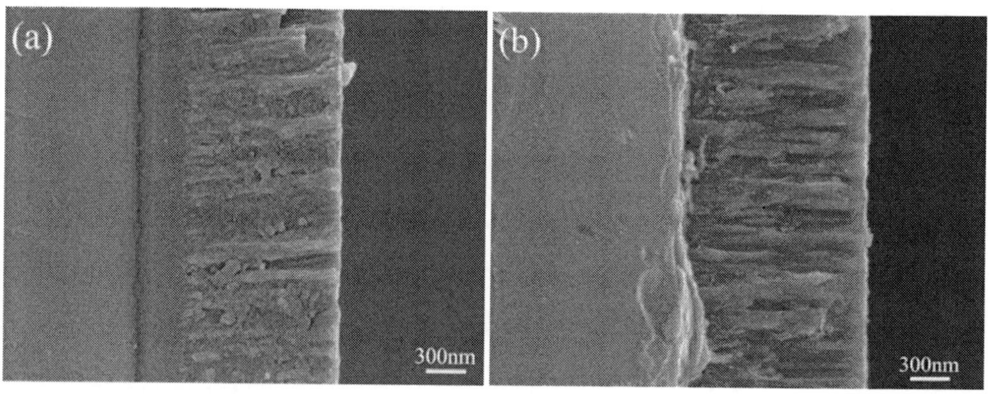


Figure 6.11. Cross-sectional SEM morphologies of two annealed samples, (a) TiN, 800 °C, (b) $TiB_{0.79}N_{1.09}$, 1000 °C

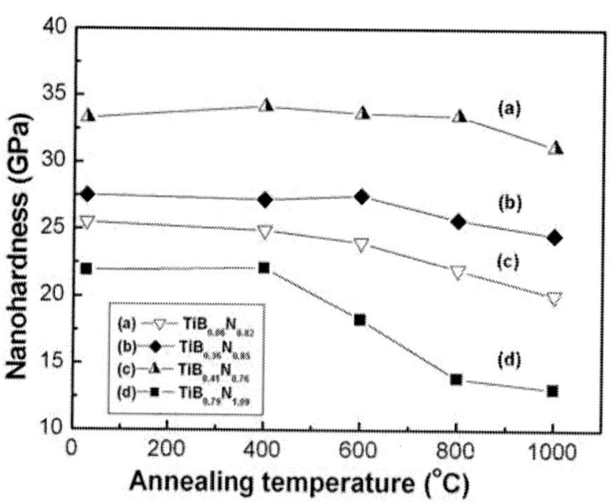


Figure 6.12. Nanohardness values of (a) $TiB_{0.41}N_{0.76}$ (18.9 at.% B), (b) $TiB_{0.36}N_{0.85}$ (16.3 at.% B), (c) $TiB_{0.06}N_{0.82}$ (3.2 at.% B), (d) $TiB_{0.79}N_{1.09}$ (27.4 at.% B) as a function of annealing temperature.

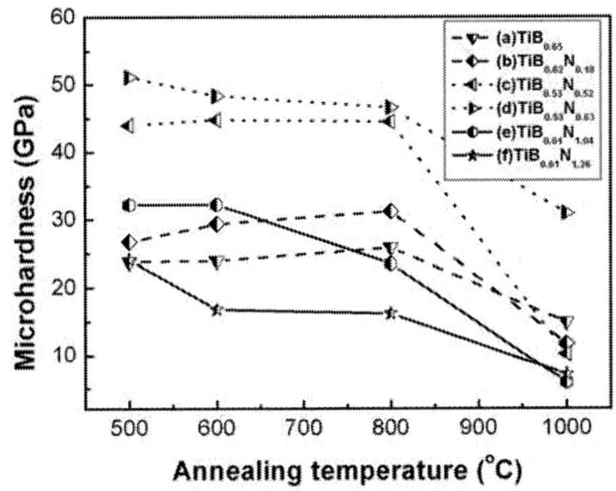


Figure 6.13. Microhardness values of thin films with different N contents as a function of annealing temperature. These films were deposited at 500 °C.

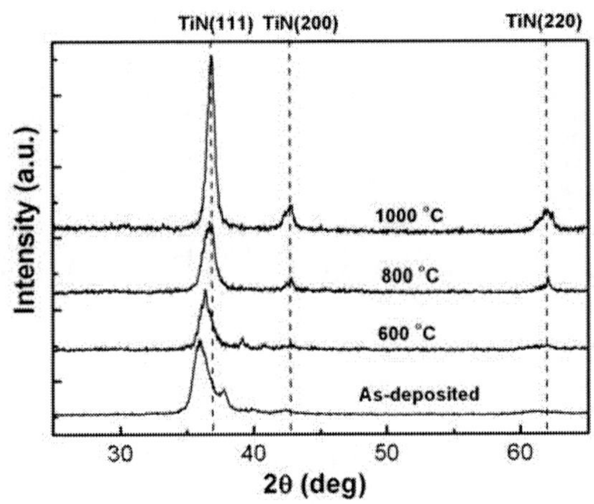


Figure 6.14. XRD θ-2θ spectra of $TiC_{0.24}N_{0.71}$ after annealing at different temperatures.

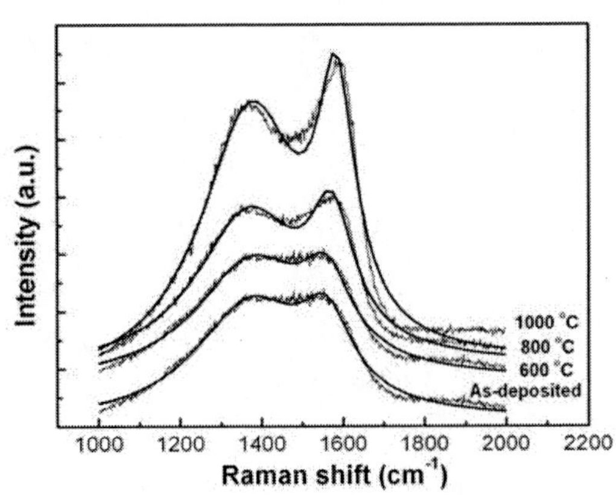


Figure 6.15. Raman spectra of $TiC_{1.61}N_{0.73}$ thin films with different annealing temperatures. The spectra with D and G peaks have been fitted with one Lorentzian line shape for the D peak and one BWF line shape for the G peak using a NLSF program.

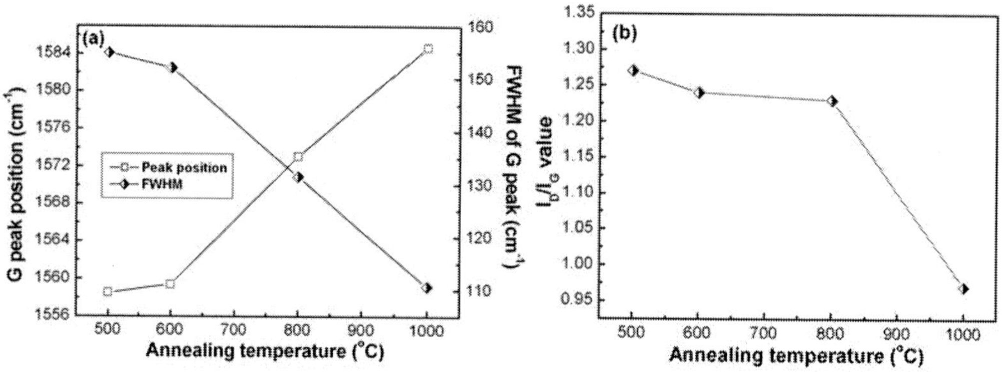


Figure 6.16. (a) The position and FWHM value of G peak as well as (b) $I_D/I_G$ value as a function of annealing temperature in $TiC_{1.61}N_{0.73}$ thin films.

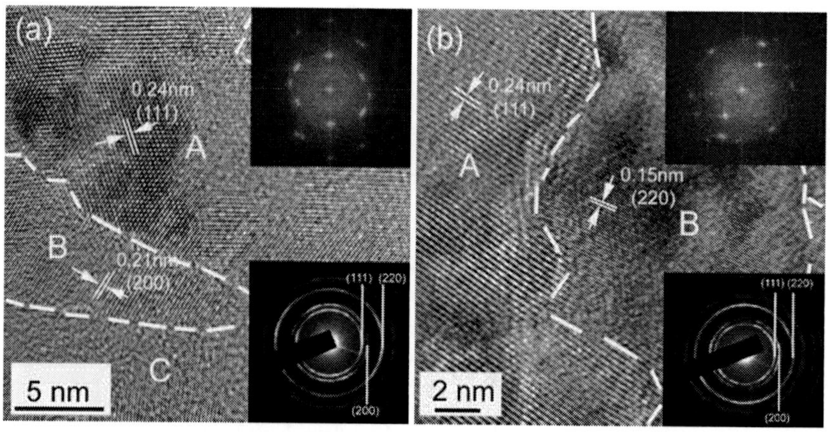


Figure 6.17. HRTEM images of (a) as-deposited and (b) 1000°C-annealed $TiC_{0.24}N_{0.71}$ thin films.

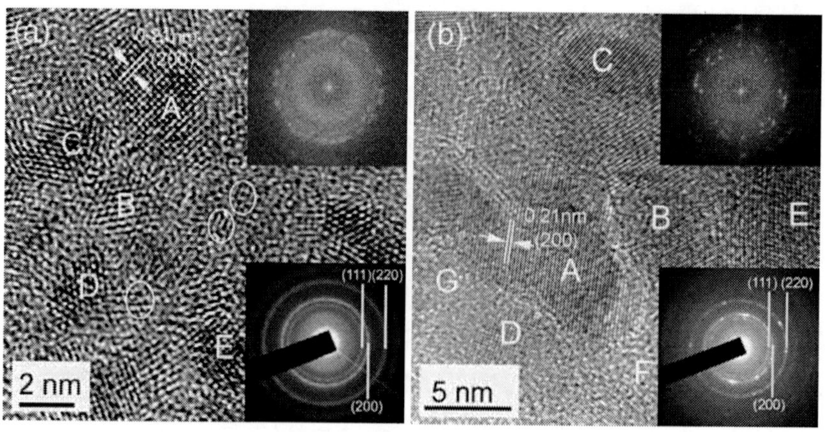


Figure 6.18. HRTEM images of (a) as-deposited and (b) 1000°C-annealed $TiC_{1.61}N_{0.73}$ thin films.

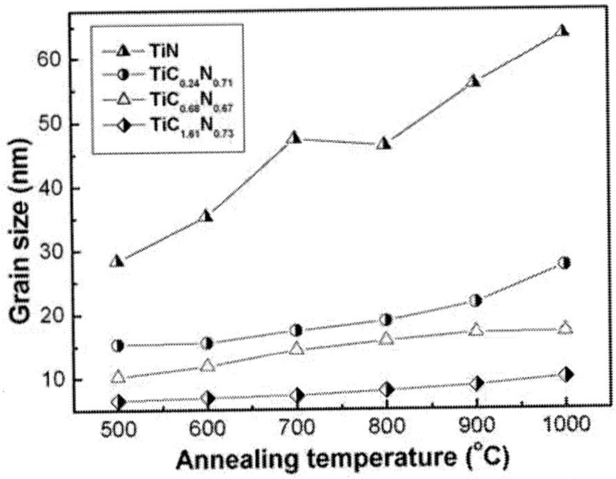


Figure 6.19. Grain sizes as a function of annealing temperature in TiN, $TiC_{0.24}N_{0.71}$, $TiC_{0.68}N_{0.67}$ and $TiC_{1.61}N_{0.73}$.

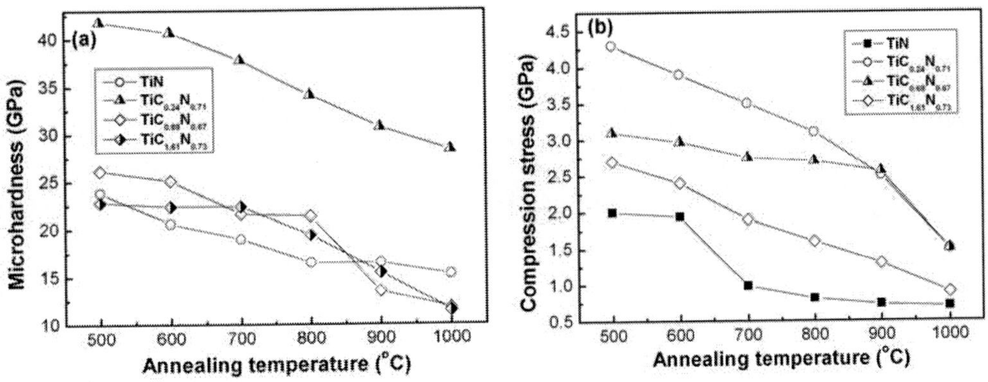


Figure 6.20. The dependences of (a) microhardness and (b) residual stress on annealing temperature.

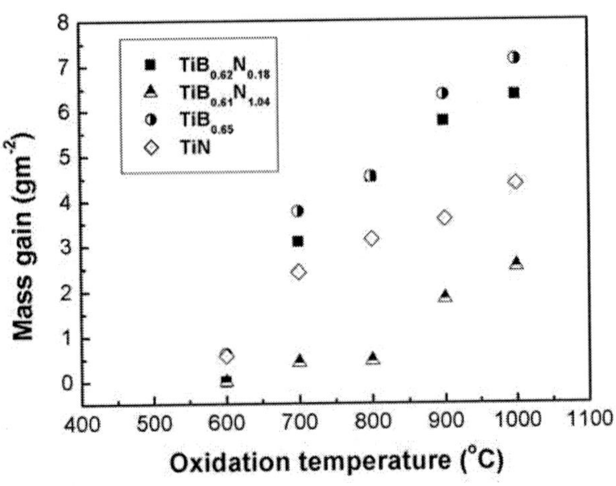


Figure 6.21. The mass gain as a function of oxidation temperature.

Figure 6.22. Plan-view SEM morphologies of $TiB_{0.61}N_{1.04}$ oxidized at (a) 600 °C, (b) 800 °C, (c) 900 °C and (d) 1000 °C.

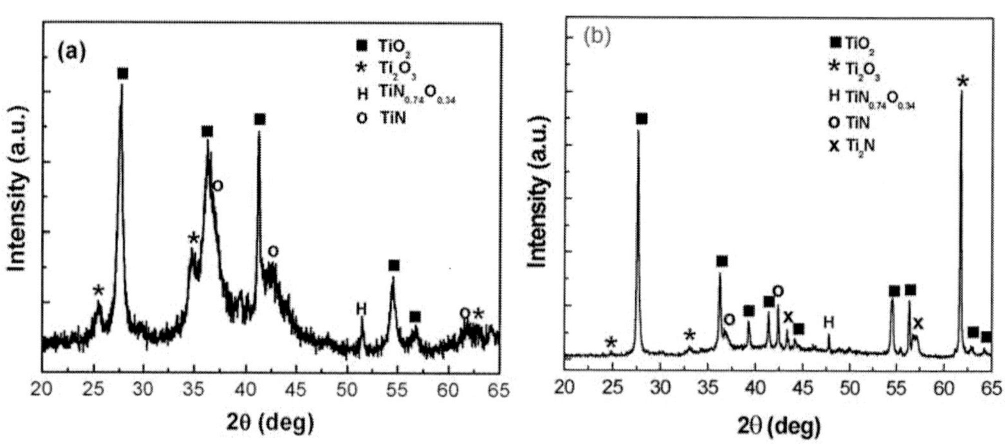


Figure 6.23. Continued on next page.

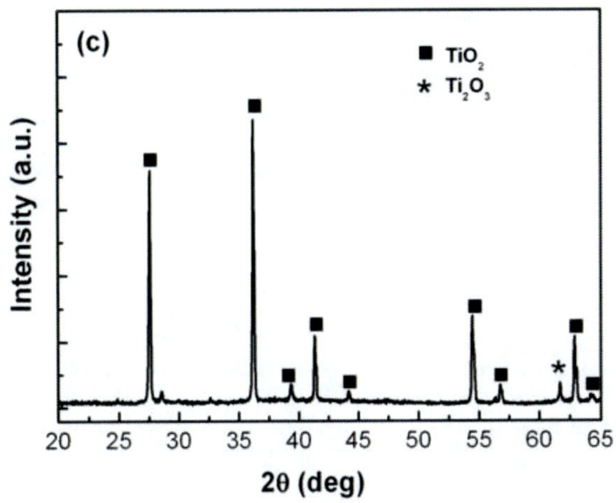


Figure 6.23. XRD spectra of (a) 600 °C-, (b) 800 °C-, (c) 1000 °C-oxidized $TiB_{0.61}N_{1.04}$.

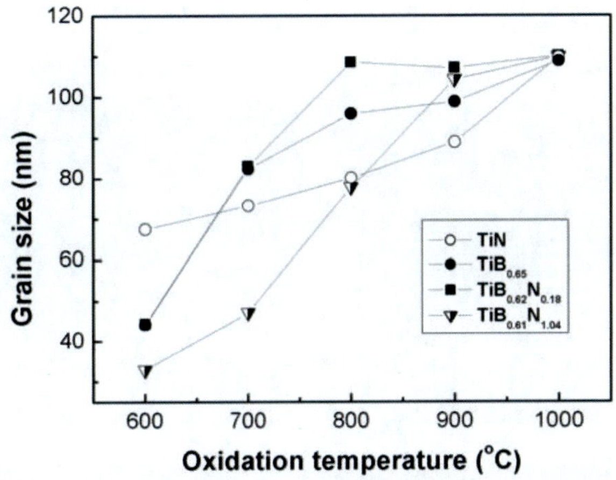


Figure 6.24. Grain size of oxide as a function of oxidation temperature.

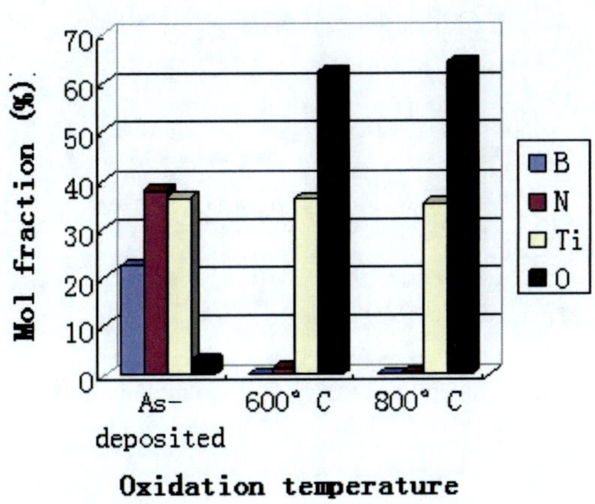


Figure 6.25. Elemental concentrations on film surface of $TiB_{0.61}N_{1.04}$ as a function of oxidation temperature.

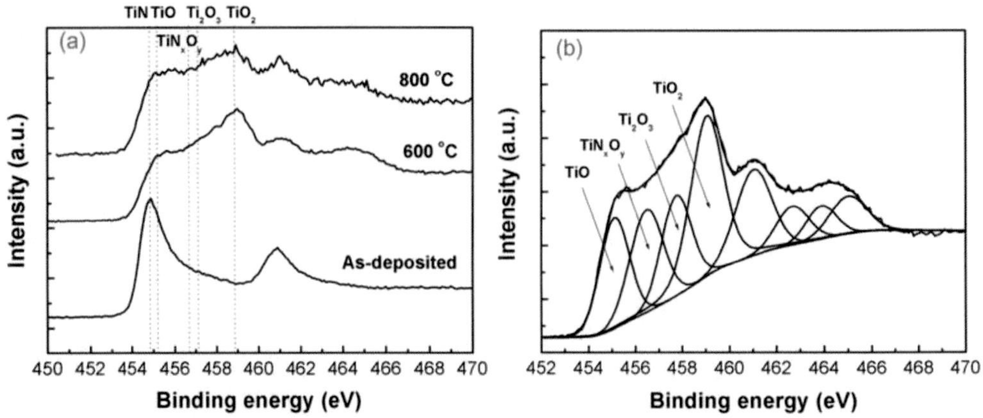


Figure 6.26. (a) Ti 2p spectra of $TiB_{0.61}N_{1.04}$ oxidized at different temperatures, and (b) a fitted Ti 2p spectrum of 600 °C-oxidized $TiB_{0.61}N_{1.04}$.

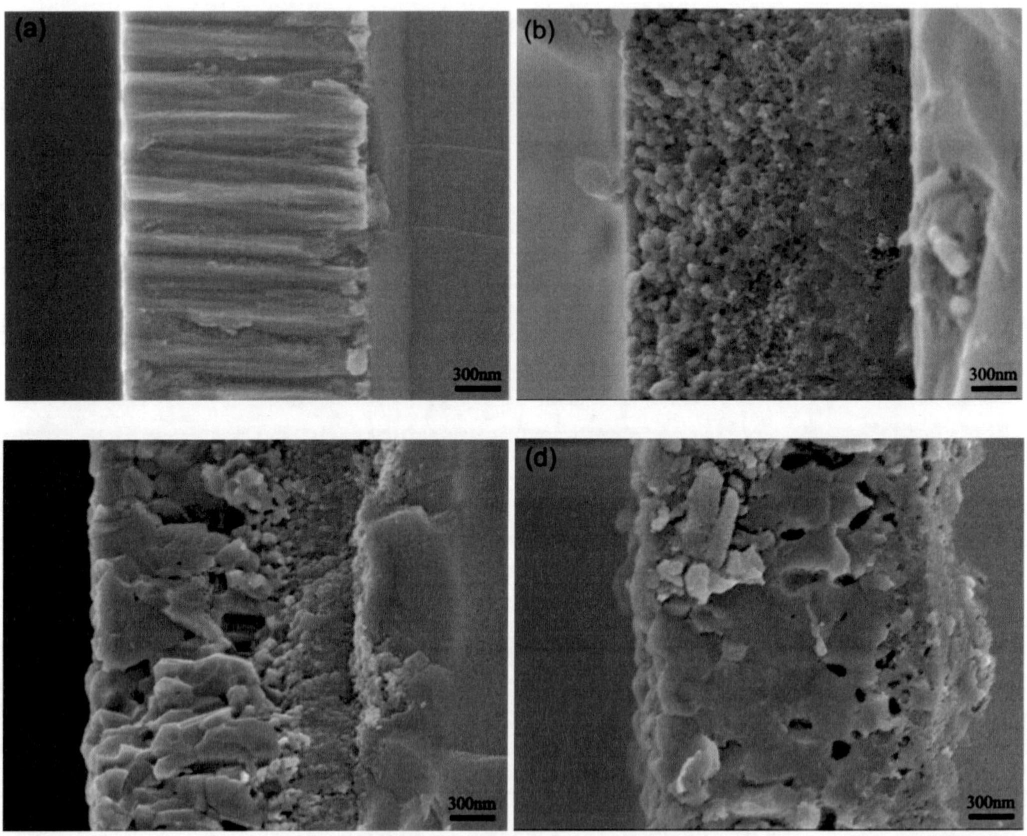


Figure 6.27. Cross-sectional SEM morphologies of (a) as-deposited, (b) 800°C-, (c) 900°C- and (d) 1000°C-oxidized $TiB_{0.61}N_{1.04}$.

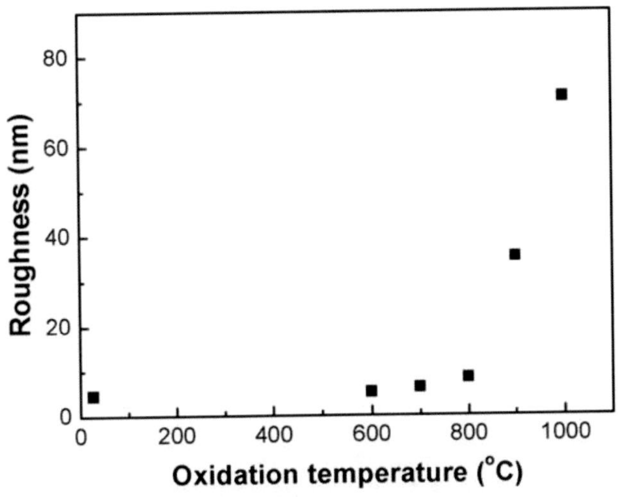


Figure 6.28. Surface roughness value as a function of oxidation temperature in $TiB_{0.61}N_{1.04}$.

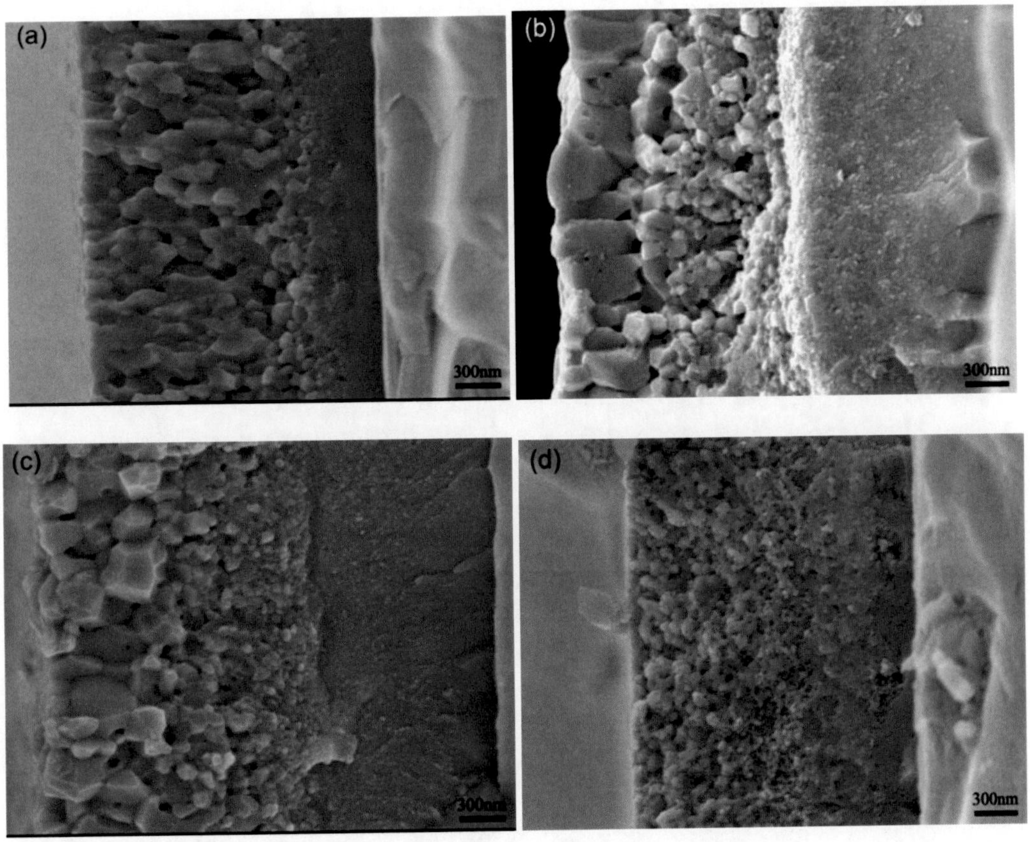


Figure 6.29. Cross-sectional SEM morphologies of (a) TiN, (b) $TiB_{0.65}$, (c) $TiB_{0.62}N_{0.18}$ and (d) $TiB_{0.61}N_{1.04}$ after oxidation at 800°C.

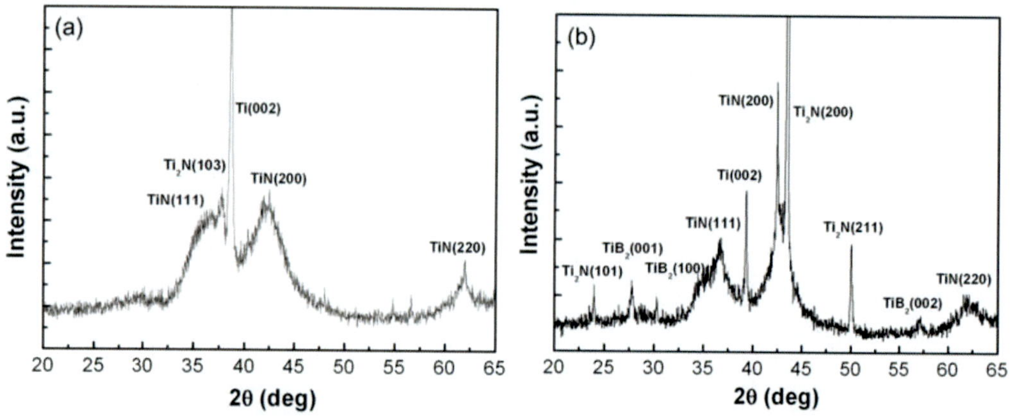


Figure 6.30. XRD spectra of (a) as-deposited and 800°C-annealed $TiB_{0.61}N_{1.04}$.

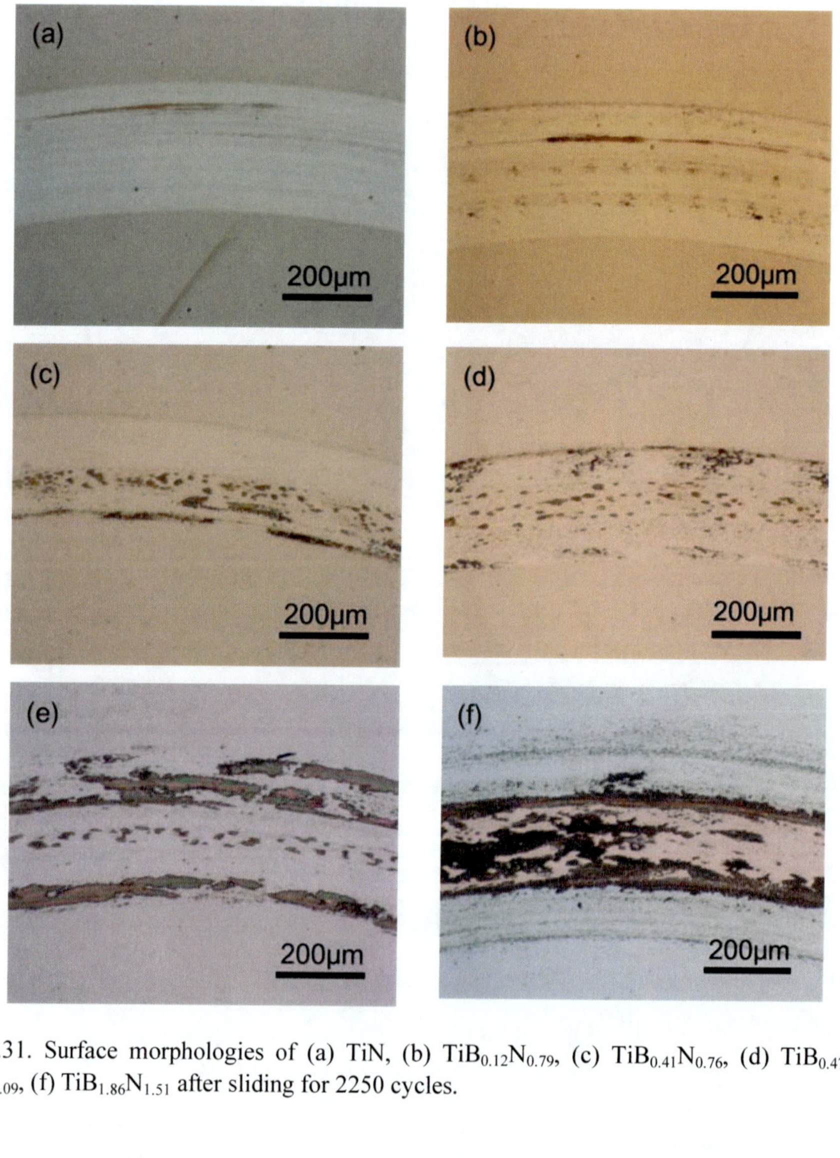


Figure 6.31. Surface morphologies of (a) TiN, (b) $TiB_{0.12}N_{0.79}$, (c) $TiB_{0.41}N_{0.76}$, (d) $TiB_{0.47}N_{0.76}$, (e) $TiB_{0.79}N_{1.09}$, (f) $TiB_{1.86}N_{1.51}$ after sliding for 2250 cycles.

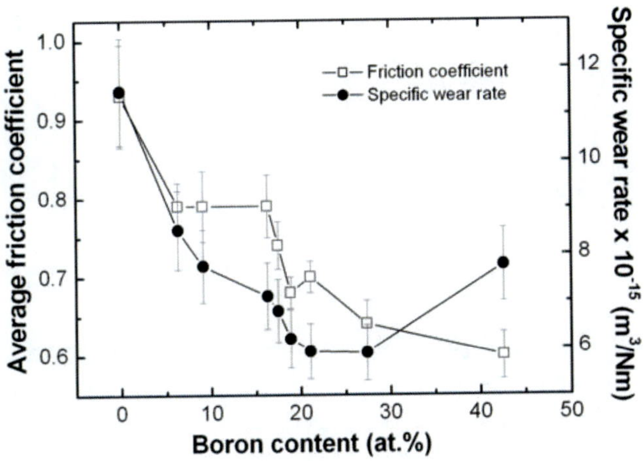


Figure 6.32. Average friction coefficient and specific wear rate of Ti-B-N thin films with increasing boron content.

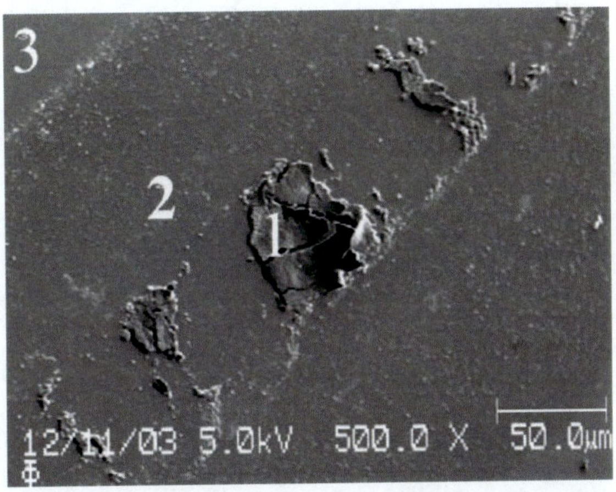


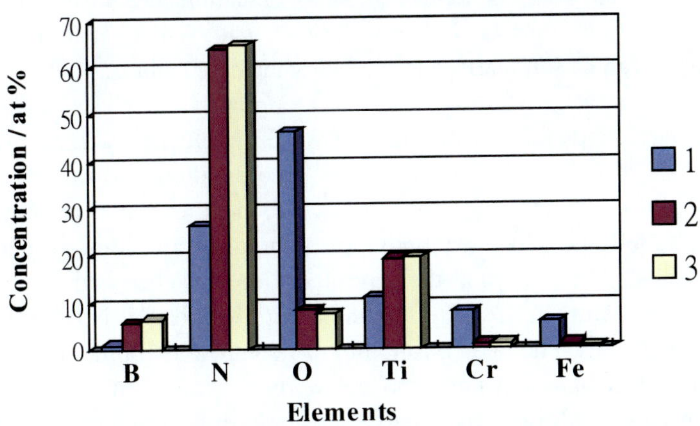


Figure 6.33. AES analysis of wear debris (region 1), worn surface (region 2) and film surface (region 3) in $TiB_{0.47}N_{0.76}$.

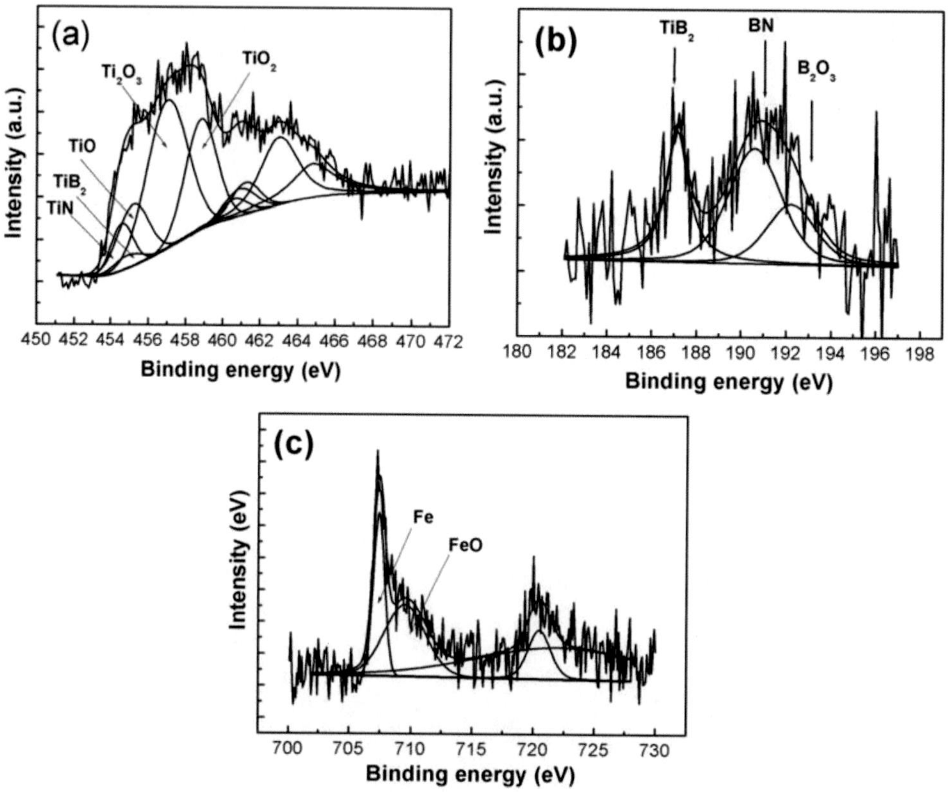


Figure 6.34. XPS spectra of (a) Ti 2p, (b) B 1s and (c) Fe 2p of debris in $TiB_{1.86}N_{1.51}$.

# 7. Summary

Significant progress has been made in understanding phase configuration, nanostructure, mechanical properties as well as thermal stability, oxidation and wear behaviours in TiN based nanocomposite thin films e.g. Ti-B-N, Ti-C-N and Ti-B-C-N thin films. The main results in the chapter can be summarised in the following conclusions:

(1) Present deposited pure TiN film exhibited well crystallized nanocrystallites. Incorporation of B or C, or their combined incorporation led to the formation of nanocomposite thin films of nanocrystalline grains embedded into amorphous matrices. Their nanostructure evolution and phase configuration transformation strongly depended on the miscibility of additives with TiN and their affinity with Ti. When additive (B or C) had a complete miscibility with TiN, solid solution Ti(N,C) or Ti(N,B) thin films were preferentially formed. Only when this miscibility overstepped its limitation, uncompleted and completed segregated two-phase nanocomposites were successively formed respectively with further incorporation of additives. In ternary nanocomposite thin films the nanocrystallites less than 10 nm were usually observed. Finer nanocrystallites were found with combined incorporation of B and C, which resulted in XRD-amorphous like nanostructure in quaternary Ti-B-C-N thin films. On the other hand, with increasing additive content, a gradual decrease

in size of nanocrystallites was found, accompanying with increase of lattice faults such as point vacancies and deformation twining in nanocrystallites.

(2) Nanocomposite structure played different roles in the hardness of thin films, which depended on not only phase configuration, nanostructure but also matrix type. Percolation threshold mechanism took place under immiscible hard amorphous matrices, which was sometime broken by solid solution hardening by way of incorporation of solutes (e.g. B and C) into nitrogen-deficial $TiN_x$. On the other hand, solid solution due to incorporation of solutes into $TiN_x$ increased the residual compressive stress, while formation of nanocomposite structure had different effects on the residual compressive stress.

(3) Nanocomposite structure had a positive effect on microstructure stability. The thermal stability of nanocomposite thin films strongly depended on not only phase configuration and nanostructure, but also the matrix type. Recrystallization of amorphous matrices in nanocomposite thin films greatly decreased the residual stress, even transformed from compression to tension. The hardness stability did not show consistency with that of microstructure, and it also depended on phase configuration and composition.

(4) The nanocomposite nc-TiN/a-$TiB_2$/a-BN thin films showed a superior oxidation resistance as characterized by a two-stage oxidation process, which was related to oxidation resistance of nanocrystallites and thermal stability of amorphous matrix phase.

(5) Both friction coefficient and specific wear rate decreased with incorporation of B into TiN or increasing B content, accompanying with transformation of wear modes from abrasive wear to powder formation, subsequently to plastic deformation. An inverse trend in the wear rate was followed after formation of soft h-BN phase in Ti-B-N thin films. Oxygen of $H_2O$ in air participated in and reacted with films, and formed mainly Ti oxide in debris, which was believed to have an effect on increasing wear resistance.

## References


[1] H.Holleck, in: A.Kumar, Y.W.Chung, J.Jmoore, J.E.Smugeresky (Eds.), *Surface Engineering: Science and Technology* **I,** The Mineral, Metals and Materials Society, 1999, pp. 207-231.

[2] S. Carvalho, L.Rebouts, A.Cavaleiro, L. A. Rocha, J. Gomes and E. Alves, *Thin Solid Films* **398-399** (2001) 391.

[3] W. Gissler, *Surf. Coat. Technol.* **68-69** (1994) 556.

[4] K. Kutschej, P. H. Mayrhofer, M. Kathrein, P. Polcik, and C. Mitterer, *Surf. Coat. Technol.* **188-189** (2004) 358.

[5] S. Vepřek, G. J. Vepřek-Heijman M, P. Karvankova, J. Prochazkz, *Thin Solid Films* **476** (2005)1.

[6] M. J. Demkowicz, A. S. Argon, *Phys. Rev. Lett.* **93** (2004) 025505.

[7] S. Vepřek, P. Nesládek, A. Niederhofer, F. Glatz, M. Jílek and M. Šíma, *Surf. Coat. Technol.* **108-109** (1998) 138.

[8] J. A. Thornton, *J. Vac. Sci. Technol.* **11** (1974) 66.

[9] J. A. Thornton, *J. Vac. Sci. Technol.* **A4**(6) (1986) 3059.

[10] C. L. Yeh, and H. C. Chuang, *Ceramics Int.* **30**(5) (2004) 705.

[11] S. Zhang, D. Sun, Y. Q. Fu, and H. J. Du, *Surf. Coat. Technol.* **167** (2003) 113.

[12] T. P. Mollart, J. Haupt, R. Gilmore and W. Gissler, *Surf. Coat. Technol.* **86-87** (1996) 231.
[13] C. Heau, J. P. Terrat, *Surf. Coat. Technol.* **108-109** (1998) 332.
[14] B. Matthes, E. Broszeit, K. H. Kloos, *Surf. Coat. Technol.* **57** (1993) 97.
[15] C. Mitterer, P. Losbichler, F. Hofer, P. Warbichler, P. N. Gibson, W. Gissler, *Vacuum* **50** (1998) 313.
[16] D. A. Shirley, *Phys. Rev. B* **5** (1972) 4709.
[17] E. Selbach, K. Schmidt and M. Wang, *Thin Solid Films* **188** (1990) 267.
[18] O. Knotek, F. Loffler, M. Bohmer and C. Stossel, *Surf. Coat. Technol.* **49** (1991) 263.
[19] H. Karner, J. Laimer, H. Stori and P. Rodnammer, *Surf. Coat. Technol.* **39/40** (1991) 263.
[20] M. A. Bake, A. Steiner, J. Haupt and W. Gissler, *J.Vac. Sci. Technol.* **A13** (1995) 1633.
[21] M. A. Baker, T. P. Mollart, P. N. Gibson and W. Gissler. *J.Vac.Sci. Technol.* **A15**, (1997) 284.
[22] C. Ruby, R. Ott, F. Huang, M. L. Weaver and J. A. Barnard, *Sur. Interface Anal.* **29**, (2000) 283.
[23] J. F. Pierson, V. Chapusot, A. Billard, M. Alnot, Ph. Bauer. *Sur. Coat. Technol.* **151/152** (2002) 526.
[24] R. D. Ott, C. Ruby, F. Huang, M. L .Weaver and J. A. Barnard, *Thin Solid Films* **377/378** (2000) 602.
[25] R. Wiedemann, V. Weihnacht and H. Oettel. *Surf. Coat. Technol.* **116/119** (1999) 302.
[26] P. Losbichler, C. Mitterer, P. N. Gibson, W. Gissler, F. Hofer, P. Warbichler, *Surf. Coat. Technol.* **94-95** (1997) 297.
[27] J. F. Pierson, F. Bertrn, J.P.Bauer, and J.Jolly, *Surf. Coat. Technol.* **142-144** (2001) 906.
[28] Q. Y. Liu, J. Q. Hu, K. B. Tang, B. Deng, Y. T. Qian, G. Zhou and X. M. Liu, *Chem. Phys. Lett.* **314** (1-2) 37.
[29] Y. S.Yang, S. C. Lee and C.-Y. A. Tsao, *Surf. Coat. Technol.* **141** (2001) 78.
[30] K. Narasimhan, S. Prasad Boppana, and D. G. Bhat, *Wear* **188** (1995) 123.
[31] L. A. Dobrazański, and M. Adamiak, *J. Mater. Proc. Technol.* **133** (2003) 50.
[32] R. Hauert, J. Patscheider, L. Knoblauch, and M. Diserens, *Adv. Mater.* **11**(1999) 175.
[33] D. J. Cheng, W. P. Sun and M. H. Mon, *Thin Solid Films* ***146*** (1987) 45.
[34] H. M. Gabrief and K. H. Kloos, *Thin Solid Films* **118** (1984) 243.
[35] E. Ertuerk, O. Knotek, W. Bergmer, H. -G. Prengel, H. -J. Heuvel, H.G. Dederichs and C. Stoessel, *Surf. Coat. Technol.* **46** (1991) 39.
[36] S. Vepřek, *J. Vac. Sci. Technol. A* **17** (5) (1999) 2401.
[37] S. -H. Jhi, J. Ihm, S. G. Louie, and M. L. Cohen, *Nature* **399** (1999) 132.
[38] D. Martínez-Martínez, J. C. Sánchez-Lopez, T. C. Rojas, A. Fernández, P. Eaton, and M. Belin, *Thin Solid Films* **472** (2005) 64.
[39] P. Nesládek, and S. Vepřek, *Phys. Status Solid.* **A177** (2000) 53.
[40] S. Vepřek, A. Niederhofer, K. Moto, T. Bolom, H. -D. Mannling, P. Nesladek, G. Dollinger and A. Bergmaier, *Surf. Coat. Technol.* **133-134** (2000) 152.
[41] S. Vepřek, S. Christiansen, M. Albrecht , and H. P. Strunk, *Mater. Res. Soc. Symp. Proc.* **457** (1997) 407.
[42] Y. H. Lu, Z. -J. Liu, and Y. G. Shen, *Acta mater.* **54** (2006) 2897.
[43] D. A. Shirley, *Phys. Rev. B* **5** (1972) 4709.

[44] G. B. Harris, *Phil. Mag.* **43** (1952) 113.
[45] C. A. Carrasco, V. Vergara S., R. Benavente G., N. Mingolo, J. C. Rios, *Mater. Character.* **48** (2002) 81.
[46] Z. L. Wang, *Part. Part. Syst. Charact.* **18** (2001) 142.
[47] W. C. Oliver, G. M. A Pharr, *J Mater Res* **7** (1992) 1564.
[48] J. M. Antunes, A. Cavaleiro, L. F. Menezes, M. I. Simoes and J. V. Fernandes, Surf. Coat. Technol. 149 (2002) 27.
[49] H. Oettel, R. Wiedemann, *Surf. Coat. Technol.* **76-77** (1995) 265.
[50] J. O. Carneiro, V. Teixeira, A. Portinha, F. Vaz and J. A. Ferreira, *Rev. Adv. Mater. Sci.* **7** (2004) 32.
[51] J. Almer, and M. Odén, *J. Vac. Sci. Technol.* **A18**(1) (2000) 121.
[52] B. Rauschenbach, J. W. Gerlach, *Cryst. Res. Technol.* **35** (2000) 675.
[53] S.Vepřek, S.Mukherjee, P. Karvankova, H.-D.Manning, J.L.He, K.Moto, and J.Prochazka, *J. Vac. Sci. Technol.* **A21**(3) (2003) 532.
[54] P. H. Mayrhofe, M. Stoiber and C. Mitterer, Script. Mater. 53 (2005) 241.
[55] Y. H. Lu, Z. K. Xu, Haydn Chen, Y. G. Shen and K. Y. Li, *Appl. Phys. Lett.* **87** (2005) 151902.
[56] P. Karvankova, M. G. J. Veprek-Heijman, O. Zindulka, A. Bergmaier, S. Vepřek, *Surf Coat Technol* **163-164** (2003) 149.
[57] X. Z. Liao, J. Y. Huang, Y. T. Zhu, F. Zhou and E. J. Lavernia, *Phil. Mag.* **83**(26) (2003) 3065.
[58] G. T. Gray, *Acta mater.* **46** (1988) 1827.
[59] S. Mahajan and D. F. Williams, *Int. Metall. Rev.* **18** (1973) 43.
[60] J. O. Stiegler and C. J. Mchargue, Deformation twinning, edited by Reed-Hill R E, J. Hirth P and Rogers H C, (New York, 1964) 209.
[61] L. Hultman, *Vacuum* **57**, 1 (2000).
[62] Q. Q.Yang, L. H. Zhao, and L. S. Wen, *Surf. Coat. Technol.* **78** (1996) 27.
[63] P. H. Mayrhofe, C. Mitterer, J. G. Wen, I. Petrov and J. E. Greene, *J. Appl. Phys.***100** (2006) 044301.
[64] A. Niederhofer, P. Nesládek, H. -D. Mánnling, K. Moto, S. Vepřek and M. Jílek, *Surf. Coat. Technol.* **120-121** (1999) 173.
[65] H. Novotny, F. Benesovsky, C. Brukl and O. Schob, *Mh. Chem.* **92** (1961) 403.
[66] J. E. Krzanowski, *Surf. Coat. Technol.* **188-189** (2004) 376.
[67] J. G. Naeini, B. M. Way, J. R. Dahn and J. C. Irwin, *Phys. Rev. B* **54**, 144 (1996).
[68] A .C. Ferrari, *Diamond Relat. Mater.* **11** (2002) 1053.
[69] D. G. McCulloch, S. Prawer and A. Hoffman, *Phys. Rev. B* **50**(9) (1994) 5905.
[70] R. O. Dillon and J. A. Woollan, *Physics Rev. B* **29**(6) (1984) 3482.
[71] M. DiDomenico, Jr. S. H. Wemple, S. P. S. Porto, and R. P. Bauman, *Phys. Rev.* **174** (1968) 522.
[72] L.G. Cancado, K. Takai and T. Enoki, *Appl. Phys. Lett.* **88** (2006) 163106.
[73] P. B. Barna, M. Adamik, *Thin Solid Films* **317** (1998) 27.
[74] L. S. Yu, J. M. E. Harper, J. J. Cuomo, D. A. Smith, *Appl. Phys. Lett.* **47** (1985) 932.
[75] D. H. Kuo, K. W. Huang, *Thin Solid Films* **394** (2001) 81.
[76] Y. H. Lu and Y.G. Shen, *Appl. Phys. Lett.* **90** (2007) 221913.
[77] L. A. Dobrazański, M. Adamiak, *J. Mater. Proc. Technol.* **133** (2003) 50.

[78] C. Morant, P. Prieto, A. Forn, J. A. Picas, E. Elizalde, J. M. Sanz, *Surf. Coat. Technol.* **180-181** (2004) 512.
[79] Y. H. Lu, Y. G. Shen, Z. F. Zhou, K. Y. Li, and H. Chen, *Mater. Sci. Technol.* **22** (10) (2002) 1255.
[80] S.Vepřek, *Thin Solid Films* **317** (1998) 449.
[81] S. Abraham, J. E. Ok and K. H. Kim, *J. Mater. Process. Technol.* **187-188** (2007) 571.
[82] D. Zhang, E. Sutter, J. J. Moore, G. G. W. Mustoe, E. A. Levashov, J. Disam, *Thin Sold Films*, **398-399** (2001) 320.
[83] C. Mitterer, P. H. Mayrhofer, M. Beschliesser, P. Losbichler, P. Warbichler, F. Hofer, P.N. Gibson, W. Gissler, H. Hruby, J. Musil, J. Vlček, *Surf. Coat. Technol.* **120-121** (1999) 405.
[84] A. D. John, Lang's Handbook of Chemistry, 15th edition, McGRAW-Hill. INC. New York, 1999.
[85] D. H. Jung, H. Kim, G. R. Lee, B. Park, J. J. Lee and J. H. Joo, *Surf. Coat. Technol.* **174-175** (2003) 638.
[86] P. Holubář, and M. Jílek, M. Šíma, *Surf. Coat. Technol.* **120-121** (1999) 184.
[87] W. Dworschak, K. Jung, H. Ehrhadt, *Thin Solid Films*, **254** (1995) 65.
[88] S. M. Aouadi, F. Namavar, T. Z. Gorishnyy, S. L. Rohde, *Surf. Coat. Technol.* **160** (2002) 145.
[89] R. A. Andrievski, *J. Solid State Chem.***133** (1997) 249.
[90] S. Vepř ek, S. Reiprich, and S. Z. Li, *Appl. Phys. Lett.* **66** (20) 2641.
[91] J.A. Thornton and D.W. Hoffman, *Thin Solid Films* **171** (1989) 5.
[92] A. Inoue, *Acta mater* **48** (2000) 279.
[93] P. H. Mayrhofer, H. Willmann and C. Mitterer, *Thin Solid Films* **440** (2003) 174.
[94] P. Karvankova, M. G. J. Veprek-Heijman, O. Zindulka, A. Bergmaier and S. Vepřek, *Surf. Coat. Technol.* **163/164** (2003) 149.
[95] S. Vepřek, M. Haussmann, and S. Reiprich, *J. Vac. Sci. Technol. A* **14**(1) (1996) 46.
[96] J. W. Cahn, *Acta Metall.* **11** (1963) 1275.
[97] F. Li, J. S. Lannin, *Appl. Phys. Lett.* **61** (17) 2116.
[98] M. F. Hochella, J. R. Lindsay, V. G. Mossotti, C. M. Eggleston, *Amer. Miner.* **73** (1988) 1449.
[99] J. E. Naeini, B. M. Way, J. R. Dahn, and J. C. Irwin, *Phys. Rev. B*, **54**(1) (1996) 144.
[100] B. Rother, and H. Kappl, *Surf. Coat. Technol.* **96** (1997) 163.
[101] D. McIntyre and J. E. Greene, *J. Appl. Phys.* **67** (3) (1990) 1542.
[102] M. Diserens, J. Patscheider and F. Lévy, *Surf. Coat. Technol.***120-121** (1999) 158.
[103] G. Llauro, F. Gourbilleau, F. Sibieude and R. Hillel, *Thin Solid Films* **315** (1998) 336.
[104] J. F. Pierson, E. Tomasella and Ph. Bauer, *Thin Solid Films* **408** (2002) 26.
[105] H. A. Jehn, Surf. Coat. *Technol.* **131** (2000) 433.
[106] M. Tamura, H. Kubo, *Surf. Coat. Technol.* **54-55** (1992) 255.
[107] E. Broszeit, B.Matthes B, W. Herr and K. H. Kloos, *Surf. Coat. Technol.* **58** (1993) 29.
[108] R. Kullmer, C. Lugmair, A. Figueras, J. Bassas, M. Stoiber and C. Mitterer, *Surf. Coat.* Tech*n*ol. **174-175** (2003) 1229.

[109] C. Rebholz, A. Leyland P. Larour, C. Charitidis and S. Logothetidis, *Surf. Coat. Technol.* **119** (1999) 648.
[110] Z. P. Huang, Y. Sun, and T. Bell, *Wear* **173** (1994) 13.
[111] Y. H. Lu, P. Sit, T. F. Hung, Haydn Chen, Z. F. Zhou, K. Y. Li, and Y. G. Shen, *J. Vac. Sci. Technol.* **B23** (2005) 449.
[112] Q. Yang, L.R. Zhao, *Surf. Coat. Technol.* **173** (2003) 58.
[113] E. Sirrvio, M. Sulonen, H. Sundquist, *Thin Solid Films* **96** (1982) 93.
[114] Y. H. Lu, Z. F. Zhou, P. Sit, Y. G. Shen, K. Y. Li and Haydn Chen, *Surf. Coat. Technol.* **187** (2004) 98.
[115] Y. Kimura, T. Wakabayashi, K. Okada, T. Wada, H. Nishikawa, *Wear* **232** (1999) 199.
[116] O. O. Ajayi and K. C. Ludema, *wear*, **140** (1990) 191.
[117] H. Mohrbacher, B. Blanpain, J. P. Celis, J. R. Roos, *Wear* **180** (1995) 43.
[118] K. Holmberg, A. Matthews, *Coatings Tribology*, Tribology Series: vol. 28, Elsevier Science, Amsterdam, 1994, pp.118.

In: New Research on Nanocomposites
Editors: Luis M. Krause et al, pp. 223-246
ISBN: 978-1-60456-799-1


*Chapter 7*

# CARBON – METAL (NI, TI) NANOCOMPOSITE FILMS AS PROTECTIVE COATINGS

***Katarína Sedlácková[1,2], Róbert Grasin[1] and György Radnóczi[1]***
[1]Thin Film Physics, Research Institute for Technical Physics and Materials Science, Hungarian Academy of Sciences, Konkoly-Thege M. út 29-33, 1121 Budapest, Hungary
[2]Institute of Electrical Engineering, Slovak Academy of Sciences, Dúbravská cesta 9, 84104 Bratislava, Slovakia

## Abstract

Carbon - metal (Ni, Ti) nanocomposites were prepared by dc magnetron sputtering. The films were deposited at different deposition temperatures (25 - 800°C) on silicon (001) substrates covered with 300 nm thick siliconoxide. The input power of the magnetrons with Ni target was between 5 W and 25 W, and the magnetrons with Ti target between 10 W and 40 W. The input power of carbon target was 150 W.

We found correlation between structural, electrical and mechanical properties of films, and developed the optimal structure of films for protective applications. Transmission electron microscopy and high resolution electron microscopy were used to determine the structure of films. The standard four probe technique was used for electrical measurements. The mechanical properties of films, e.g. hardness, modulus of elasticity were measured by nanoindentor.

In both cases (C-Ni, C-Ti), the highest hardness (C-Ni ~ 11 GPa, C-Ti ~ 18 GPa) and highest modulus of elasticity (C-Ni ~ 114 GPa, C-Ti ~ 205 GPa) were obtained for the coating preparet at 200°C consisting of columnar ($Ni_3C$ and TiC) crystallites and 2 nm thin amorphous and graphite-like carbon matrix. The hardness of the coating is independent on the metal content but is influenced by the structure of thin carbon matrix.

## Introduction

Carbon-based nanocomposite thin films have large application potential because they possess unique mechanical properties, espesially high hardness, high elasticity, and a low widely adjustable friction coefficient [1-4]. Various carbon-based composite films are formed as combinations of thin films based on pure carbon [5].

Pure carbon-based thin films come out in several modifications determined by deposition technology. Carbon-based nanocomposite thin films include different carbon phases: amorphous phase (a-C), graphite like carbon (GLC), fullerene-like carbon (FLC) , and diamond-like carbon (DLC) [1,4,6].

Modern methods of vacuum deposition provide great flexibility for manipulating material chemistry and structure, leading to films and coatings with special properties. These new special properties of nanocomposites are often unachievable in bulk materials.

A combination of pure carbon-based thin films with metallic nanoparticles can enhance certain physical properties in a nanocomposite. For example, hardness can be improved by the introduction of boundaries or particles of hard phases for stoping deformation [3]. Carbon films are alternatively modified by the addition of metals, e.g. Ni [7,8], Cu and Ag [3], Cr [9, 10], and Ti [10]. This approach is used to form composite layers.

Nanocomposite structures composed of crystalline and (or) amorphous phases are of rapidly growing interest. With the real-time control of the deposition parameters e.g. pressure, temperature, functionally graded thin films can be produced. Nanocomposite coatings have enhanced mechanical, magnetic and interesting electric properties compared to conventional single-phase coatings. Tailored physical properties can be produced and reproduced only if detailed knowledge of the structure and the atomic processes, taking place during film formation is avialable.

## Growth of One-Component Films

In a series of reviews [17-20], to explain the achievements made in the Movchan´ and Demshichin´s work the following main phenomena were considered: nucleation, crystal growth coalescence and grain growth. The corresponding atomic processes are condensation, adsorption, surface diffusion, agglomeration for nucleation and bulk self-diffusion for grain growth. The corresponding main driving forces are the decrease of bulk free energy for nucleation and crystal growth, and minimization of interface/surface.

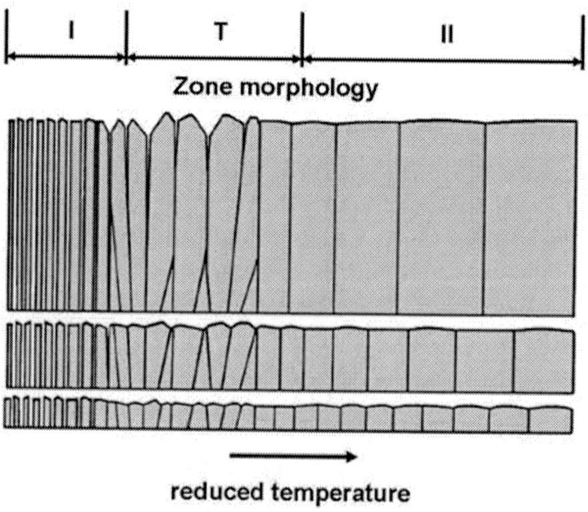


Figure 1. The schematic structure zone diagram of a one-component system as a function of the reduced temperature (Ts/Tm). Ts is the deposition temperature, Tm is the melting temperature of the material.

Fig. 1 shows the typical Structure Zone Diagram (SZD) of a one-component system. If the substrate temperature $T_s$ is below 0.2 of the melting temperature ($T_s < 0.2\ T_m$), nucleation occurs at random orientation and high density and coalescence is hindered.

The nucleated crystals grow independently, leading to a columnar structure of parallel crystals without any texture (Zone I).

In an intermediate temperature region ($0.2\ T_m < T_s < 0.5\ T_m$) the grains show a competing growth after nucleation. Surface diffusion plays an important role in this, and the driving force is the decrease of surface energy. V-shape grains are the result of this competing growth (Zone T). In the case, at temperatures when coalescence is effective ( $T_s > 0.5\ T_m$) the width of columns reaches their height (Zone II).

For all temperatures and conditions, nucleation occurs on the substrate in one-component films. As a result, all one-component films must be columnar in their nature.

## Growth of Multi-component Films

The growth of individual crystals of the first phase in two component systems can be stopped by a covering second phase material.

If atoms of two different chemical species arrive at the substrate surface, a two dimensional alloy layer forms from the adsorbed atoms after accomodation.

The schematic possible SZD for two-component system is shown in Fig. 2. In such system the first component grow in three dimensional way and the oxide growes in two dimensional way in the sence that a continuous film develops at very small thickness. The scheme established in the case of Al/Al-oxide systems [20] has been utilized in growing nanocomposite carbon-metal films.

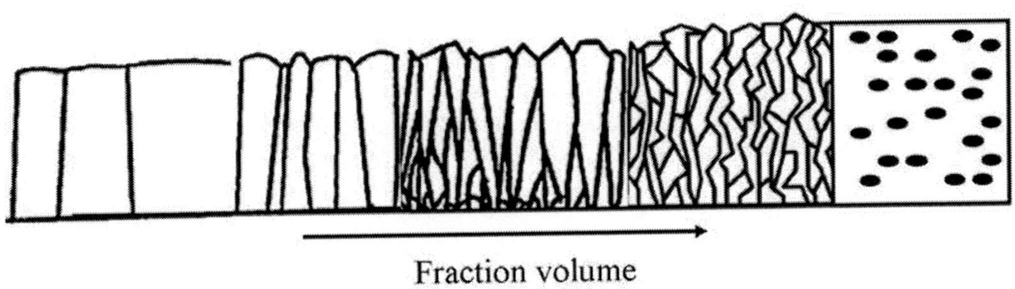


Figure 2. The schematic structure zone diagram of two-component system as a function of the incident first component to second component fraction volume.

At very low oxide (in our films the carbon has the same function) fraction the film grows in Zone II morphology, but the surface structure of the growing crystals is modified [20]. The structural changes in the films as a function of increasing carbon concentration are shown in Fig. 1. Carbon-containing Ni or Ti films deposited at room temperature correspond with restructuration texture II zone in pure metal zone diagram SZM. At low carbon / metal arrival rate ratios, carbon is incorporated into the grain boundaries and continues to accumulate during grain boundaries migration. The resulting texture remains Zone II with columns extending through the film. The structure has a lower degree of preferred orientation and a smaller column size as shown in Fig. 1.

If the carbon content is slightly higher, coarsening during coalescence is severely suppressed. Than the columns have the random orientation. At still higher carbon concentrations, the carbon covers the metal islands which have different orientations and coarsening during coalescence is blocked. The film is consisting of globular grains and no columns form. If further increasing the carbon / metal flux ratio, the grain size decreases. In this case, the grain size can reach the nanometers and by exploiting these phenomena, new micro- and nanostructures can be controllably formed.

## Growth of Carbon-Metal Nanocomposite Thin Films

Nanoscale composite films are often produced by physical deposition methods. In these films the nanosize of the crystalline grains is ensured by the formation of a continuous covering layer of the (amorphous) matrix during growth, blocking the growth of the crystallites.

Several techniques have been proposed for the deposition of metal-containing carbon films, which include sputter deposition [7, 8, 11, 12], plasma assisted chemical vapor deposition (CVD) [13], ion beam-assisted deposition (IBAD) [14], and plasma source ion implantation (PSII) [15].

Application oriented preparation of thin films needs reproducible technologies. As long as setting up a reproducible production characterised by a given set of preparation parameters is possible by an experimental trial and error technique, to understand the effect of these parameters on the properties of films is indispensable. Therefore the knowledge of the driving forces and available kinetics of the structure forming processes for each material system is essential. Furthermore, the relation of the structure and property have to be clarified as well. When avialable, this type of knowledge is usually summarized in Structure Zone Diagrams (SZD), first suggested by Movchan and Demshuchin for micro-structural morphology of thin films as the function of the growth temperature [16].

A „modern" SZD shows how the morphology, phase composition, crystallographic texture and other structural properties depend on the preparation conditions and specifies the main atomic processes leading to the described structure.

# Preparation of Films

Carbon-metal composite thin films were prepared by dc magnetron sputtering of metal (Ni, Ti) and C targets in argon at 0.25 Pa (Fig. 3.). The films were deposited on 300 nm thick oxidized (001) silicon substrates and freshly clevead NaCl single crystals as substrates.

The input power of the carbon target was kept constant at 150 W while the input power of the metal target was changed in a range of 5÷40 W. These values corresponds to a deposition rate between 0.05÷0.07 nm/s. The target-substrate distance was 70 mm. The nominal tickness of the films was 20÷600 nm. The deposition temperature dependence of film growth at different temperature range 25 ÷800 °C was also studied.

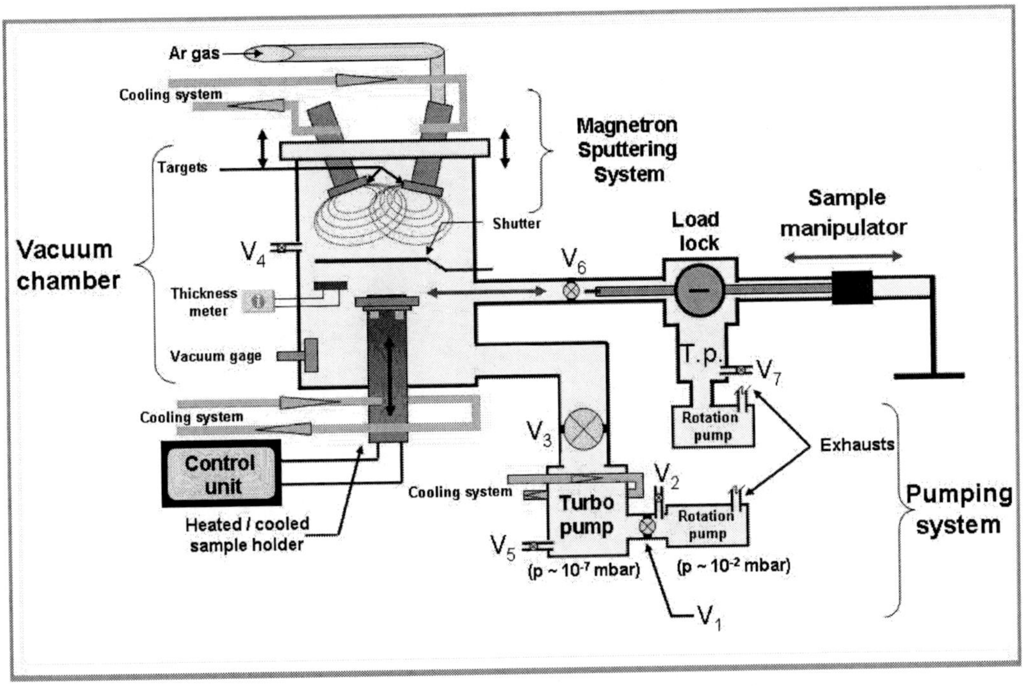


Figure 3. Schematic figure of dc magnetron sputtering.

The structural investigations were perfomed on a Philips CM-20 Microscope using a 200 kV accelerating voltage. The elemental analysis of film composition was performed also in this microscope which is equipped with a NORAN EDS (Energy Dispersive Spectrometer), with a HP-Ge detector. For high resolution electron microscopy (HREM) investigations were used a JEOL 3010 microscope at 300 kV accelerating voltage .

Dynamic nanohardness and modulus of elasticity were measured by NanoTest 600 nanomechanical tester (Micro Materials Ltd., UK) [21] The Berkovich type indentor-head was applied for hardness measurements in depth control mode with 50 nm maximum penetration. The Oliver-Pharr formalism was used in the evaluation of the load-penetration curves [22].

The AC bridge LR-700 and a four-probe technique were used for the electrical resistivity measurement in a temperature range of 4.2 ÷ 340 K. The current-voltage dependences were measured by a standard technique at several temperatures. The I-V curves measured were then numerically differentiated. The magnetoresistivity in parallel and perpendicular fields to the film surface was measured in magnetic fields up to ±0.5 T.

Figure 4 shows a dependence of the content of Ni and C (at.%) in the thin films as a function of Ni- target power [23].

The C-target power set to 150 W gives an optimal deposition rate [24]. The 1:1 atomic ratio of Ni to C was achieved at 25 W of Ni-target power. A lower power of the Ni source resulted in a decrease of the Ni content.

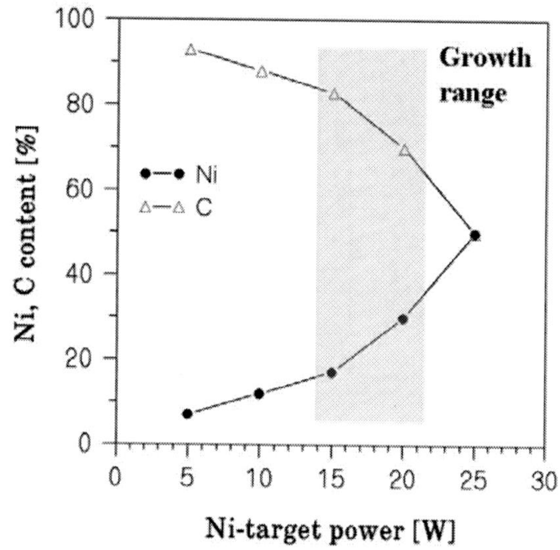


Figure 4. The dependence of Ni and C content in the films (in at.%) as a function of Ni-target power.

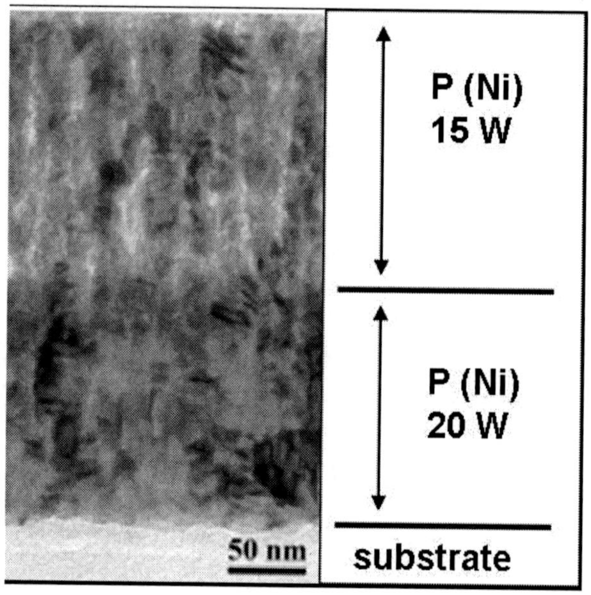


Figure 5. Cross-sectional TEM image of C-Ni nanocomposite thin film prepared at 20°C with different Ni-target power showing columnar character of crystalline grains (dark) and thin carbon matrix between columns (light).

Figure 5 shows a thin film deposited at two different Ni source powers. The lower part was deposited at 20 W of Ni target. The content of Ni was 30 at%. The upper half of the film was deposited at 15 W of Ni source. The EDS analysis showed 82 at% of C and 18 at% of Ni. The upper layer appears to be a more regular nanocomposite with a better continuity of a thicker matrix phase and less defined metallic crystallites. The column width is in a range of 10 ÷ 30 nm.

# Properties of Carbon – Metal (Ni, Ti) Thin Films

## Structure of Carbon-Nickel Thin Films

TEM analysis of the films prepared between 25°C and 400 °C revealed a columnar structure of nanocomposite (Fig. 6 a-c). These thin films were composed of columnar $Ni_3C$ crystalline grains and a disordered carbon matrix (width of 1 ÷ 3 nm), localised between the columnar crystallites. In the case of the film deposited at 200°C the column length is practically equal to the film thickness. With increasing the C content the column length decreases slightly (Fig. 6b). The increase of deposition temperature to 400°C caused partial ordering in the carbon matrix. This structure, called as graphite-like carbon, is built from graphene layers of ~ 0.35 nm spacing (Fig.6c and Fig.7).

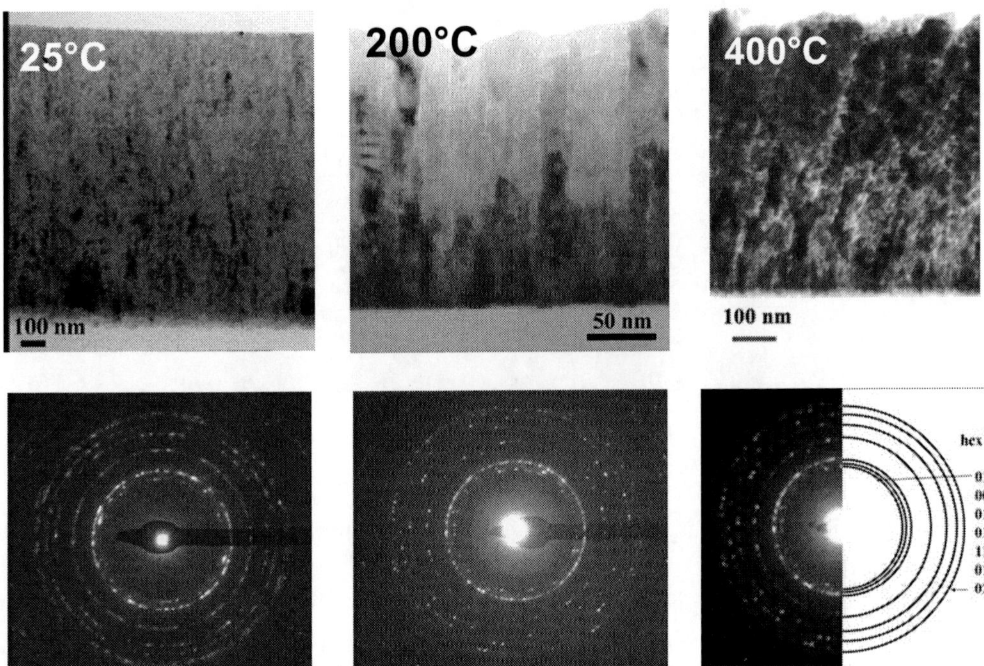


a) b) c)

Figure 6. Cross section TEM images of C-Ni nanocomposite films prepared by dc magnetron sputtering at different deposition temperatures and their electron diffractions. a) 25°C, b) 200°C and c) 400°C. The diffraction pattern shows the presence of crystalline phase, which can be either hexagonal $Ni_3C$.

The crystallites are of hexagonal symmetry fitting $Ni_3C$ (JCPDS 04-0853) unit cell. The similarity of these unit cells does not allow us to clearly distinguish between these two phases by electron diffraction. Due to the small grain size, the phases cannot be resolved either by XRD analysis. However, as discussed in [23], we assume that the crystalline phase observed in films prepared below 400°C can be identified as the metastable $Ni_3C$ phase. The existence of the carbide phase is supported by the ESCA measurements reported in [24].

In the samples deposited at room temperature the carbon matrix can be considered practically amorphous.

The dissolution of C in the $Ni_3C$ type lattice is deduced from the volume ratio of the crystalline and inter-grain carbon phases exceeding substantially the Ni:C ratio (being about 1:4) corresponding to the composition of the film. HREM analysis of the thin films revealed that carbon was present in the matrix as amorphous (a-C) and/or graphite-like carbon (GLC). Gradual ordering into GLC displaying 5 ÷ 10 graphene layers of 0.35 nm spacing between two crystalline grains (Fig. 7) was observed at deposition temperature 200°C.

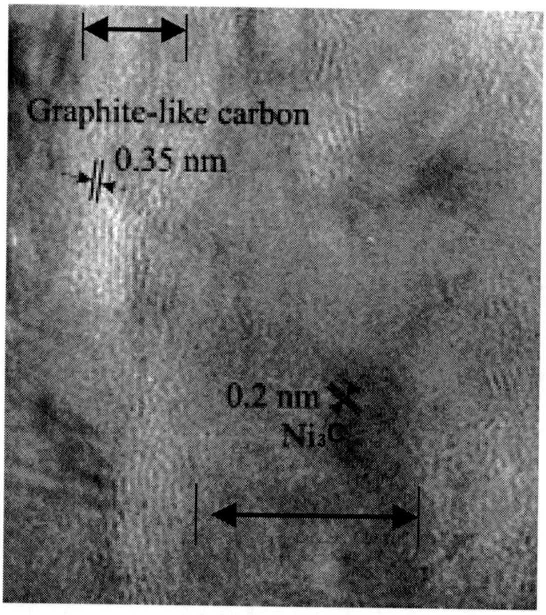


Figure 7. HREM image of C(Ni) nanocomposite film prepared at 200°C shows graphene layers of GLC between the $Ni_3C$ crystalline grains.

The deposition temperature 200°C is a typical temperature for the inset of the growth of the GLC phase [4].

In the films prepared at 600 and 800°C the columnar morphology of crystalline grains changed to globular one (Fig. 8a-b). The stable fcc Ni phase formed (Fig. 8) [25].

The innermost diffraction ring belongs to carbon, corresponding to the (002) reflection of graphite, originating from a relatively large volume of ordered graphene layers. From the asymmetry of intensity in this ring, one can conclude that the graphene layers are oriented mainly parallel to the surface normal. The HREM cross-section image revealed that the carbon matrix separating adjacent Ni grains consisted of graphene layers that encapsulated the globular Ni grains. Figure 10 shows that the small crystalline Ni grains are encapsulated into a ordered GLC matrix.

The phase change from hexagonal $Ni_3C$ to fcc Ni in the film occured simultaneously with the change of columnar structure to globular. The stable Ni phase does not dissolve carbon, and carbon atoms are segregated to form the carbon phase. As a result, the number of carbon atoms forming the matrix increases compared to the case of $Ni_3C$ formation. The forming carbon matrix encapsulates the Ni grains preventing their columnar growth. Therefore, the globular morphology is due to C segregation and to repeated nucleation of the Ni phase [26]. The size of Ni grains is between 10 ÷ 100 nm.

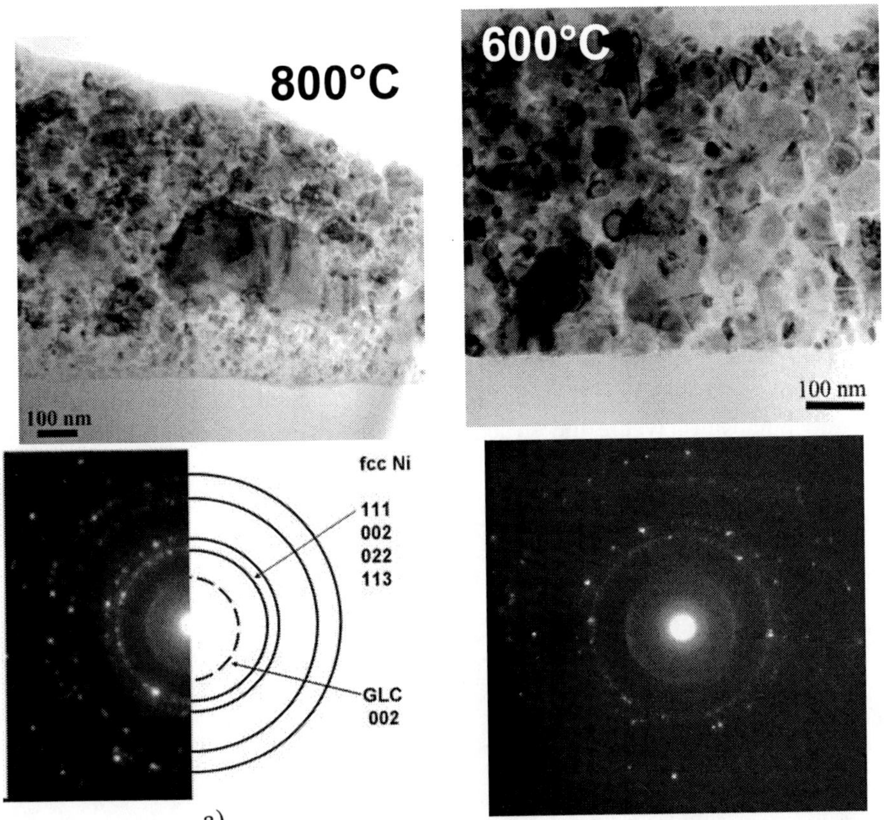


a) b)

Figure 8. Cross-section TEM image of C-Ni nanocomposite films prepared at 600°C (a) and 800°C (b) and their electron diffractions. The images show the globular structure of Ni grains. The electron diffraction reveals the presence of the GLC phase and fcc Ni.

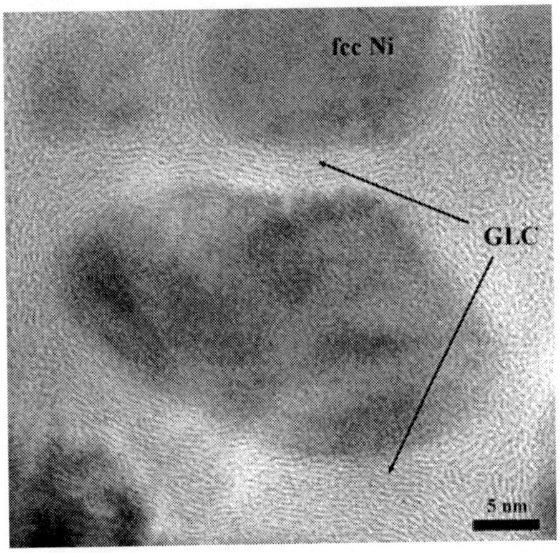


Figure 9. HREM image of the C-Ni nanocomposite film prepared at 600°C shows ordered graphene layers of GLC between fcc Ni grains. The atomic distance between grapheme layers is 0.35 nm.

As could be expected, we found that deposition from two separate magnetrons (C and Ni) at somewhat oblique deposition or from one composite C-Ni target, as was described in [23], did not drastically influence the morphology and phases of the C-Ni nanocomposite films. Even varying of the composition in the range shown in Fig. 3 leads just to quantitative changes in the ratio of the crystalline (Ni or $Ni_3C$) to disordered (carbon) phases, however, the overall morphology and phases remain unchanged (Fig. 5).

## Structure of Carbon-Titanium Thin Films

As revealed by TEM analysis, all the films prepared at temperatures ranging from 25 ˚C to 800 ˚C had a columnar structure (Fig. 10 – Fig. 13). This situation is different from the case of C-Ni nanocomposites, where the structure of films depends on deposition temperatures (Fig. 6-9).

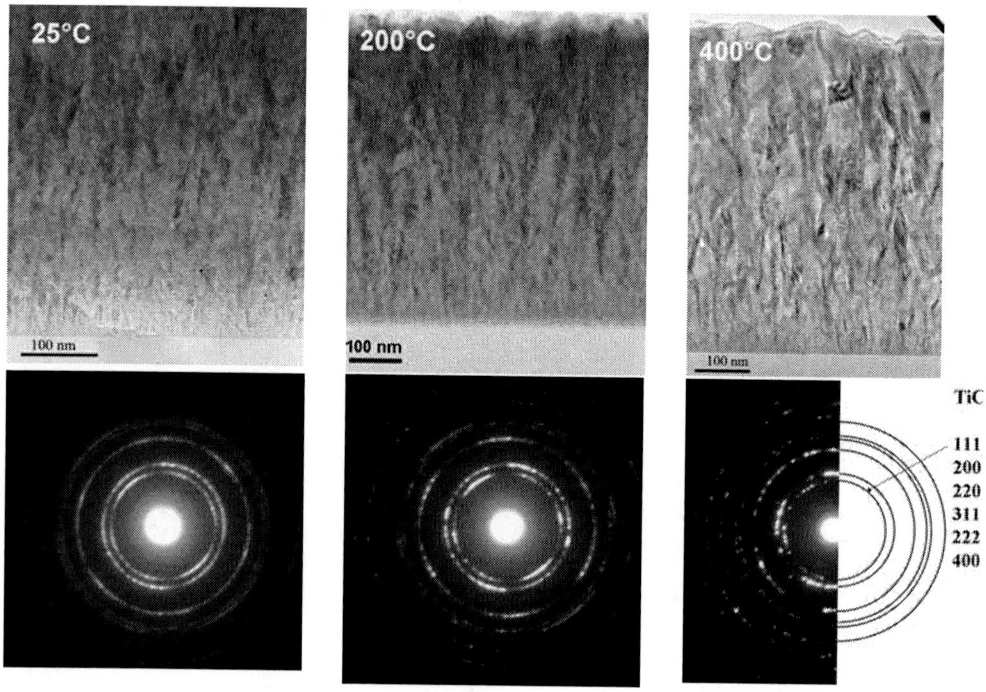


Figure 10. Cross section TEM images of C-Ti nanocomposite films prepared by dc magnetron sputtering at different deposition temperatures and their electron diffractions. a) 25°C, b) 200°C and c) 400°C. The diffraction pattern shows the presence of crystalline phase, which is cubic TiC.

EDS elemental analysis of film composition showed that all the films deposited between 25 °C and 800 °C contained ~ 80 at% C and 20 at% Ti. According to Balden et al. [27], the optimal content from structure (crystallites separeted by thin carbon matrix) of the doped C films is one with a low metal concentration within 1 – 20 at%.

The cross section TEM images of C-Ti nanocomposites deposited between 25°C and 400 ˚C are shown in Fig. 9. The collective peculiarity of these films is a columnar crystalls separated by thin carbon matrix.

The increasing of deposition temperature caused the flare of crystalline columns from ~ 10 nm (25°C) to ~ 40-50 nm (400°C). The electron diffraction showed the TiC phase of crystallites (card no. JCPDFWIN 32_1383) (Fig. 10).

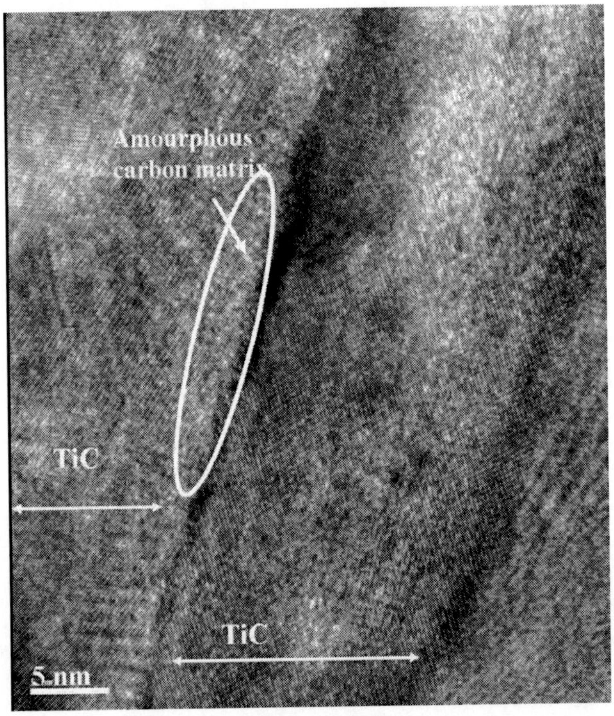


Figure 11. Cross section HREM image of C-Ti film prepared at 200°C shows the colmnar structure of TiC crystalls and thin amorphous carbon matrix between columns.

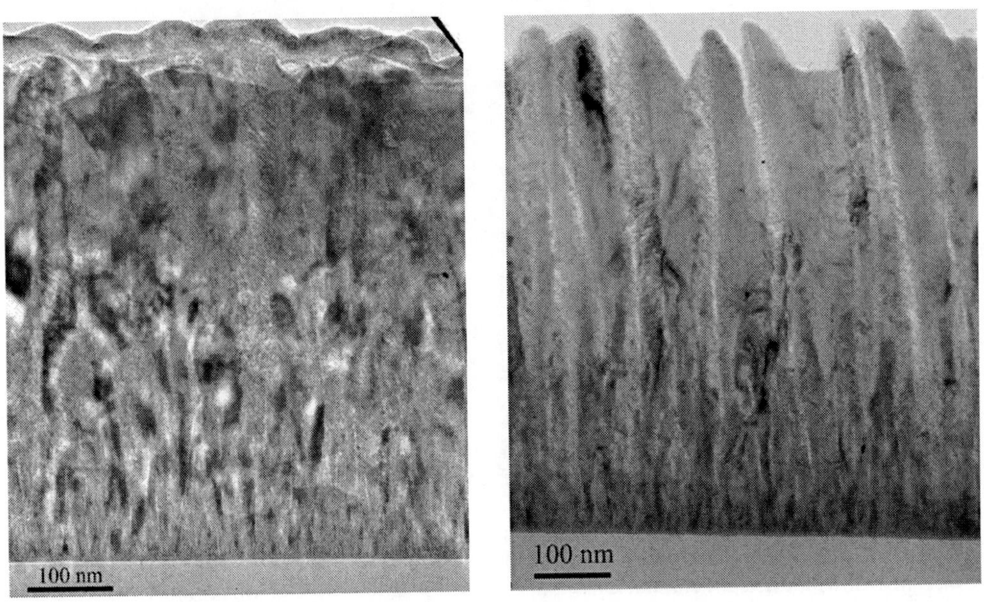


Figure 12. Continued on next page.

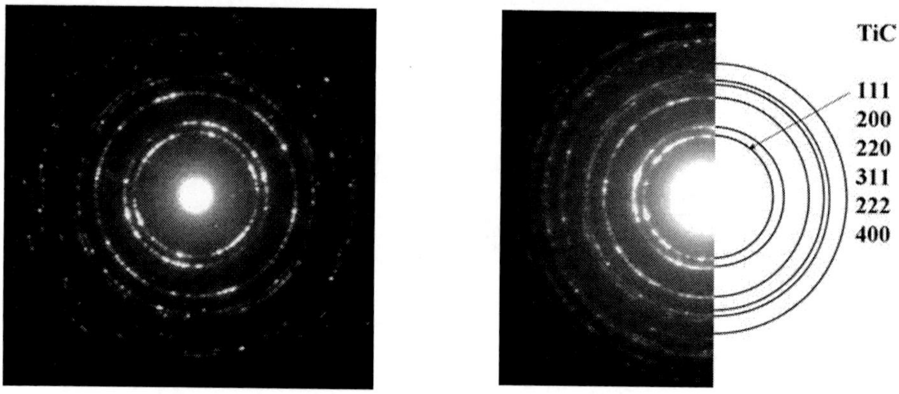


Figure 12. Cross section TEM images of C-Ti nanocomposite films prepared by dc magnetron sputtering at different deposition temperatures and their electron diffractions. a) 600°C and b) 800°C. The diffraction pattern shows the presence of crystalline phase, which is cubic TiC.

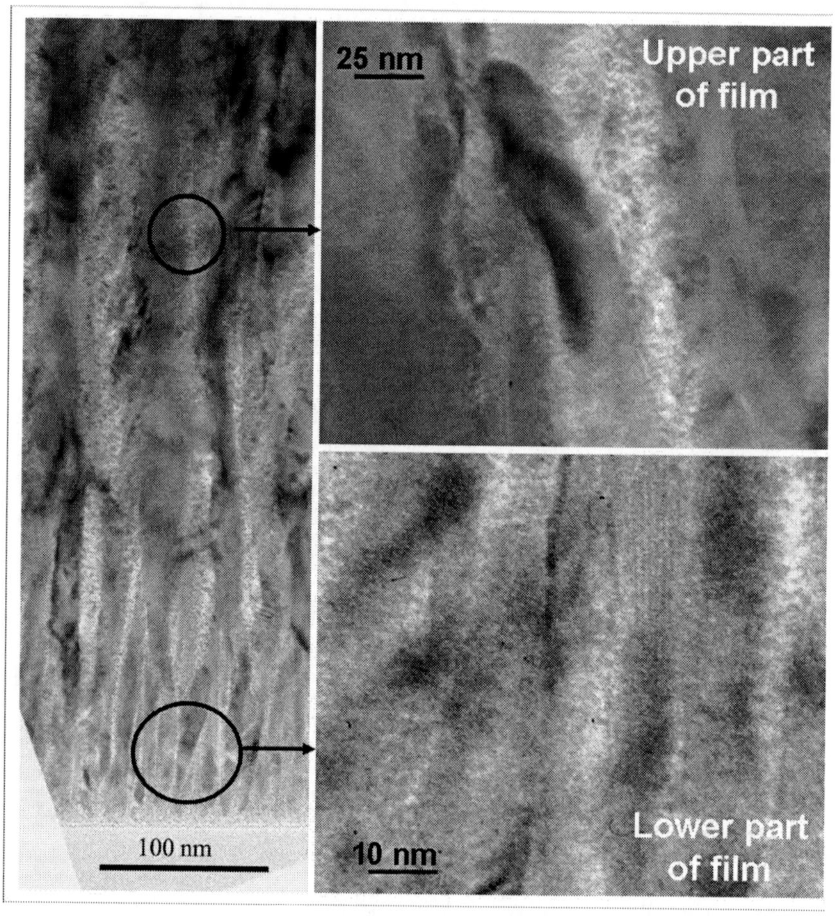


Figure 13. Cross section TEM image of C-Ti nanocomposite prepared at 800°C. The film shows columnar growth with different column and matrix thickness at bottom and top regions.

Cross section HREM image of nanocomposite prepared at 200°C shows the atomic layers of TiC column which atomic distance is 0.25 nm (Fig. 11) and ~ 2 nm thin amorphous carbon matrix between columns.

Further increasing of deposition temperature resulted in growth of the TiC columns (Fig. 12-13) close to 100 nm width near the surface of the films. The thickness of the carbon matrix between adjacent TiC columnar crystallites was ~ 10-20 nm (Fig. 13).

The electron diffractions of films prepared up to 600°C confirm the cubic TiC phase of crystallites (Fig. 12). In all films were attributed mainly to (111), (200) and (220) diffraction maximum from TEM electron diffraction. This result is in the correlation with work of Zhang et. al [28], they were observed no dominant texturing in TiC, too.

## Mechanical Properties of Carbon-Nickel Thin Films

The mechanical properties, hardness (H) and reduced modulus of elasticity (E) of C-Ni films strongly depends on the deposition temperature [29]. The hardness of nanocomposites ranges from 2 GPa to 11±2 GPa (Fig. 14). The highest hardness ~11 ± 2 GPa was measured for thin films prepared at 200°C.

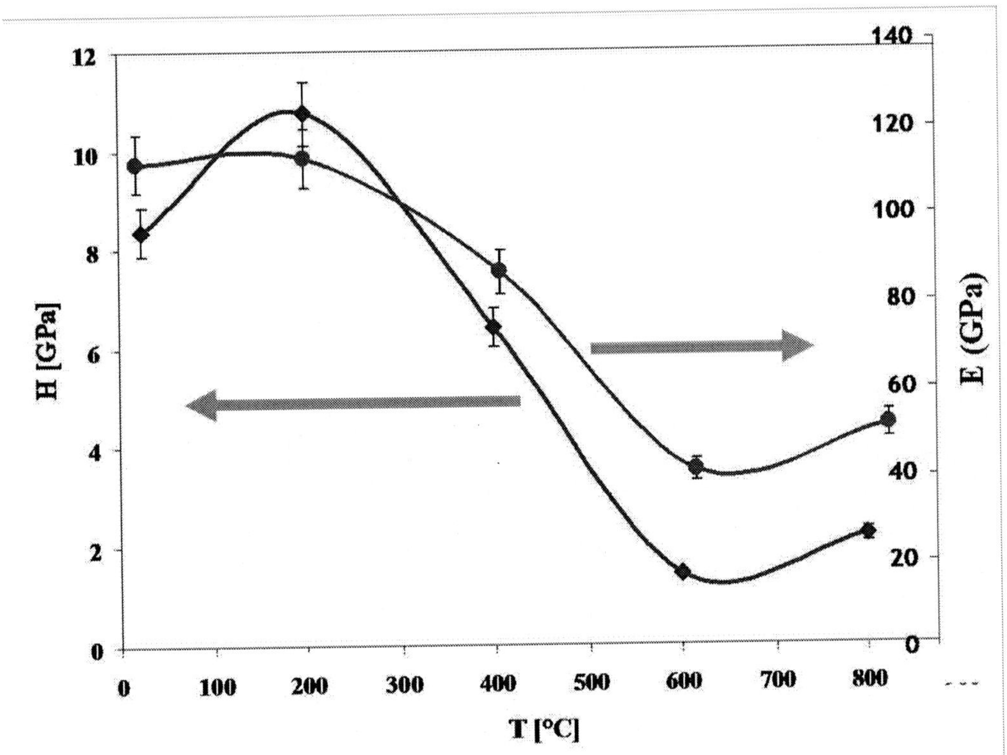


Figure 14. Hardness (H) and reduced modulus of elasticity (E) of C-Ni nanocomposites as a function of depostion temperature.

The dependence of the reduced modulus of elasticity on the deposition temperature is similar to that of hardness (Fig. 14). The values of the elasticity modulus ranged from 40-50 GPa to ~110 GPa. The highest elasticity modulus was determined in the nanocomposite prepared at 200°C, alike to the hardness. With the increase of deposition temperature above 200°C the elasticity modulus of the film decreases.

The H/E ratio is used to describe the failure mechanism of the materials. If the deformation of the films is dominated by plastic or elastic deformation, it shows H/E > 0.1 [30]. The ratio H/E of C-Ni composites was measured between 0.04 and 0.09 (Table 1). The elastic recoveries of C-Ni films ranging between 0.18 and 0.47 were also determined in the nanoindentation tests. The highest H/E ratio ~ 0.1 and elastic recovery ~ 0.5 indicate that the deformation of the composites arises mainly from elastic deformation of the matrix which determines the elasticity of the asperities in a tribological contact [30].

**Table 1. H/E ratio and the elastic recovery (ERP) of C-Ni composite thin films deposited at 15 W Ni magnetron power with 17 at.% Ni content at different deposition temperature.**

| $T_{substrate}$ (°C) | 25 | 200 | 400 | 600 | 800 |
|---|---|---|---|---|---|
| H / E | 0.074 | 0.094 | 0.074 | 0.033 | 0.042 |
| ERP | 0.36 ± 0.04 | 0.47 ± 0.08 | 0.36 ± 0.06 | 0.15 ± 0.03 | 0.18 ± 0.05 |

Plan view HREM images of films deposited at 200°C are shown in Fig 15. These films were prepared at the same temperature but at different input power of the magnetron with Ni target (5 – 25W). The aim was the verification of the effect of Ni content on the structural and mechanical properties.

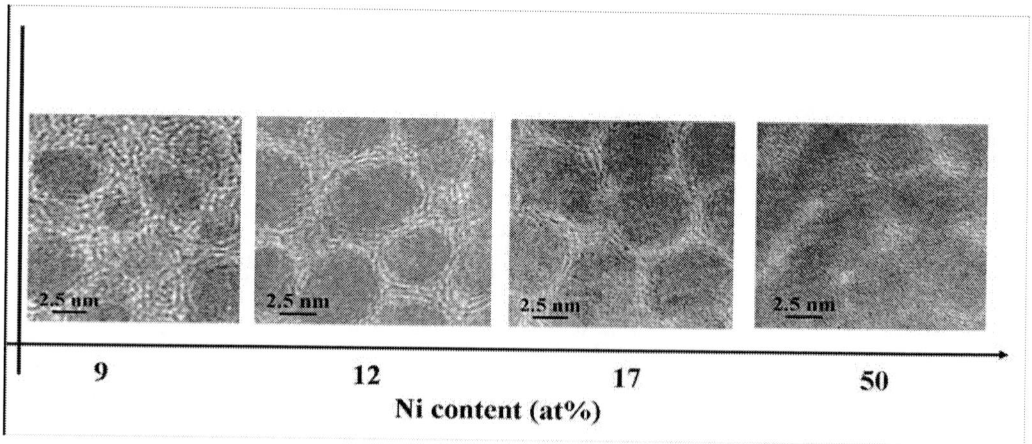


Figure 15. Plan view HREM images of 20 nm thick C-Ni nanocomposite films deposited at 200°C as a function of the Ni content.

The size of crystalline grains increases with Ni content in the films. The films of 9 at% Ni content consist of 3-5 nm $Ni_3C$ grains embedded in 2 nm thick carbon matrix. This carbon matrix consists of ordered regions with 3 – 5 graphene layers with spacing ~ 0.35 nm and also disordered regions.

The increase of Ni content resulted in increase of $Ni_3C$ grain size ( ~ 5 - 6 nm), when at 50 at% of Ni content coalescence of the crystallites started and a network of $Ni_3C$ crystallites formed. The $Ni_3C$ serves as matrix and C is the embedded phase. On the other hand, the increase of Ni content in the films and growth of average size of crystalline grains resulted in the decrease of carbon matrix thickness. In the films with Ni contents between 12 at% and 17 at% , the carbon matrix consists only of ordered thin carbon phase. The columnar structure was characteristic for all these thin films.

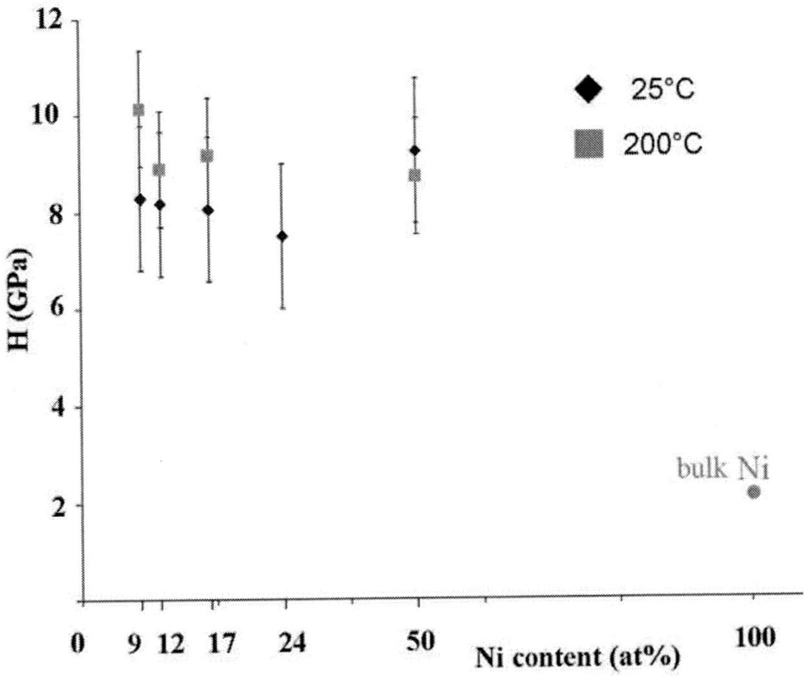


Figure 16. Hardness of C-Ni composite thin films vs. Ni content (at.%). The hardness of pure bulk nickel is 2 GPa.

As it is shown in Fig. 16, the hardness of these films is not influenced by the Ni content. With increasing Ni content from 9 to 50 at%, the hardness of films was practically constant. From this fact and the application of the rule of mixtures [31] to these results we can conclude, that the hardness of the matrix and embedded $Ni_3C$ crystallites must be practically the same. As the grain size of the crystalline phase changes at least by a factor of two, no Hall-Petch behavior of the composite is observed [32]. This involves a deformation mechanism, activating sliding of grains at the interfaces. This latter observation supports also the statement, i.e. the hardness of the studied C-Ni films was determined by the hardness of the carbon matrix. This thin carbon matrix (2- 3 nm) which isolates the columnar crystallites consists of disordered and graphite-like carbon (Fig. 7).

The situation in the case of the dependence of the modulus of elasticity vs. Ni content is different from hardness as showsn in Fig. 17. The reduced modulus of elasticity increases with increasing Ni content giving an extrapolation value equal to that of bulk Ni (E(Ni) = 205

GPa). The application of the rule of mixtures to this situation gives a result for the modulus of the carbon matrix around ~ 100 GPa, which corresponds to measurements on $CN_X$ and DLC films [4].

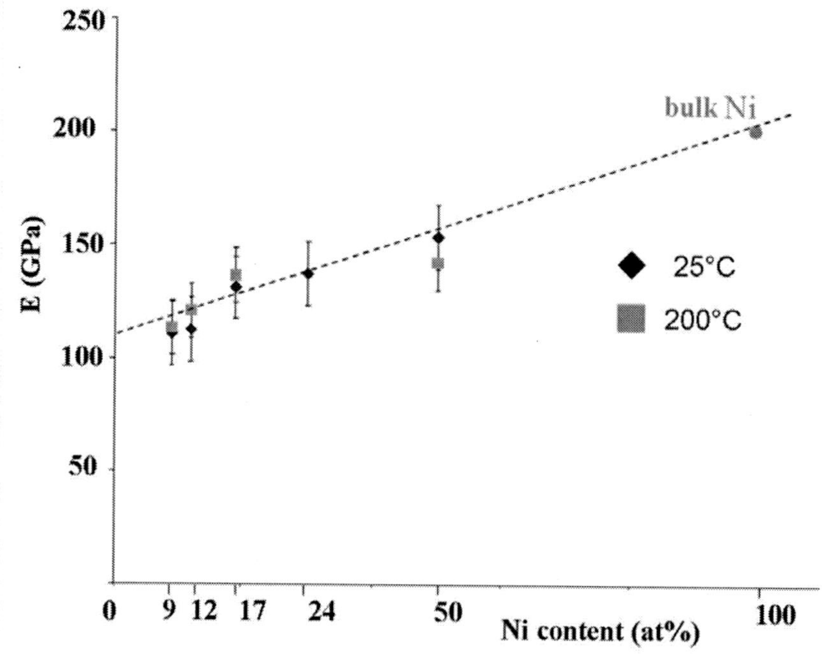


Figure 17. Reduced modulus of elasticity of C-Ni composite thin films vs. Ni content (at%). The modulus of elasticity of pure bulk nickel is 205 GPa.

## Mechanical Properties of Carbon- Titanium Thin Films

The hardness and the reduced modulus of elasticity of C-Ti nanocomposite thin films showed a distinct variation depending on the deposition temperature (Fig. 18). The H and E values are quite high in the interval of deposition temperatures from 25 °C to 400 °C. But the film deposited at 200 °C had the highest hardness ~ 18 GPa and the highest modulus of elasticity ~ 205 GPa. At 600 °C these values drop abruptly stepwise. The similar behaviour was found also in the C-Ni nanocomposites [8].

**Table 2. H/E ratio and the elastic recovery parameter (ERP) of C-T composite thin films deposited at 40 T magnetron power with 17 at.% Ti content at different deposition temperature.**

| $T_{substrate}$ (°C) | 25 | 200 | 400 | 600 | 800 |
|---|---|---|---|---|---|
| H/E | 0.079 | 0.086 | 0.085 | 0.052 | 0.052 |
| ERP | 0.41 ± 0.07 | 0.47 ± 0.09 | 0.4 ± 0.14 | 0.25 ± 0.07 | 0.26 ± 0.06 |

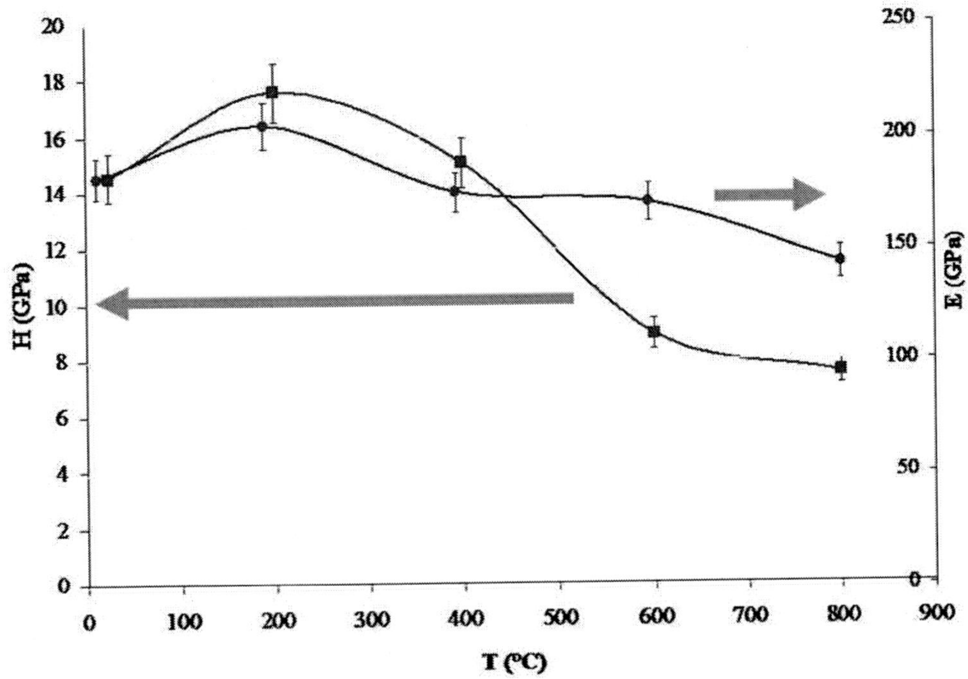


Figure 18. Hardness (H) and reduced modulus of elasticity (E) of C-Ti nanocomposites as a function of depostion temperature.

The H/E ratio is used to describe the failure mechanism of the materials. The ratio H/E of C-Ti nanocomposites was measured between 0.052 and 0.086 (Tab. 2). The elastic recoveries of C-Ti films ranging between 0.25 and 0.47 were also determined in the nanoindentation tests. The highest H/E ratio ~ 0.1 and elastic recovery ~ 0.5 indicate that the deformation of the nanocomposites arises mainly from elastic deformation of the C matrix which determines the elasticity of the asperities in a tribological contact [30]. The large H/E and ERP (Elastic Recovery Parameter) values predict good wear-resistant properties of the C-Ti nanocomposite films.

## Electrical Properties of Carbon-Nickel Thin Films

The structural studies revealed four constituents of different electrical conductivity: Amorphous carbon, graphite-like carbon, $Ni_3C$, and Ni. They are ranked according to the increasing conductivity. Depending on deposition temperature their content in the corresponding composite film varies and so does the type of conductivity.

The results of R(T) measurements normalized to R(300 K) are shown in Fig. 18. We found that all the values of room-temperature resistivities lie within the interval given by the resistivities of the nanocomposite constituents, i.e. graphite (or amorphous carbon) and nickel. In table 3, the values of the temperature coefficient of resistivity (TCR) determined from the R(T) dependences, together with relevant structural characteristics are also shown. Though the variation in the resitivities or TCR are not large, we show in the following that even these minor differences provide useful information, and they can be clearly correlated with the structural data.

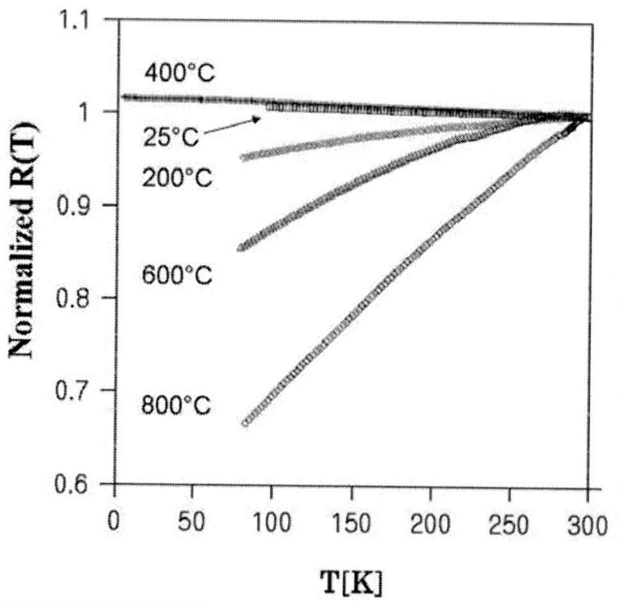


Figure 19. Temperature dependence of the resistivity of C-Ni nanocomposite thin films prepared at substrate temperature from 25 to 800°C.

The films deposited at 600 and 800°C show positive TCR, which implies a metallic type of conductivity (Tab. 3). It means that in a film there must exist at least a single percolation conduction path formed by particles with metallic character.

The nanocomposite thin films prepared between 25 and 400°C have a slightly negative TCR of about (-4 ÷ -6).$10^{-5}$ $K^{-1}$, which can be explained by the fact that the carbon matrix surrounding metallic particles is disordered (a-C was identified by TEM), and thus much less-conductive than GLC found mainly in films prepared above to 600°C. Since the amorphous carbon barriers are very thin (1 ÷ 3 nm) the electrons can tunnel across them. If the carbon separator were thick, one would expect an R(T) dependence similar to that of semiconductors – a steep increase of resistivity with decreasing temperature. The presence of the tunneling effect is clearly proved by the G(V) = dI/dV vs bias voltage curves obtained at low temperatures. They are of parabolic shape (Fig. 20) for both films showing TCR < 0. The presence of the tunneling effect is in accord once with TCR < 0 because at higher temperatures some electrons are thermally activated to tunnel, which enhances the probability of a tunneling event and results in a decrease in the resistance. These films can be considered to be a 2D network of tunnel junctions of random barrier parameters. To be more precise, we cannot exclude that some small regions in the film have eventually metallic type of conductivity (because TCR is smaller than that in the Nb/Si junctions), but we can argue that these small regions do not form a percolation path.

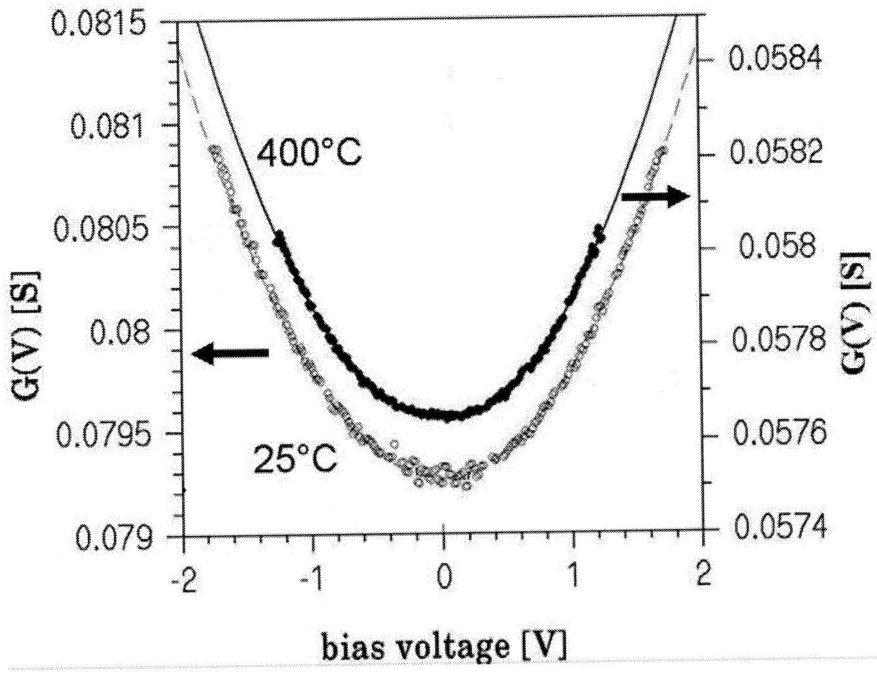


Figure 20. G(V) dependence obtained at 79K for C(Ni) nanocomposite prepared at 400°C (25°C) respectively, is a shallow parabola, providing the evidence for the tunneling of electrons across thin amorphous carbon barrier.

If we take into account that in the matrix, the ratio of GLC to be a-C content increases generally with increasing deposition temperature, we can consider the behaviour of the sample prepared at 400°C rather non-consequent because, according to the tendency mentioned, one would expect even higher positive TCR compared with that of the sample prepared at 200°C.

A mixture of GLC and amorphous carbon in the matrix was found also in the sample prepared at 200°C, but in this case the well-conducting GLC regions connected directly the adjacent $Ni_3C$ nanoparticles (or better said nanowires in this sample). Therefore, it is supposed that the GLC "interconnections" are responsible for the slightly positive TCR of the film prepared at 200°C. Since the aspect ratio is so high (1:60) we assume that even a direct metallic contact between some adjacent nanowires is likely, because of the half-resistivity of this sample (see Tab. 3).

**Table 3. Estimated resistivity (ρ) and temperature coefficient of resistivity (TCR) of C-Ni nanocomposite thin films as a function of deposition temperature ($T_{substrate}$).**

| $T_{substrate}$ [°C] | 25 | 200 | 400 | 600 | 800 |
|---|---|---|---|---|---|
| TCR | $-0.4\ 10^{-5}$ | $3.0\ 10^{-4}$ | $-0.66\ 10^{-4}$ | $10\ 10^{-4}$ | $15\ 10^{-4}$ |

The values for pure bulk nickel are ρ = 6.8 μΩcm, TCR = 0.006 and for graphite ρ = 3.6 mΩcm, TCR = -0.0005.

The nanocomposites prepared at 25 and 400°C were also tested for the presence of tunneling magnetoresistivity (TMR effect [33]) since the tunneling effect was proved by G(V) curves shown in Fig. 8. However, we obtained straight R(B) lines within the available external field $B_{max} = \pm 0.5$ T. The explanation why these films have not shown the expected negative magnetoresitivity was found by electron diffraction: the $Ni_3C$ type crystalline nanograins were of hcp phase (Fig. 6), which is a nonmagnetic form of nickel.

Since Ni crystallites in the C(Ni) prepared between 600°C and 800°C were found by electron diffraction to be of fcc phase, which is magnetic, we decided to measure magnetoresistivity in the film prepared at 600°C. Indeed, the shape of the R(B) curves in parallel and in perpendicular fields (Fig. 21) indicates the so-called anisotropic magnetoresistivity (AMR), which is encountered in materials containing transient elements [34]. The AMR value is very low ~ 0.1% when compared with AMR ~ 3% commonly achievable in permalloy sensors [34]. The ferromagnetism in these samples was also confirmed by Magneto-Optical Kerr Effect (MOKE) measurements.

The bell-like shape of the R(B) curve in the perpendicular field can be fitted with the common formula [35]: $R(B)=R_o + \Delta R[1-(B/B_o)^2]$, where $R_o$ is the resistance in zero applied field, $B_o$ is the sum of demagnetizing and anisotropy fields, and $\Delta R$ is the peak-to-peak value obtained from the angular dependence of the resistivity on applied field. The fit is shown in Fig. 21.

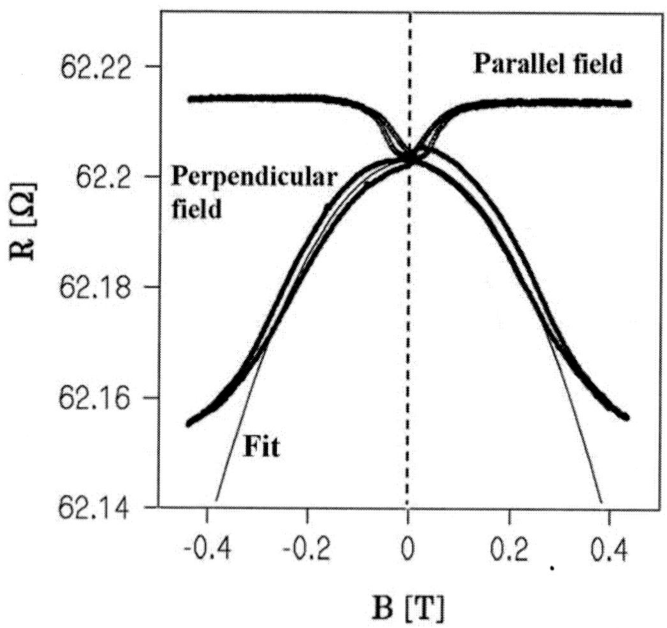


Figure 21. Dependence of the resistivity on parallel and perpendicular fields for C(Ni) nanocomposite prepared at 600°C. The shape of the R(B) curve indicates the so-called anisotropic magnetoresistivity (AMR).

## Electrical Properties of Carbon-Titanium Thin Films

The resistance dependence on the temperature was measured in the interval of 80-330 °K for all the films deposited at different temperature.

The results of R(T) measurements, normalized to R(300 °K) are shown in Fig. 22. The C-Ti nanocomposite shows different values of the temperature coefficient of resistivity (TCR) for various deposition temperatures. There is a relation between the sign of TCR and structure: the films deposited at temperatures ranging from 25 to 600 °C have a negative TCR, which can be explained by the structural observation that the thin carbon matrix is disordered (amorphous). Recalling our recent paper [8], we suppose that in transport a tunnelling current can play some role since the amorphous carbon can be considered a barrier between adjacent metallic columns.

The film prepared at 800 °C has a slightly positive TCR, which implies different mechanisms of electrical transport. The film shows a metallic type of conductivity which implies that in the film at least a percolation conduction path formed by metallic particles TiC exists.

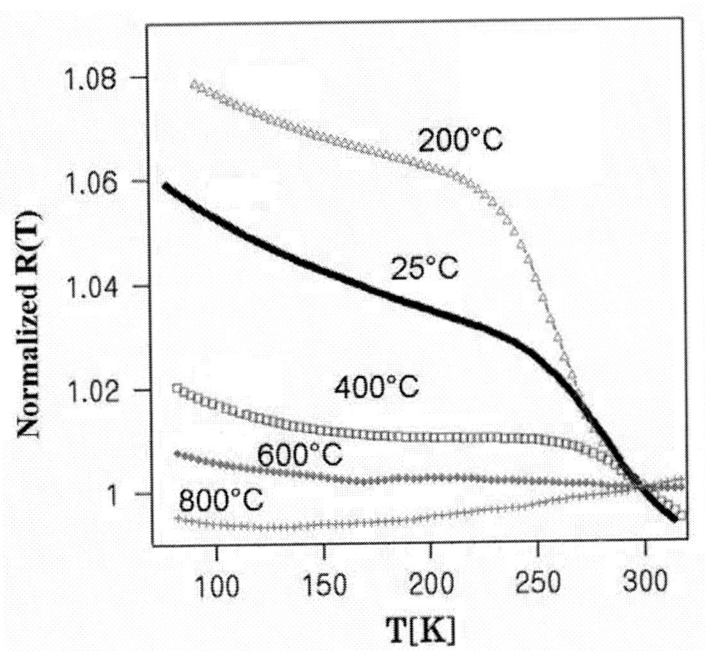


Figure 22. Temperature dependence of the normalized resistivity R(T) of C-Ti nanocomposite thin films prepared at substrate temperature from 25 to 800 °C.

The most peculiar R(T) dependence was seen for the film deposited at 200 °C, which has also extraordinary mechanical properties. The high slope of the R(T) curve in the interval of 250 – 300 °K can be related to strong non-linearity of I(V) (current-voltage dependence) curves (not shown in this paper), the origin of which is not clear at the moment.

# Conclusion

Carbon- metal (Ni, Ti) nanocomposite thin films were prepared by dc magnetron sputtering at temperatures between 25 and 800°C. The C-Ni films deposited below 400°C show a columnar structure of hexagonal $Ni_3C$ type crystalline grains embedded in a matrix consisting of an amorphous and/or graphite-like carbon. Above 400°C the nanocomposite consists of globular fcc Ni grains sized between 50 and 100 nm that were separated by the ordered GLC phase. Above 400°C the metastable $Ni_3C$ phase becomes unstable and the stable fcc Ni phase forms.

In the case of C-Ti nanocomposites the situation is different. All C-Ti films (except for the one prepared at 800 °C) consisted of TiC columns 10 – 100 nm wide that were embedded in a carbon matrix. The thickness of the carbon matrix between adjacent TiC columns was ~ 2 – 5 nm.

The studies of C-Ni and C-Ti nanocomposites films showed that the high hardness and modulus of elasticity are consequences of columnar structure and the combination of graphite-like and amorphous carbon phase in the matrix. The best mechanical properties were observed in both cases, in the films composed of columnar $Ni_3C$ crystallites (C-Ni) or TiC columns (C-Ti), with column length practically equal to the film thickness, separated by thin carbon matrix. The carbon matrix consisted of disordered and ordered graphite-like carbon phase.

The H/E ratio of ~ 0.09 ÷ 0.1 and elastic recovery of 0.5 (in C-Ni and C-Ti) indicate that the deformations of the films arise mainly from deformation of the matrix which determines the elasticity of the asperities in a tribological contact. As was shown on film series prepared at low deposition temperature (25°C and 200°C) at various power of Ni target, the hardness of films was independent of Ni content in nanocomposite, suggesting an interface sliding deformation mechanism. On the contrary, the elasticity modulus shows a linear dependence on Ni content of the films corresponding to the rule of mixtures [36].

A correlation was found between the structure and electrical properties of the C-metal nanocomposites. The films prepared at room temperature, in which the prevailing matrix material is the disordered carbon exhibit the tunneling effect, and they can be considered an insulator-metal composite with a 2D random network of tunnel junctions consisting of metallic $Ni_3C$ or TiC nanoparticles separated by thin amorphous carbon barriers. In the films grown at high deposition temperature (600 and 800°C), where fcc Ni globular crystallites and TiC non-linear columns are separated by well-conductive GLC no tunneling effect was revealed by differential conductance measurements, and the nanocomposite shows metallic behaviour with positive TCR.

## Acknowledgements

This work was supported in part by the Hungarian National Science Foundation (OTKA T-048699 and T-043359) and by the European Community´s Human Potential Program under contract HPRN-CT-2002-00209, (New Fullerene-like Materials). Katarína Sedláčková and Róbert O. Grasin acknowledge the financial support of the same EU project.

# References

[1] Yang, S.; Teer, D.G.; *Surf. Coat. Technol.* 2000, 131, 412-416.
[2] Laidani, N.; Miotello, A.; Perrière, J.; *J Appl. Surf. Sci.* 1996, 99, 273-284.
[3] Patscheider, J.; Nanocomposite hard coatings for wear protection, *MRS Bull* 2003, 28 (3), 180-184.
[4] Hultman, L.; Neidhardt, J.; Hellgren, N.; Sundgren, J.E.; Fullerene-like carbon nitride: A resilient coating material. *MRS Bulletin* 2003, 28 (3), 194-201.
[5] Karabutov, A.V.; Godeev, S.K.; Ralchenko, V.G.; Korchagina, S.B.; Lavrishev, S.V.; Vlasov, I.I.; *Diamond and Relat. Mater.* 2003, 10-11, 1698-1704.
[6] Vepřek, S.; *J. Vac. Sci. Technol. A* 1999, 17(5), 2401 - 2420.
[7] Radnóczi, G.; Kovács, G.J.; Pécz, B.; Geszti, O.; Sáfrán, G.; *Vacuum* 2003, 71, 185-91
[8] Sedláčková, K.; Lobotka, P.; Vávra, I.; Radnóczi, G. ; *Carbon* 2005, 43, 2192 - 2198.
[9] Yang, S.; Jones, A.H.S.; Teer, D.G.; *Surf. Coat. Technol.* 2000, 133-134, 369-375.
[10] Zhang, S.; Bui, X.L.; Jiang, J.; Li, X.; *Surf. Coat. Techol.* 2005, 198, 206 - 211.
[11] Bewilogua, K.; Cooper, C.V.; Specht, C.; Schroder, J.; Wittorf, R.; Grischke, M.; *Surf. Coat. Technol.* 2000, 127, 224- 231.
[12] Monaghan, D.P.; Teer, D.G.; Logan, P.A.; Efeoglu, I.; Arnell, R.D.; *Surf. Coat. Technol.* 1993, 60, 525- 530.
[13] Oguri, K.; Arai, T.; *Surf. Coat. Technol* 2001, 47, 710 - 721 .
[14] Gago, R.; Abrasonis, G.; Mücklich, A.; Möller, W.; Czigany, Z.; Radnoczi, G.; *App. Phys. Lett.* 2005, 87, 7190 – 7191.
[15] Baba, K.; Hatada, R.; *Surf. Coat. Technol.* 2002 ,158-159, 373-376.
[16] Movchan, B. A., Demschichin, A.V., *Fiz. Met. Metalloved.*, 1969, 28, 83-89.
[17] Thornton, J. A. ; *Ann. Rev. Mater. Sci.*, 1977, 7, 239 - 260.
[18] Messier, R.; Giri, A. P.; Roy, A. R.; *J. Vac. Sci. Technol.*, 1984, A2, 500-503.
[19] Grovenor, C. R. M.; Hentzell, H. T. G.; Smith, D. A.; *Acta Metall.*, 1984, 32, 773-781.
[20] Petrov, I.; Barna, P. B.; Hultman, L.; Greene, J. R.; *J. Vac. Sci. Technol.*, 2003, A21, 117 - 127.
[21] Kusano, Y.; Hutchings, I.M.; *Surf Coat Technol* 2003,169-170,739-742.
[22] Oliver, W.C.; and Pharr, G.M.; *J. Mater. Res.* 1992, 7, 1564-1583.
[23] Radnóczi, G.; Kovács, G.J.; Sáfrán, G.; Sedláčková, K.; Geszti, O.; Ujvári, T.; et al. In: Senkov ON, et al, editor. *Metallic Materials with High Structural efficiency*, Netherlans; Dordrecht: Kluwer Academic Publisher, 2004, 101-112.
[24] Kovács, G.J.; Sáfrán, G.; Geszti, O.; Ujvári, T.; Bertóti, I.; Radnóczi, G.; Surface and *Coating Technology* 2004, 180-181, 331-334.
[25] Shi, J.; Nittono, O.; J. of. Mat. Science Lett. 1996, 15, 928-930.
[26] Barna, P.B.; and Adamik, A.; Thin Solid Films 1998, 317, 27-33.
[27] Balden, M.; Cieciwa, B.T.; Quintana, I.; de Juan Pardo, E.; et al. *Surface & Coatings Technology* 2005, 200, 413- 417.
[28] Zhang, S.; Bui, X.L.; Jiang, J.; Li, X.; *Surface & Coatings Techn.* 2005, 198, 206-211.
[29] Ujvári, T.; Tóth, A.; Kovács, G.; Sáfrán, G.; Geszti, O.; Radnóczi, G.; et.al. *Surf Interface Anal* 2004, 36, 760-764.
[30] Neidhadrt, J.; Czigány, Z.; Brunell, I.F.; Hultman, L. *J.; Appl. Phys.* 2003, 93, 3002-3014.

[31] Fan, Z.; Tsakiropoulos, P.; Miodownik, A.P.; *J. of Mat. Science*, 1994, 29, 141-150.

[32] Hahn, H.; Padmanabhan, K.A.; *Philos. Mag.* 1997, B76, 559–571.

[33] Mitani, S.; Fujimori, H.; Takanashi, K.; Yakushiji, K.; Ha, J.G.; et al. *J Magn Magn Mater* 1999, 198-199, 179-84.

[34] Zhao, B.; Yah, X.; Pakhomov, A.B.; *J Appl Phys* 1997, 81(8), 5527-5529.

[35] Hauser, H.; Stangl, G.; Hochreiter, J.; Chabicovsky, R.; Aigner, P.; et al. Magneto-Resistive Sensors; in: "Sensor & packaging". Wien Austria: ÖVE. 2003, 49–56.

[36] Sedlácková, K.; Czigány, Zs.; Ujvári, T.; Bertóti, I.; Grasin, R.; Kovács, Gy.J.; Radnóczi, G.; *Nanotechnology* 2007, 18, 445604-445609.

In: New Research on Nanocomposites
Editors: Luis M. Krause et al, pp. 247-280
ISBN: 978-1-60456-799-1


*Chapter 8*

# Periodic Non-Oxide Semiconductor-Based Organic-Inorganic Nanocomposites

***Ningzhong Bao*[*]*, Liming Shen and Arunava Gupta***
Center for Materials for Information Technology (MINT), The University of Alabama, Tuscaloosa, AL 35487, USA

## Abstract

The area of periodically arranged organic-inorganic nanocomposites (OINCs) prepared by self-assembly in the presence of organic molecules underwent dramatic growth. Compared with large amounts of heavy research on the oxide-based materials, such as silicate-base materials, non-oxide-based OINCs, metal-chalcogenide-based OINCs, are particularly much more attractive. Although the considerable challenge of the construction of non-oxide analogues which is caused by the unstable precursor anions and poor understanding of their condensation behavior in solution, metal-chalcogenide-based OINCs offer new potential opportunities in materials sciences, photo-electric conversion, photocatalysis, separation, and environmental remediation. All the reported metal-chalcogenide-based OINCs can be regarded as periodically arranged organic-inorganic nanocomposites, and divided into two categories according to the size of the periodically repeated units: (1) mesostructured (2-50 nm) surfactant-chalcogenide OINCs, and (2) microstructured (< 2 nm) open-framework organic molecule-chalcogenide OINCs. This article reviews the current state of the art in chemical synthesis, characterization, and applications of the mesostructured (2-50 nm) surfactant-chalcogenide OINCs.

## 1. Introduction

The conception of organic-inorganic nanocomposites (OINCs) has exploded since the 1980s' with the expansion of soft inorganic chemistry processes. [1-3] The field of OINCs has been widely recognized as one of the most promising and rapidly emerging research areas in many fields such as chemistry, physics, materials, etc. The OINCs can be best described as being

* E-mail address: nzhbao@mint.ua.edu; nzh_bao@yahoo.com. phone: 205-348-5041, fax: 205-348-9104. Dr. Ningzhong Bao, Center for Materials for Information Technology, 205 Bevill Building, Box 870209, University of Alabama, Tuscaloosa, AL 35487

organized into spatially identifiable domains of an organic and an inorganic component. The major driving forces behind the intense activities in this area are the new and different properties of the OINCs arising from the unique properties in reduced dimension and the great tunability at the interface of organic and inorganic realms. A greater understanding of the manipulation of the OINCs at nanoscale has led to a number of advances both in fundamental scientific research and technological applications of such nanocomposites. Thus, scientific and technical advances of the OINCs have offered the OINCs a wide range of possibilities to elaborate tailor-made materials in terms of chemical and physical properties, including novel optical, electronic, magnetic, catalytic, etc., and the high versatility for a broad range of applications in various fields. [4-6] Promising applications are expected or have already been realized in many fields of optical and electronic materials, solid electrolytes, coating technology, sensor, catalysis, and separation science. [4-6] Harnessing the potential of OINCs requires fine-tuning of the sizes, topologies, and spatial assembly of individual domains and their interfaces. This, in turn, relies on perfecting chemical routes of covalent attachment coupled with an improved fundamental understanding of self-assembly mechanisms, particularly those involving weak intermolecular forces.

In fact, both the organic and the inorganic species in the OINCs have been well studied either in the bulk form or the nanoparticle form in their respective fields. But additional novel functions in reducing dimensions at nanoscale appear as the nanophases of the organic and inorganic species, and are organized to form the OINCs. The performance of the OINCs in current applications is determined by the organization way, the interface interaction, and characteristic nature of the organic and inorganic components. The target OINCs are expected to maintain or enhance the best properties of each of the components while eliminating or reducing their particular limitations, which are capable of simultaneously satisfying several requirements. [7]

The main challenge is the development of effective methods for preparing uniformly-organized OINCs with fine tunable properties. The OINCs are generally produced using various methodologies of "soft" inorganic chemistry in liquid or sol/gel media, exploiting either self-assembly mechanisms under the influence of structure-directing agents, controlled polymerization processes in such media, or directed assembly of nano building blocks in solution or at liquid/solid interfaces. [8] A general classification is (I) the OINCs in which the inorganic and organic components interact only weakly through hydrogen bonding, van der Waals contacts, or electrostatic forces, and (II) the OINCs in which the constituents are more strongly linked through ionic/covalent bond formation. The spatial organization of dissimilar and commonly incompatible components in these OINCs produces a wealth of novel structural features, physical properties, and complex functions, which arise from the synergistic interaction of the individual constituents. [9] Most of the investigated OINCs materials are zeolites, pillared clays, mesostructured solids, nanocrystal-contained gels, glass composites, etc., which can be divided into five sorts (see Figure 1) according to the size of the periodic units of the OINCs. [10-13] The organic templating agents can be organic amines for zeolites and pillared clays, surfactant molecules for mesostructured solids, and polymer molecules for nanocrystal-contained gels and glass composites. The inorganic species are silicates, phosphates, aluminosilicates, some non-siliceous oxides, and chalcogenide clusters, metals, etc.

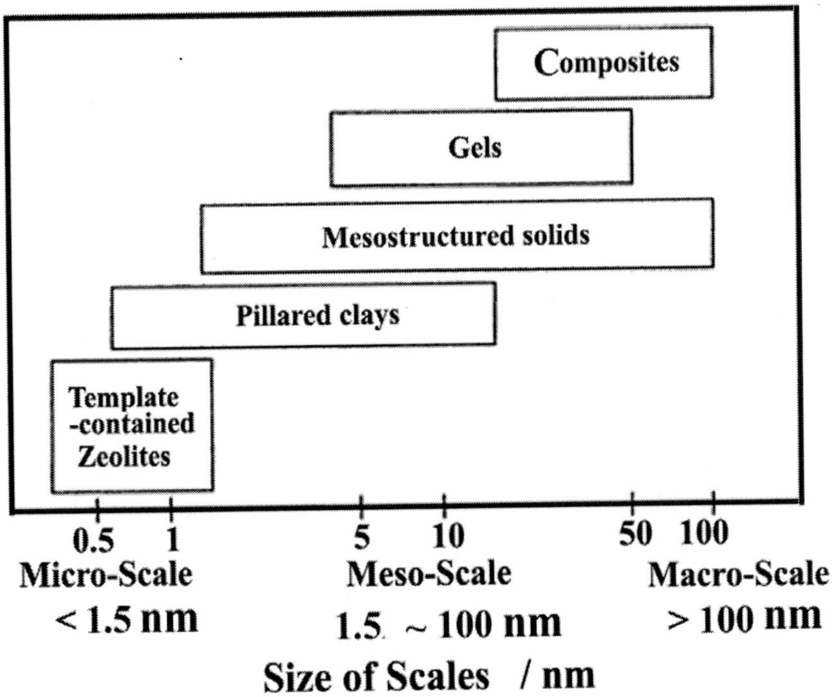


Figure 1. Size distributions of inorganic phases in different micro-, meso-, and macro-structured organic-inorganic nanocomposites with ordered nanostructures.

Metal-chalcogenides have been known well due to their distinguished optical, electric, magnetic, and catalytic properties. [14-16] Chalcogenide clusters represent the lower limit of nanoparticles of which the unique size-dependent properties have been recognized as key to future technological advancement. This area of research has become increasingly important because it provides a valuable opportunity to fundamentally explore the synthetic and structural chemistry, and the synthetic design of colloidal nanostructures. [17-19] On the other hand, applications of oxide-based OINCs have been hindered by lack of electronic, optical, or electrooptic properties. The OINCs derived from the organization of chalcogenide clusters could serve as a unique type of materials that integrate uniform porosity with classical semiconductor type solid-state properties. [20]

Compared with oxygen, sulfur has far richer coordination chemistry due to its low-lying d orbitals. However, the lack of thermodynamic stability of sulfides in the presence of oxygen has made the synthesis of mesostructured sulfides more difficult. Only two types of the mesostructured surfactant-chalcogenide OINCs have been reported. They are: (1) framework-disconnected mesostructured OINCs, such as CdS-, ZnS-, and CdTe-based OINCs, prepared via direct true-liquid-crystal templating from nonionic surfactants. [21-28] These OINCs are typically made of individual nanocrystals of non-continuous frameworks of sulfide nanocrystals; and (2) framework-connected mesostructured OINCs, such as tin sulfides, thiogermanates, and various crystalline open-framework metal sulfides and selenides, etc. The inorganic phases are chalcogenide clusters connected by the metal ion linkers. [29] Various aspects of non-oxide mesostructured OINCs have been reviewed by Kanatzidis, [29] Soler-IIIa et al., [30] Schuth, [31] and He et al. [32] However, the chemistry of metal-chalcogenides-based OINCs is much more diverse than that of the oxide-based OINCs, thus, the synthetic strategies for their production are also more mutative. We review the general pathways which can lead to mesostructured metal-chalcogenides-based OINCs. A more detailed description of selected examples will then be given in the subsequent sections, which are organized according to the chemical composition of the resulting materials.

## 2. Self-assembly of Surfactant Template and Inorganic Species

A number of models have been proposed to explain the formation of mesostructured OINCs and to provide a rational basis for the various synthesis routes. On the most common level, these models are predicated upon the presence of surfactants in a solution to guide the formation of the inorganic mesostructure from the solubilized inorganic precursors (see Figure 2). Surfactants contain a hydrophilic head group and a long hydrophobic tail group within the same molecule and will self-organize in such a way as to minimize contact between the incompatible ends. How the inorganic precursor interacts with the surfactant is the issue whereby those models diverge; the type of interaction between the surfactant and the inorganic precursor will be seen as a significant difference among the various synthesis routes, the formation models, and the resulting classes of mesostructured OINCs.

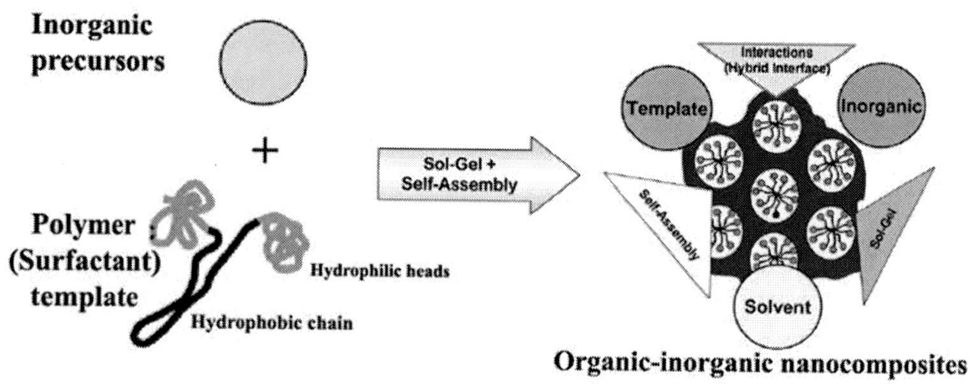


Figure 2. Schematic representation of the general formation of organic-inorganic nanocomposites from inorganic precursors and organic surfactants by sol-gel self-assembly.

### 2.1. Self-assembling Templates

The key feature in the synthesis of mesostructured OINCs materials is to achieve a well-defined segregation of organic (hydrophobic) and inorganic (hydrophilic) domains at the nanoscale. The nature of the organic-inorganic hybrid interface plays a fundamental role. The most relevant thermodynamic factors affecting the formation of a hybrid interface have been first proposed by Monnier et al. [33] and discussed in depth by Huo et al. [34] in their description of the "charge matching" model. The free energy of mesostructure formation ($\Delta G_{ms}$) is composed of four main terms, which represent, respectively, the contribution of the inorganic-organic interface ($\Delta G_{inter}$), the inorganic framework ($\Delta G_{inorg}$), the self-assembly of the organic molecules ($\Delta G_{org}$), and the contribution of the solution ($\Delta G_{sol}$). In the classical liquid crystal templating, the contribution due to the organization of the amphiphilic molecules prevails over the other interactions. In the cooperative assembly route, template concentrations may be well below those necessary for obtaining liquid crystalline assemblies or even micelles. Thus, the creation of a well-defined and compatible hybrid interface between the inorganic walls and the organic templates (i.e., $\Delta G_{inter}$) is central to the generation of a well-ordered hybrid structure with adequate curvature.

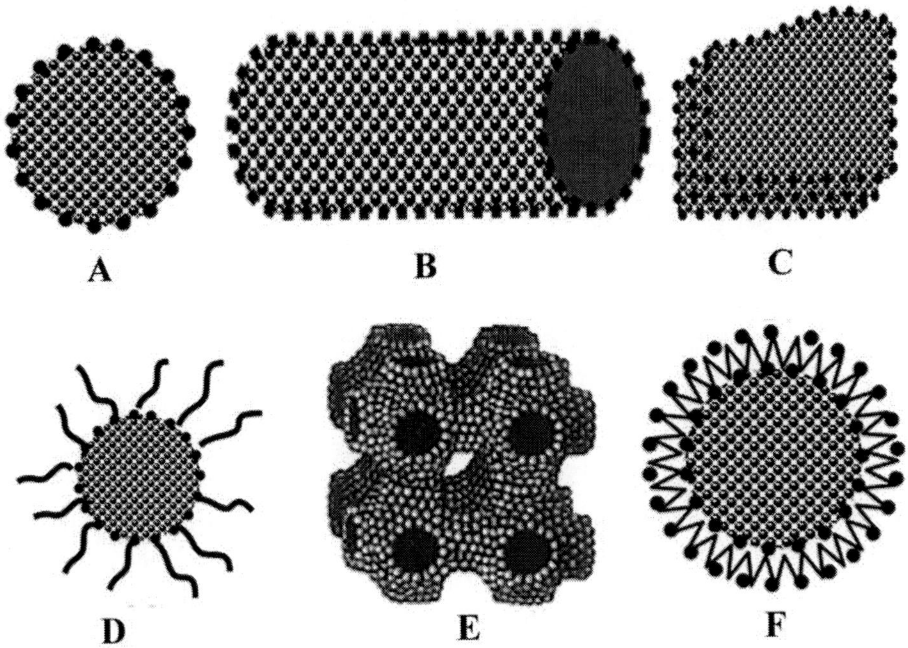


Figure 3. Micellar structures (A) sphere; (B) cylinder; (C) planar bilayer; (D) reverse micelles; (E) bicontinuous phase; and (F) liposomes.

The structures of the OINCs are determined by the shape and organization of the micellar structures (see Figure 3) [35] and the micellar aggregates (see Figure 4). Micellar aggregates organize to different shapes (such as spherical or cylindrical micelles, lamellae, etc.), permitting the coexistence of two incompatible phases. Some typical micellar structures are presented in Figure 3. In some colloidal systems, a more complex behavior has been evidenced, and other arrangements, such as the sponge-like structure (see Figure 3E), are possible. A combination of molecular geometry and intermolecular (solvophilic/phobic, Coulombic, H-bonding) and entropic interactions drives these solutions to self-assemble into colloidal systems, presenting different microstructures: spherical, cylindrical, planar, cellular, etc. The assembly depends on the nature and morphology of the discrete molecules. The numerous micellar assemblies and aggregates, constituted by the association of amphiphilic molecules, linked by weak forces (van der Waals, H-bonds, electrostatic, etc.) rather than by covalent bonds, are collectively known as association colloids or molecular organized systems. The architecture of the final OINCs directly rely on the nature of the surfactant molecules, that is, the morphology of the micellar aggregates and the interactions at the organic-inorganic interface. A variety of mesostructured OINCs are shown in Figure 4.

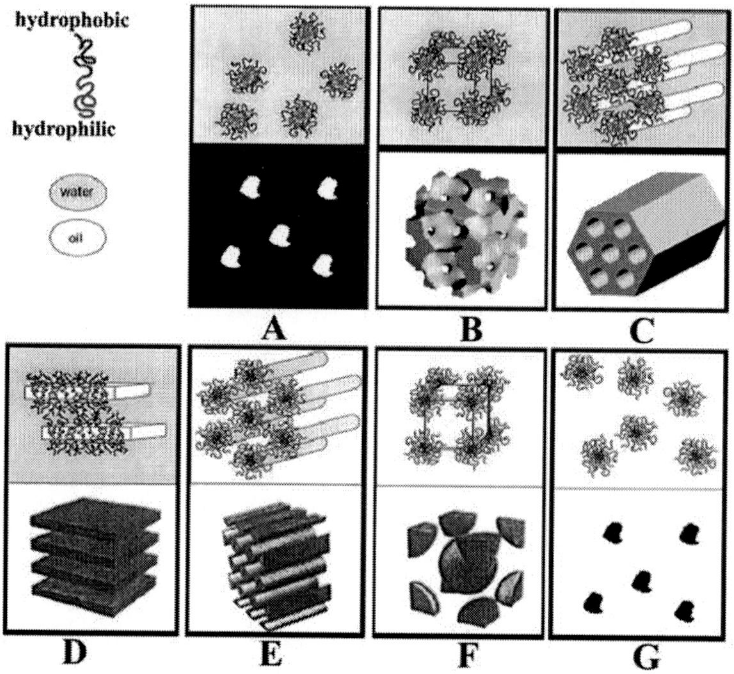


Figure 4. Schematic of different modes of self-organization of the surfactants in the presence of solvents ("water" and "oil") selective for the two blocks. (A) normal spherical micellar solution; (B) normal micellar cubic lyotropic liquid crystal (LLC); (C) normal hexagonal LLC; (D) lamellar LLC; (E) reverse hexagonal LLC; (F) reverse micellar cubic LLC; and (G) reverse micellar solution. The amphiphiles are localized at the interfaces between the water and oil domains. For reasons of clarity, only the copolymer moleculesat the cross section of the (spherical, cylindical, and planar) assemblies are depicted, and only sections of the LLC structures are shown. Corresponding organic-inorganic nanocomposites of (A) surfactant micell-embeded in inorganic phase; (B) cubic OINCs; (C) hexagonal (*p6mm*) OINCS; (D)lamellar OINCs; (E) inorganic nanorod-based OINCs; (F) cubically-arraied inorganic nanocrystal-based OINCs; and (G) inorganic nanocrystal-embaded OINCs are also shown.

## 2.2. Role of Surfactant Geometry

Upon progressive increase of surfactant concentration in the aqueous solution, a number of phases appear, always following the same order: "direct" spheres, "direct" cylinders, lamellae, "inverse" cylinders, and "inverse" spheres; this order corresponds to a monotonic variation of the interfacial curvature. Different models have been proposed to explain these experimental facts, the main parameters taken into account are (1) the hydrophobic interactions between organic chains, (2) geometric restrictions due to molecular packing, (3) molecule exchange between aggregates, (4) enthalpy and entropy of packing, and (5) electrostatic repulsion between polar heads. [36]

A relatively simple model proposed by Israelachvili and colleagues, based on geometrical considerations, explains and predicts the resulting self-assembled structures of each type. [37-38] This model considers a hydrophobic liquid-like core, the contribution of which is merely due to geometrical constraints. These geometrical considerations rely on the ratio of the polar head surface to the hydrophobic volume. The amphiphilic molecules are thus modeled like a conical fragment (the hydrophobic part) attached to a spherical (hydrophilic) head. Two main shapes (see Figure 5) are possible: direct conical (or ice-cream cone) for one hydrophobic chain, and inverse conical (or champagne cork) for two hydrophobic chains.[30]

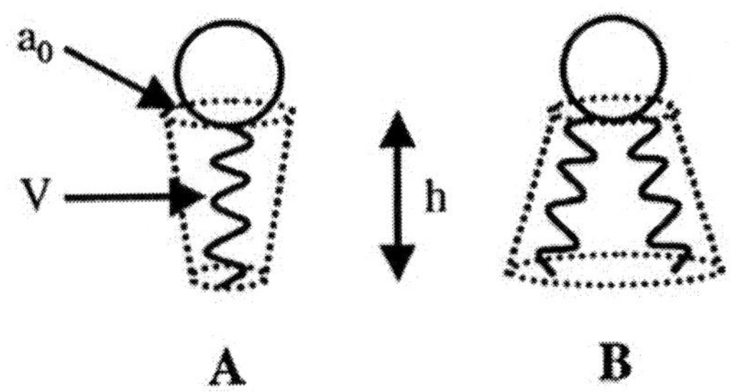


Figure 5. Schematic representation of amphiphilic molecules, adopting conical shape (icecream cone, A) or inverse conical (champagne cork, B). [Reprinted with permission from Ref. 30; Soler-Illia, G. J. de A. A. et al. Chem. Rev. 102, 4093, 2002 Copyright @ American Chemical Society (2002)].

The steric hindrance of the hydrophobic chain is characterized by the ratio v/l, where v is the chain volume and l is the chain length; in the case of the polar head, its contribution is given by the effective optimal surface, $a_0$. To ensure chain fluidity, l must verify that $l < l_c$, where $l_c$ is the length of the fully extended chain; $l_c$ can be easily estimated as a function of the number of C atoms in the chain, n. The value of the packing parameter, $g = v/l_c a_0$, links the molecular structure of the amphiphilic molecule to the architecture of the aggregates. The value of the packing parameter, $g = v/l_c a_0$, links the molecular structure of the amphiphilic molecule to the architecture of the aggregates. The relation between g parameter and different micellar structures is spherical micelles for $g < 0.33$, cylindrical micelles for $g = 0.33$-$0.5$, bilayer (vesicules) for $g = 0.5$-$1$, bilayer (membranes) for $g = 1$-$2$, inverse cylindrical micelles for $g = 2$-$3$, and inverse spherical micelles for $g > 3$. The limiting values of g can be easily calculated for an aggregate of known geometry, by using the condition of chain fluidity ($l < l_c$) and an estimation of the aggregation number (number of molecules forming the micelle). The latter can be obtained from two relationships: the aggregate surface to $a_0$ and the aggregate volume to v. Huo et al. firstly took into account the g parameter to explain the formation of different surfactant-templated oxide mesostructures. [39] In principle, the structure of the mesophase depends on the packing properties of the surfactant molecules, hence, on the value of g. The validity of this concept has been illustrated by a complete study, which took into account various parameters, such as surfactant nature, pH, presence of co-solvents or co-surfactants, and their influence on the observed phase transitions, for fixed synthesis conditions.

## 2.3. Interaction of the Hydrophilic Head

Both surfactant and inorganic soluble species direct the synthesis of mesostructured OINCs, which are strongly dependent on the interaction between surfactants and inorganic precursors. In the case of ionic surfactants, the formation of the mesostructured OINCs is mainly governed by electrostatic interactions. [40] In the simplest case, the charges of the surfactant (S) and the mineral species (I) are opposite, in the synthesis conditions. Two main direct synthesis routes have been identified: $S^+I^-$ and $S^-I^+$. Two other synthesis paths, considered to be indirect, also yield hybrid mesophases from the self-assembly of inorganic and surfactant species bearing the same charge: counterions get involved as charge compensating species. The $S^+X^-I^+$ path takes place under acidic conditions, in the presence of halogenide anions ($X^-$: $Cl^-$, $Br^-$); the $S^-M^+I^-$ route is characteristic of basic media, in the presence of alkaline cations ($M^+ = Na^+$, $K^+$). The different possible hybrid organic-inorganic interfaces are schematized in Figure 6. [30] Other synthesis routes rely on nonionic surfactants, where the main interactions between the template and the inorganic species are H-bonding or dipolar, giving birth to the so-called neutral path: $S^0I^0$, $N^0I^0$, and $N^0F^-I^+$. [41-45]

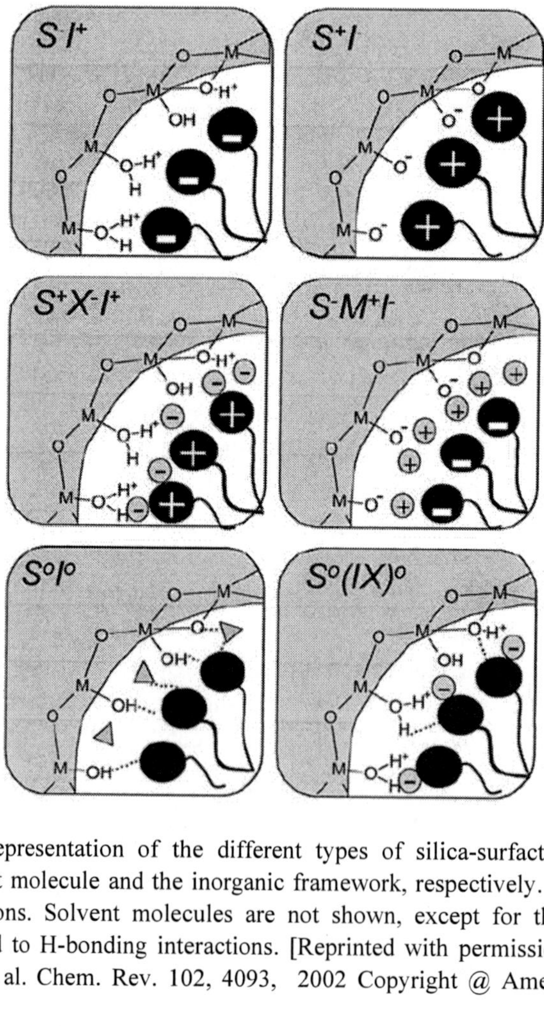


Figure 6. Schematic representation of the different types of silica-surfactant interfaces. S and I represents the surfactant molecule and the inorganic framework, respectively. $M^+$ and X represent the corresponding counterions. Solvent molecules are not shown, except for the $I^0S^0$ case (triangles); dashed lines correspond to H-bonding interactions. [Reprinted with permission from Ref. 30; Soler-Illia, G. J. de A. A. et al. Chem. Rev. 102, 4093, 2002 Copyright @ American Chemical Society (2002)]

## 3. Formation Mechanisns

Figure 7 presents an illustration of two main general synthesis strategies used to construct the mesostructured OINCs. In both of the synthesis strategies, the chemical, spatial, and structural properties of the texturing agent must be thoroughly adjusted by controlling the rates of chemical reactions, the nature of the interfaces, and the encapsulation of the growing inorganic phase. An adequate tailoring of the organomineral interface is of utmost importance for obtaining well-defined textured phases. The chemical, spatial, and temporal control of this "hybrid interface" is a major task in the challenge of developing cooperatively assembled inorganic-organic integrated systems. [31]

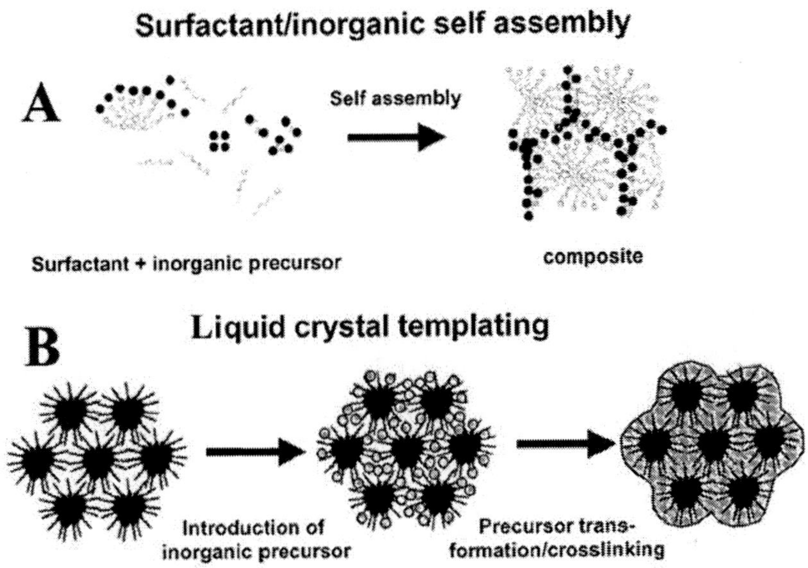


Figure 7. Schematic representation of the general pathways leading to mesostructured organic-inorganic nanocomposites. (A) cooperative self-assembly; and (B) liquid crystal templating. [Reprinted with permission from Ref. 31; Schuth, F. Chem Mater. 13, 3184, 2001 Copyright @ American Chemical Society (2001)].

As shown in Figure 7, the molecular/supramolecular templates are present in the synthesis media from the beginning; the self-assembly process of the templates is followed by the formation of the mineral network, deposited around the "self-assembled substrate". "Inorganic replication" occurs at accessible interfaces built by preorganized or self-assembled molecular or supramolecular templates, which create the mesostructure in the material. These templates can be organic compounds (surfactant molecules, amphiphilic block copolymers, dendrimers, etc.) or biomolecules, forming micellar assemblies (Route A) or liquid crystal mesophases (Route B).

### 3.1. Cooperative Self-assembly

The cooperative "templating" perspective for the mechanism of formation of the MCM-41 and MCM-48 silicate phases is that the electrostatic interface interaction between soluble anionic inorganic molecular species and cationic monotailed surfactant molecules is responsible for initiating the composite synthesis and that the pre-organized organic array is not necessary, or if present not necessarily related to the morphology of the final composite product. [46] A generalized mechanism (see Figure 8) of formation based on the specific type of electrostatic interaction between a given inorganic precursor I and surfactant head group S was proposed by Huo and co-workers. [34] Since that, a variety of surfactants, including cationic, anionic, two-tailed, with coupled head groups, fluorinated and zwitterionic species, and protonic neutral copolymers have been studied, in which the surfactant concentrations which are too low for liquid-crystal phase formation in pure water solutions.

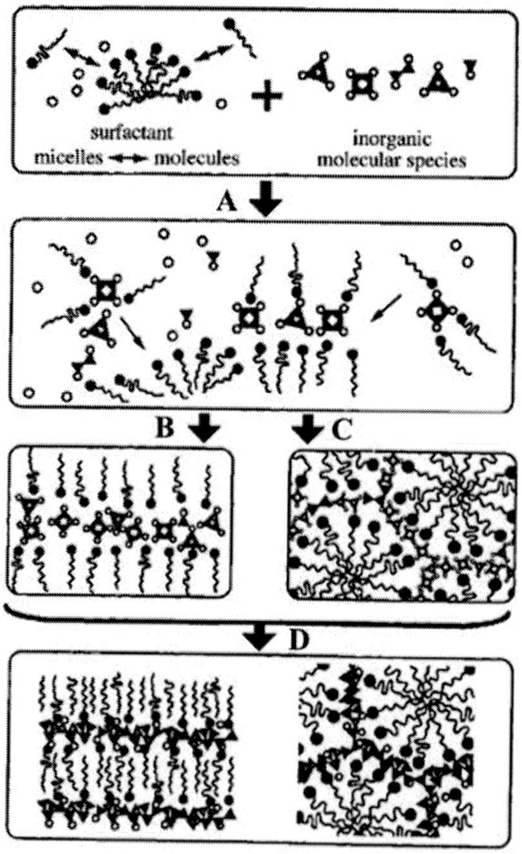


Figure 8. Cooperative templating model for OINCs synthesis. (A) Single-chain surfactant molecules react prefer-entially with inorganic polyanions. Micelles serve as a surfactant molecule source or are rearranged according to the anion charge density and shape requirement; (B and C) Nucleation and rapid recipitation of organized arrays takes place with configurations determined by the cooperative interactions of ion-pair charges, geometries, and organic van der Waals forces; and (D) Inorganic mineralization and condensation. [Reprinted with permission from Ref. 34; Huo, Q. et al. Chem Mater. 6, 1176 1994 Copyright @ American Chemical Society (1994)].

The organization of surfactant molecules into arrays at concentrations below the liquid crystal-phase region is characterized by the critical micelle concentration (CMC) which is denoted by the temperature and concentration at which finite organized ordered arrays can first be detected. The spherical or rodlike micelles are formed at the very low concentrations of surfactants. The sphere-to-rod transformation is strongly dependent on temperature, anion, and surfactant chain length. The addition of poly-dentate charged inorganic species coordinate electrostatically with surfactant molecules which may originate from loosely structured micelles or unassociated organic molecules. The poly-dentate inorganic ions in solution can be expected to be an important factor in determining organic array morphology through multidentate bonding and charge density matching of organic surfactant with the inorganic molecular species. The charge, geometry, and association of the molecular inorganic species in solution are able to be adjusted through the use of pH, cosolvent, counterions, and temperature.

The charge density of the inorganic species determines how many surfactant molecules are associated with a given inorganic molecular unit as well as the preferred orientation of the surfactant head group relative to the molecular inorganic species (see Figure 8A). This "charge density matching" determines the average intermolecular spacing between surfactant head groups and within that constraint and the energetics of the intermolecular interactions of the inorganic species, the molecular ion pairs adapt the preferred liquid crystal array morphology (see Figure 8B and C). In its simplest kinetic form at low temperatures the synthesis is a three-stage process in which the first step can be viewed as the net ionic exchange of a polycharged anionic species for a monoanion. These ion pairs can then organize into a new liquid-crystal-like array. The final step is condensation of the inorganic phase.

## 3.2. True-Liquid-Crystal Templating

As the concentration of a surfactant in a solvent is larger than the CMC, the surfactant micellar solution turns to form lyotropic liquid crystals (of either “oil-in-water” or “water-in-oil”), which are called “True-liquid-crystal mesophases”. Typically, with increasing the concentaration of a surfactant in solvents above CMC at the temperature above critical micellization temperature (CMT), lyotropic liquid crystals with continuously changed mesophases are formed. [47]The concentration-temperature phase diagram for the binary surfactant/solvent system is presented in Figure 9. The effects of temperature on the surfactant self-assembly were explored over a wide temperature range. The surfactant molecules can self-organize into different thermodynamically stable microstructures, depending on both the polymer concentration (lyotropic behavior) and the temperature (thermotropic behavior). A total of six different one phase regions have been identified. Four “gel” regions with different lyotropic liquid crystalline structures have been characterized: micellar cubic (I1), hexagonal (cylindrical) (H1), bicontinuous cubic (V1), and lamellar (smectic) (LR).

The true-liquid-crystal mesophases and different assemblies are involved in the surfactant templating assembly to synthesize ordered mesostructured nanocomposites using high concentrations of surfactants as templates. The condensation of inorganic precursors is improved owing to the confined growth around the surfactants and thus ceramic-like frameworks are formed. After the condensation, the organic templates can be removed by calcination, extraction, etc. The inorganic materials "cast" the mesostructures, pore sizes, and symmetries from the liquid-crystal scaffolds. Direct templating of microemulsion liquid-crystal mesophases were used to synthesize a variety of mesostructured sufactant-chalcogenide OINCs.

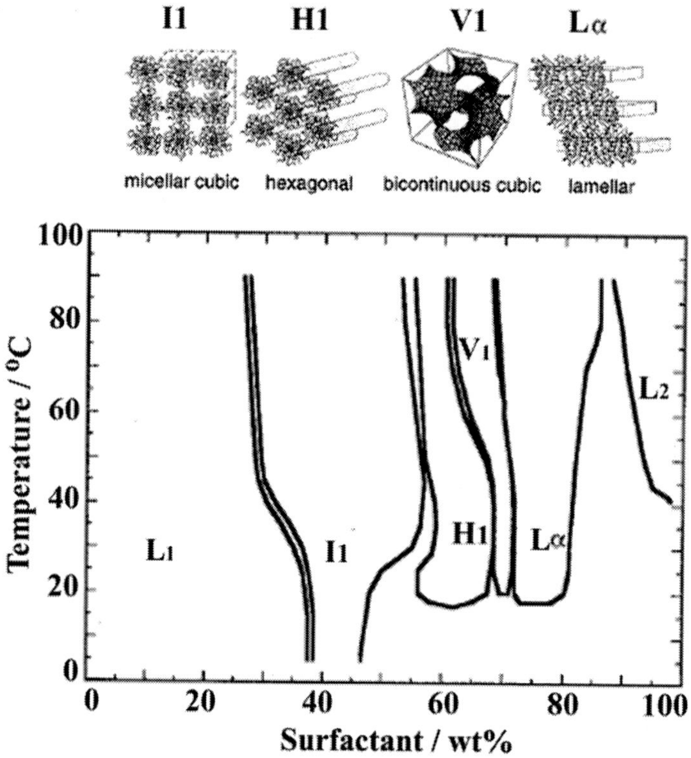


Figure 9. The concentration-temperature phase diagram of the surfactant-solvent binary system. The concentrations are expressed in wt %. The phase boundaries of the one-phase regions are drawn with solid lines. I1, H1, V1, and LR denote micellar cubic, hexagonal (cylindrical), bicontinuous cubic, and lamellar (smectic) lyotropic liquid crystalline phases, respectively, while L1 and L2 denote solvent-rich and surfactant-rich solutions. The samples whose compositions fall outside the one-phase regions are dispersions of two different phases. The tie lines in the two-phase regions are parallel to the concentration axis (isothermal).

# 4. Surfactant/Metal-Chalcogenide-Based Mesostructured OINCs

## 4.1. Framework-Disconnected Mesostructured OINCs

Various groups have reported mesostructured chalcogenide-based OINCs, such as CdS, CdSe, ZnS and $MoS_4^{2-}$. [25, 26,48-50] The representative synthetic strategy was developed by Stupp and co-workers.[49-50] This method involves passing $H_2S$ or $H_2Se$ over a hexagonal mesophase containing the ether $C_{18}H_{35}(OCH_2CH_2)_{10}OH$, water, and $Cd^{2+}$ or $Zn^{2+}$ salts. As shown in Figure 10, the synthesis pathway uses the true liquid-crystal templating (TLTC) that is formed in the polymer aqueous solution; this phase is loaded with a metal ion precursor solution and then exposed to hydrogen sulfide gas. This results in the formation of metal sulfides in the liquid crystal by which the metal sulfide is mesostructured. As a typical example, the structure and properties of the hexagonal CdS-nanocrystal/$C_{18}H_{35}(OCH_2CH_2)_{10}OH$ OINCs has been shown in Figure 11. Instead of hydrogen sulfide, hydrogen selenide can be used which results in the formation of hexagonally ordered metal selenide nanocrystals. Following this route, mesostructured CdS, CdSe, and ZnS-based OINCs have been obtained, while attempts to synthesize mesostructured $Ag_2S$, CuS, HgS, and PbS failed. This led the authors to the conclusion that a specific interaction between the metal ions and/or the precipitate and the liquid crystal is necessary.

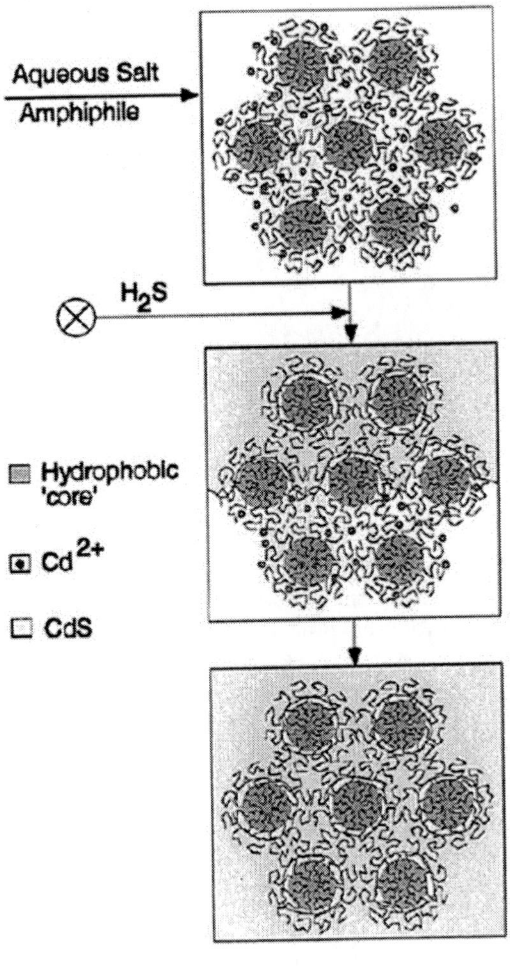


Figure 10. Schematic representation of the process used to synthesize hexagonally ordered metal chalcogenide nanocrystals. [Reprinted with permission from Ref. 50; Tohver, V. et al. Chem Mater. 9, 1495, 1997 Copyright @ American Chemical Society (1997)].

In this approach, both chemical nature and structure of the amphiphile in direct templating are very important. Moreover, the order of the mesostructures in the obtained

OINCs was even found to be dependent on the anion of the metal salts, and not all of the chalcogenides can form ordered mesostructured OINCs because of the disconnected chalcogenide nanocrystals frameworks. This first direct templating process opens up a variety of synthetic avenues to the formation of novel nanostructures. There are a large number of amphiphilic liquid crystals, with lattice constants ranging from a few nanometers to tens of nanometers, which include lamellar, hexagonal, cubic, and bicontinuous phases. Potentially, many of these systems could be mineralized, generating materials with an array of novel structures and properties.

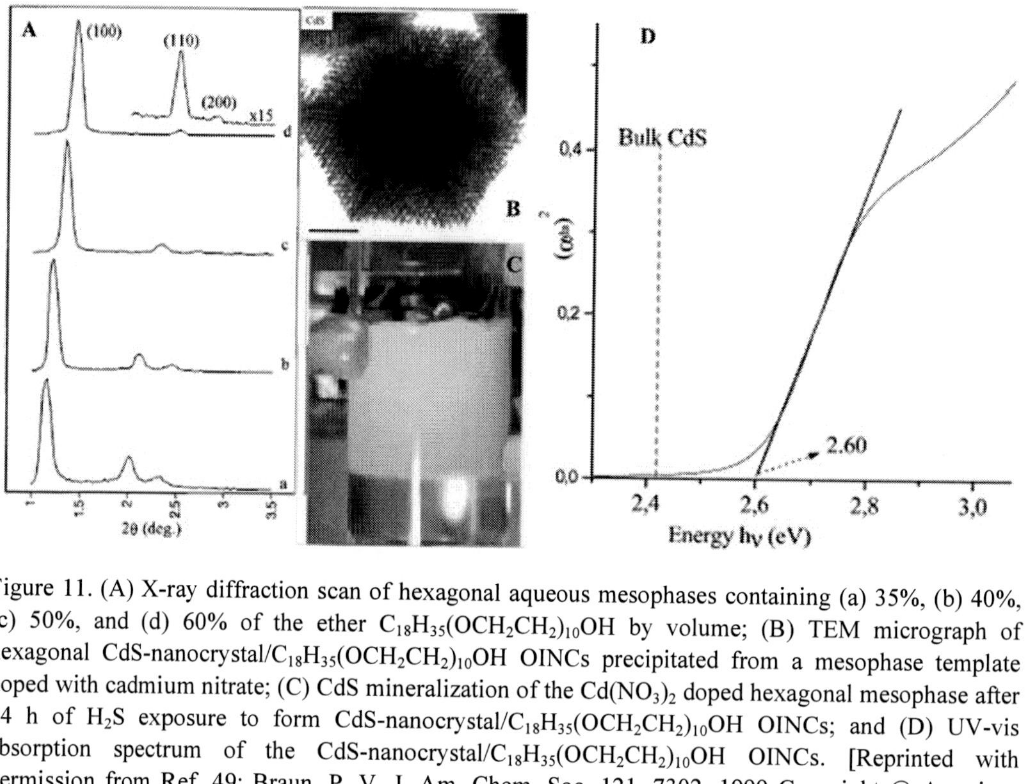


Figure 11. (A) X-ray diffraction scan of hexagonal aqueous mesophases containing (a) 35%, (b) 40%, (c) 50%, and (d) 60% of the ether $C_{18}H_{35}(OCH_2CH_2)_{10}OH$ by volume; (B) TEM micrograph of hexagonal CdS-nanocrystal/$C_{18}H_{35}(OCH_2CH_2)_{10}OH$ OINCs precipitated from a mesophase template doped with cadmium nitrate; (C) CdS mineralization of the $Cd(NO_3)_2$ doped hexagonal mesophase after 24 h of $H_2S$ exposure to form CdS-nanocrystal/$C_{18}H_{35}(OCH_2CH_2)_{10}OH$ OINCs; and (D) UV-vis absorption spectrum of the CdS-nanocrystal/$C_{18}H_{35}(OCH_2CH_2)_{10}OH$ OINCs. [Reprinted with permission from Ref. 49; Braun, P. V. J. Am. Chem. Soc. 121, 7302, 1999 Copyright @ American Chemical Society (1999); and from Ref. 50; Tohver, V. et al. Chem Mater. 9, 1495, 1997 Copyright @ American Chemical Society (1997)]

## 4.2. Framework-Connected Mesostructured OINCs

As motioned above in the section "4.1 Framework-Disconnected Mesostructured OINCs", not all of the chalcogenides can form ordered mesostructured OINCs via the direct templating process because of the disconnected chalcogenide nanocrystal framworks. However, unlike the silicas which derive from the self-polymerization of silicates, chalcogenide clusters are more easily assembled via coordination reactions between cluster anions and linking metal cations, forming the connected chalcogenide frameworks. The first two chalcogenide-based OINCs with a connected framework were reported for tin, [51-52] which were both lamellar. Subsequently, also ZnS-based OINCs were described. The best investigated system, however, seems to be the germanium/sulfur system, where the first mesostructured OINC, a layered material, was reported in 1997. [53] Since that, some of the best known building blocks are the tetrahedral [$MQ_4$], adamantane [$M_4Q_{10}$] (M=Ge, Sn; Q= S, Se, Te), the dimeric [$Sn_2Q_6$]$^{4-}$ (Q = S, Se), and $(TMA)_4[M_4Q_{10}]$ (TMA tetramethylammonium) units, mainly reported by the Ozin's and Kanatzidis' groups. The generic structures of these anions are shown in Figure 12. These, the divalent metal ions ($Zn^{2+}$, $Cd^{2+}$, $Hg^{2+}$, and $Co^{2+}$, employed as charge-balancing and inorganic framework-chain-linking ions), and the charged surfactant molecules have been employed to produce structures with wormholes, hexagonal honeycomb arrays of parallel surfactant phases, and the more rare cubic architectures where the inorganic framework maps on to a 3D gyroid surface.

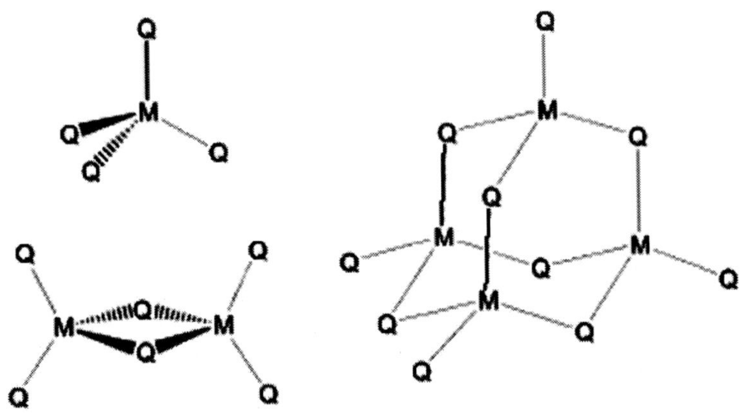


Figure 12. Molecular building blocks used for the preparation of chalcogenide mesostructures.

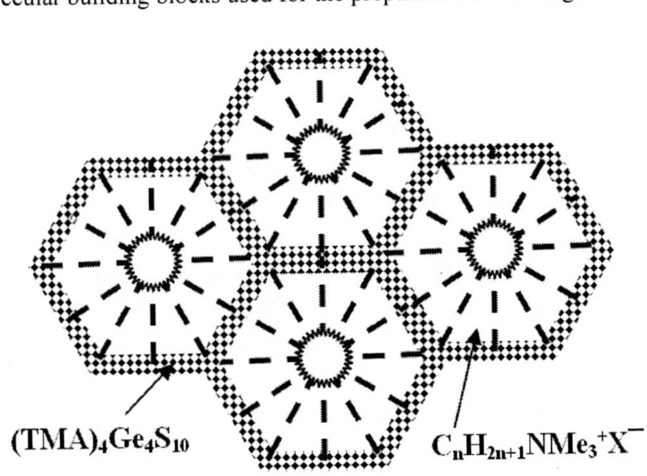


Figure 13. Proposed architecture of a mesostructured germanium sulfide material. $(TMA)_4[Ge_4S_{10}]$ and (CTA)Br were dissolved in formamide to give a viscous mesophase of the adamantoid clusters charge balanced by surfactant. The addition of a metal salt, the adamantoid clusters link to form a supramolecular, hexagonal, mesostructured material in which the adamantoid clusters are linked by metal ions in the structure.

The problem of the low solubility of the sulfide precursors was solved by using formamide as the solvent instead of water. Because of the high dielectric constant of formamide, surfactant phase diagrams in this solvent are often quite similar to the ones in water. [54] Kanatzidis, however, also succeeded in synthesizing similar compounds from an essentially aqueous system. [55] An interesting expansion of this synthetic approach was reported in a recent publication, where the $Ge_4S_{10}$ units were coassembled with biologically relevant $Fe_4S_4$ clusters, resulting in a well-ordered hexagonal mesostructure. Figure 13 is a general model of the proposed framework in which transition metals are linked to the adamantanoid ( such as $[Ge_4S_{10}]^{4-}$) clusters around a lyotropic liquid crystalline phase in formamide. The gyroid is a fascinating triply periodic minimal surface that divides all of space into two congruent regions. The occurrence, nature, and properties of the gyroid surface are of broad scientific interest. [56]

#### 4.2.1. Mesostructured $[M_4Q_{10}]^{4-}$ (M=Ge, Sn; Q= S, Se, Te)-Based OINCs

The most direct sulfur and selenium analogs of hexagonal mesoporous silica would be the thio- and seleno-alumino silicates (i.e., systems based on $[SiS_4]^{4-}$ and $[SiSe_4]^{4-}$), but these materials are unstable to hydrolysis and difficult to handle. Therefore, to assemble metal chalcogenides one needs a working analog to mimic some of the characteristics of the $[SiO_4]^{4-}$ unit. [57] A suitable building unit with a tetrahedral topology and similar charge is a group of adamantane thioanion $[M_4Q_{10}]^{4-}$ (M=Ge, Sn; Q= S, Se, Te), which is made of four cornerlinked $[M_4Q_{10}]^{4-}$ tetrahedra. [58]

Ozin's group reported a non-aqueous surfactant-templating assembly of adamantanoid $[Ge_4S_{10}]^{4-}$ cluster precursors in formamide, where $(TMA)_4Ge_4S_{10}$ (TMA: tetramethylammonium) was used as the inorganic precursor and $C_nH_{2n+1}NMe_3^+X^-$(n=12, 14, 16, and 18, and X=$Cl^-$ and $Br^-$) were used as structure-directing agents (see Figure 13). [59-60] Transition metal cations such as $Co^{2+}$, $Ni^{2+}$, $Zn^{2+}$, or $Cu^+$ were employed as charge-balancing and inorganic framework-chain-linking ions to create OINCs with well-ordered hexagonal mesostructures, as shown in Figure 14. Later, Kanatzidis' group found that the selection of solvent and surfactant seems to play a major role in these systems. Extensive experiments with $H_2O$, alcohols, or their mixtures mainly resulted in disordered pore structures with wormhole character. Only when formamide was used did the chemistry change decisively towards more ordered hexagonal pore structures, which also were stoichiometrically different from those obtained in aqueous systems. They confirmed that the selection of the solvent plays a major role in influencing the structure of these systems, and successfully prepared hexagonal frameworks with $[M_4Q_{10}]^{4-}$ (M=Ge, Sn; Q= S, Se) anions in conjunction with a wide variety of metal ions, including the tetrahedral trivalent $Ga^{3+}$, $In^{3+}$, and $Sn^{4+}$. [61] Other ions that generally do not adopt tetrahedral coordination, such as $Sb^{3+}$, $Bi^{3+}$, $Sn^{2+}$, and $Pb^{2+}$ also yield mesostructures. [62] As shown in Figure 15, the uniform hexagonal symmetry of the mesostructures in CPGaGeS and CPInGeS is readily observed by TEM. The Se analogs CPGaGeSe, CPInGeSe, CPSbGeSe, and CPSnGeSe OINCS also possess hexagonal symmetry. Optical absorption UV-vis spectra show well-defined band gaps (Eg) of 3.01 and 2.97 eV for CPGaGeS and CPInGeS, respectively. A unique feature of both materials is the intense photoluminescence (PL) observed when excited with light above the band gap. With an excitation line of 3.35 eV (370 nm) an intense green emission at 77 K with a maximum 2.35 eV (528 nm) and 2.32 eV (535 nm) was observed for CPGaGeS and CPInGeS, respectively. This is the first observation of light emission from mesostructured metal thiogermanates. Light emission from amorphous chalcogenide frameworks is rare and underscores the potential of these systems for photonic function.

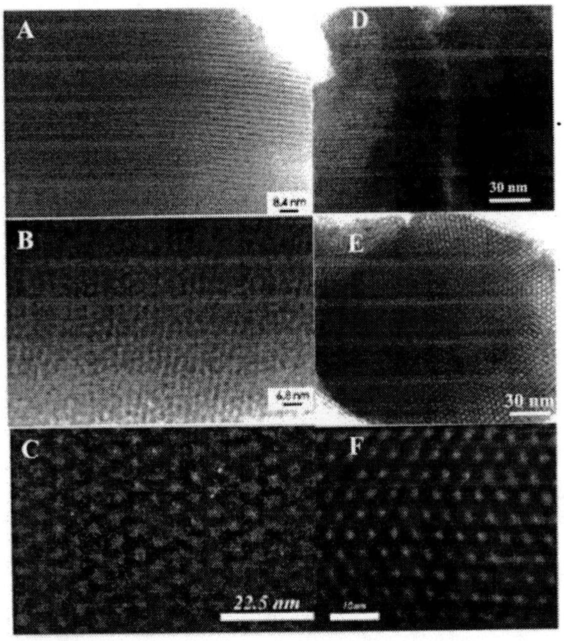


Figure 14. TEM images of (A) CTA-Ni/$Ge_4S_{10}$ parallel to the channel axis; (B) CTA-Ni/Ge4S10 perpendicular to the channel axis; (C) CTACu/$Ge_4S_{10}$ perpendicular to the channel axis; (D) CTA-Co/$Ge_4S_{10}$ parallel to the channel axis; (E) CTA-Co/$Ge_4S_{10}$ perpendicular to the channel axis; and (F) Expanded view of the hexagonal mesostructure in CTA-Co/$Ge_4S_{10}$. [Reprinted with permission from Ref. 59; MacLachlan, M. J. J. Am. Chem. Soc. 121, 12005, 1999 Copyright @ American Chemical Society (1999)]

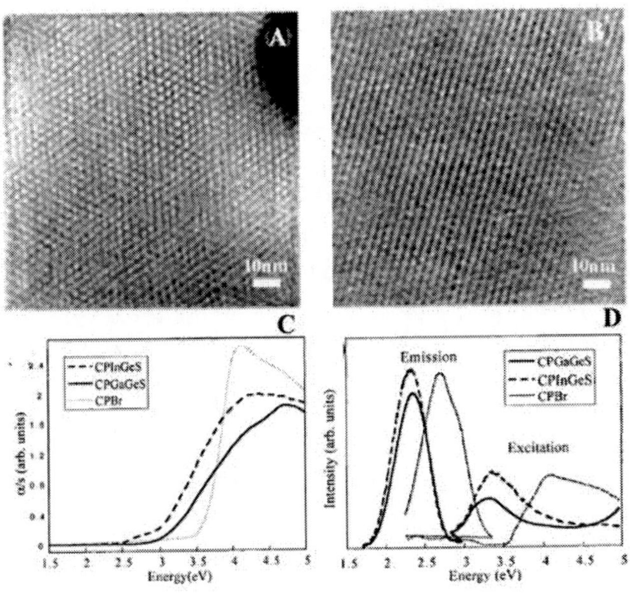


Figure 15. Representative TEM images of the indium germanium sulfide framework (CPInGeS): (A) parallel to the pore tunnels; (B) perpendicular to the pore tunnels; (C) Electronic absorption spectra; and (D) Photoluminescence and excitation spectra. [Reprinted with permission from Ref. 61; Rangan, K. K. J. Am. Chem. Soc. 122, 10230, 2000 Copyright @ American Chemical Society (2000)]

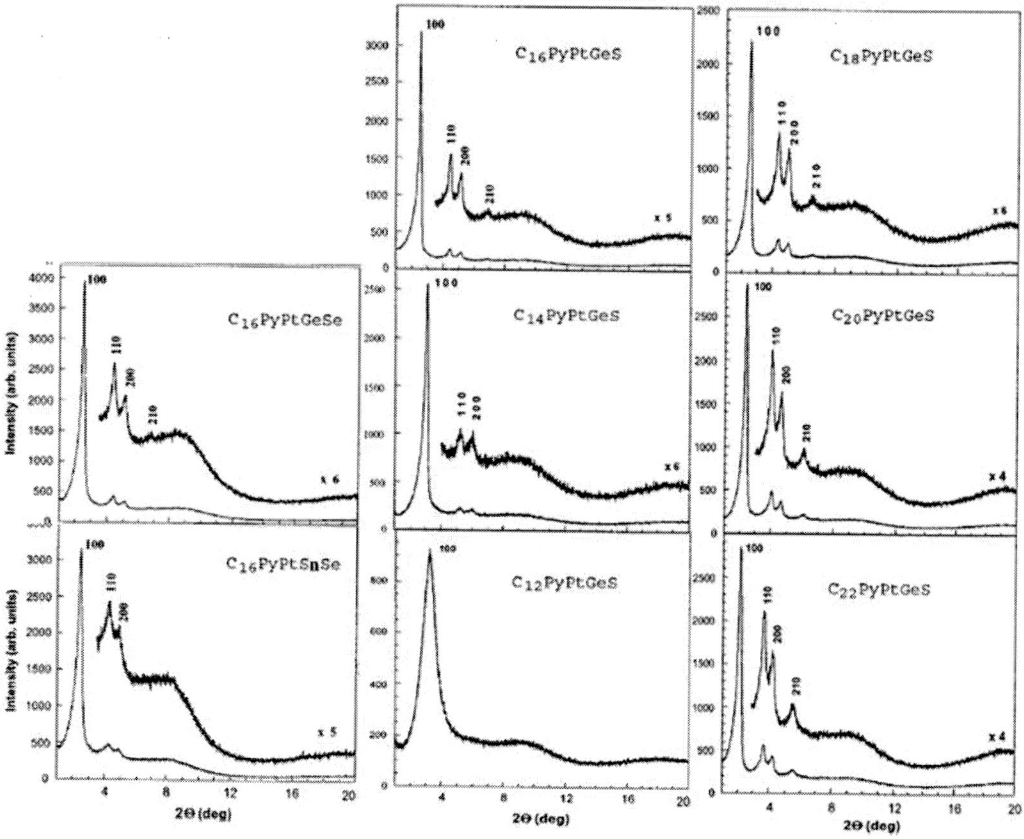


Figure 16. X-ray powder diffraction patterns of mesostructured CnPyPtGeS where n = 12, 14, 16, 18, 20, and 22. [Reprinted with permission from Ref. 66; Trikalitis, P. N. et al. J. Am. Chem. Soc. 127, 3910, 2005 Copyright @ American Chemical Society (2005)].

Kanatzidis' group also found that OINCs with highly periodic hexagonal honeycombs of platinum-germanium chalcogenide and platinum-tin selenide frameworks were able to be prepared by linking corresponding $[M_4Q_{10}]^{4-}$ (M=Ge, Sn; Q= S, Se) clusters with $Pt^{2+}$ ions. [63] The non-oxidic honeycombs designated as $C_nPyPtGeQ$ and $C_nPyPtSnSe$ were templated by the lyotropic liquid-crystalline phase of alkylpyridinium surfactant $[C_nH_{2n+1}NC_5H_5]Br$ (CnPyBr) with n= 12, 14, 16, 18, 20, and 22. Although the materials are amorphous at the microscale, they have crystalline mesoporosity with well-ordered and aligned surfactant-filled cylindrical pores. In addition to high mesoscopic order, the pore-pore separation is adjustable with the surfactant chain length (i.e., value of n) (see Figure 16). The quality of these materials, as judged by the degree of hexagonal order, rivals or exceeds that reported for the highest quality MCM-41 silicates. The materials have the lowest band gap reported so far for mesostructured chalcogenides solids, in the range $1.5 < Eg < 2.3$ eV. The CnPyPtGeS analogues show intense photoluminescence at 77 K when excited with light above the band gap.

### 4.2.2. Mesostructured Iron Sulfide-Contained OINCs

Bulk iron(II) and iron(III) sulfides have shown potential applications as cathodes, high refractive, index materials and catalysts for coal liquefaction. The catalytic formation of complex organic molecules from simple precursors is considered to take place on the surface of iron monosulfide or pyrite as a result of redox reactions between FeS and $FeS_2$. Furthermore, iron sulfides are particularly interesting since they play an important role in life's beginning. Fe–S clusters are exceptional and ubiquitous in biological systems as they provide a major part of the foundation for electron transfer, catalytic, and enzymatic processes in living organisms. Many proteins, such as ferredoxins and hydrogenases, contain Fe–S clusters to utilize their redox properties. If these clusters were to be incorporated in mesostructures, they could produce an interesting biomimetic redox functionality. The construction of mesostructured chalcogenide OINCs that incorporate biological clusters, such as $Fe_4S_4$, has been demonstrated. [64]

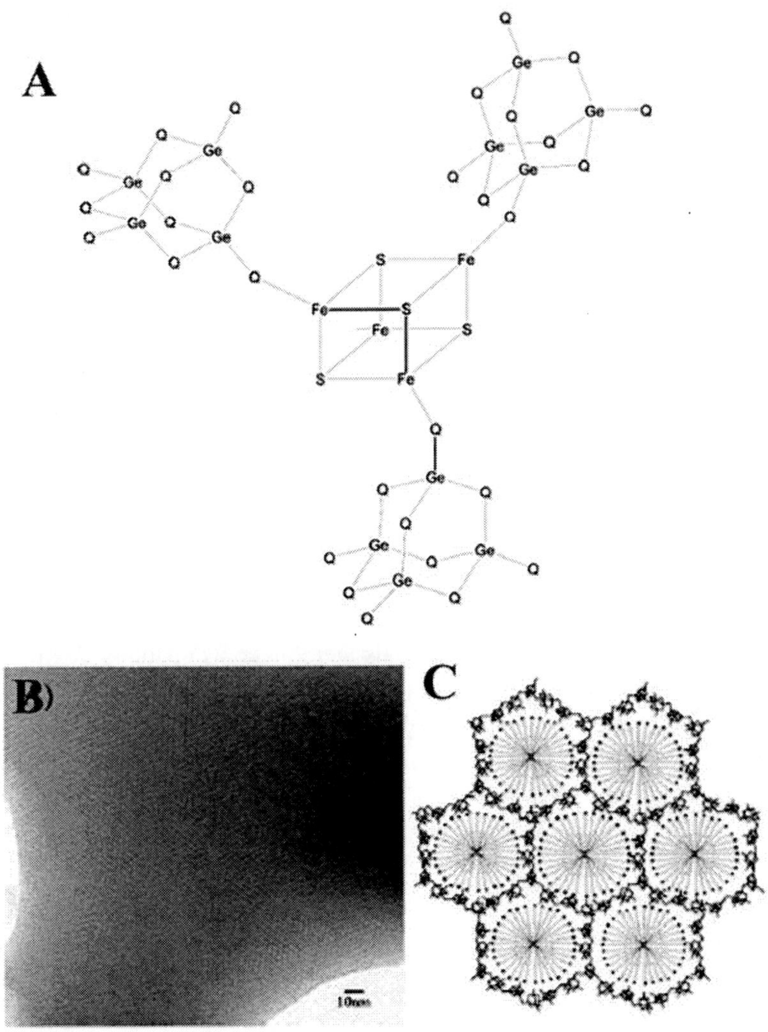


Figure 17. (A) Proposed schematic of the linking of $Fe_4S_4$ clusters with $[Ge_4Q_{10}]^{4-}$ clusters; (B) pore tunnel organization in $Fe_4S_4$–$[Ge_4Q_{10}]^{4-}$ examined with TEM; and (C) schematic of the mesostructured framework of $Fe_4S_4$–$[Ge_4Q_{10}]^{4-}$. [Reprinted with permission from Ref. 66; Trikalitis, P. N. et al. J. Am. Chem. Soc. 127, 3910, 2005 Copyright @ American Chemical Society (2005)].

The first reported mesostructured iron sulfide-contained OINCs are hexagonal mesostructured iron sulfide prepared via a neutral template route or an ionic template. Employment of an ionic surfactant hexadecyltrimethylammonium hydroxide/chloride as a structure-directing agent, and an amorphous iron(III) sulfide, as iron and sulfur sources yielded $C_{16}TMA\text{-}Fe_2S_3\text{-}M$ OINCs with ordered hexagonal mesostructures after hydrothermal synthesis at 110 °C for 16 h. [65] Kanatzidis' group also reported different surfactant-templated, mesostructured phases incorporating the biologically relevant, redox-active $Fe_4S_4$ cluster as a basic tetrahedral building block, linked by redox-inert blocks of $[M_4Q_{10}]^{4-}$ (M=Ge, Sn; Q= S, Se, Te) (see Figure 17A). [56] This methodology of supramolecular assembly of mesostructured $[M_4Q_{10}]^{4-}$ (M=Ge, Sn; Q= S, Se, Te)-based OINCs could be expanded to synthesize stable binary systems including $[Fe_4S_4]$ and $[Ge_4Q_{10}]^{4-}$ (Q=S, Se) clusters as building blocks for the inorganic frameworks. This material possessed hexagonal mesoscopic order and was prepared in the non-aqueous solvent mixture of formamide and dimethyl formamide (DMF) (see Figure 17 B and C). This general method has opened a doorway for preparing a stable hybrid system with multiple functionality.

### 4.2.3. Mesostructured $[SnSe_4]^{4-}$- and $[Sn_2Se_6]^{4-}$-Based OINCs

Other new classes of mesostructured selenide-based OINCs with an open framework based on tetrahedral $[SnSe_4]^{4-}$, $[SnTe_4]^{4-}$, and $[Sn_2Se_6]^{4-}$- ions were reported recently by the Kanatzidis group. [63-64] $[SnSe_4]^{4-}$ and $[SnTe_4]^{4-}$ possess the same tetrahedral structure as $[SiO_4]^{4-}$ and are smaller and heavier than the $[Ge_4Q_{10}]^{4-}$ (Q=S, Se) clusters. Therefore, it was expected that a similar topological structure to silicas and semiconductors with narrower band gaps than metal germanium chalcogenides would be obtained. Meanwhile, the dimeric $[Sn_2Se_6]^{4-}$ on the other hand is special, because it not only gives various mesophases but also leads to a interesting class of cubic mesostructured materials when combined with $Pt^{2+}$. All of These materials are important not only because of both their interesting electronic properties and their varied pore organization, which includes disordered worm-hole, hexagonal, and even cubic phases.

It was known that $[SnSe_4]^{4-}$ These materials were made by spatially controlled assembly of $[SnSe_4]^{4-}$ ions in formamide with various divalent metals such as $Mn^{2+}$, $Fe^{2+}$, $Co^{2+}$, $Zn^{2+}$, $Cd^{2+}$, and $Hg^{2+}$ in the presence of cetylpyridinium surfactant as the template. Their general chemical formula were described as $(CP)_{4-2x}M_xSnSe_4$ (where $1.0<x<1.3$). [63-64] XRD patterns (see Figure 18) and TEM images (see Figure 19) revealed that, under the same experimental conditions, all the materials had hexagonal mesostructures such as silica MCM-41 except for the Zn analogue, which had a cubic mesostructure such as MCM-48. The structure of the Zn materials was dependent on the surfactant concentration. Pore arrangements vary from disordered wormhole to hexagonal, with spacing of 35-40 Å. More importantly in the CP/Zn/SnSe4 system a cubic mesophase was discovered for the first time with Ia-3d symmetry. When the concentration was reduced by 50 %, the product had an hexagonally ordered mesophase with a uniform pore size. The presence of the tetrahedral $[SnSe_4]^{4-}$ ion clusters was demonstrated by $^{119}Sn$ Mossbauer spectroscopy. The UV-visible/near-IR spectroscopy showed the band-gap transition in the energy range of 1.4 ± 2.4 eV, similar to those values of CdTe (1.45 eV) and GaAs (1.54 eV) semiconductors.

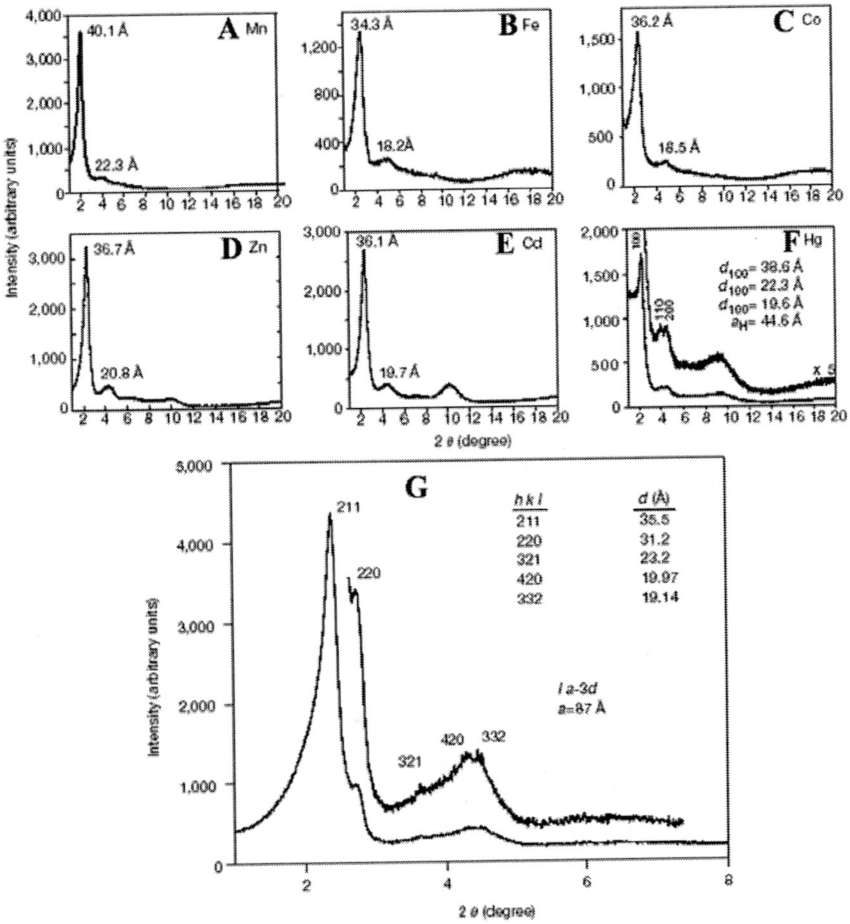


Figure 18. X-ray patterns of mesostructured chalcogenide-based OINCs. (A) CP-Mn/$SnSe_4$; (B) CP-Fe/$SnSe_4$; (C) CP-Co/$SnSe_4$; (D) h-CP-Zn/$SnSe_4$; (E) CP-Cd/$SnSe_4$; (F) CP-Hg/$SnSe_4$; and (G) c-CP-Zn/$SnSe_4$. [Reprinted with permission from Ref. 67; Korlann, S. D. et al. J. Am. Chem. Soc. 127, 12516, 2005 Copyright @ American Chemical Society (2005)].

As for $[SnTe_4]^{4-}$ ions, its solution behavior of $[SnTe_4]^{4-}$ is similar to that of $[SnSe_4]^{4-}$. Initial attempts to synthesize mesostructured materials by using $[SnTe_4]^{4-}$ as the precursor and various linkage metals, e.g., $Ni^{2+}$, $Co^{2+}$, $Zn^{2+}$, $Cd^{2+}$, $Hg^{2+}$, in formamide solvent resulted in either highly disordered or completely amorphous phases. The use of $Pt^{2+}$ with its slower reaction kinetics gave more promising results. [66] One reason might be the faster rate of $Pt^{2+}/[SnTe_4]^{4-}$ coupling compared with the $Pt^{2+}/[SnSe_4]^{4-}$ system. Mesostructured tellurides have been investigated less, because the precursor anions $[MTe_4]^{4-}$, $[M_2Te_6]^{4-}$, and $[M_4Te_{10}]^{4-}$ (M=Ge, Sn) are more difficult to prepare and handle. Whereas the $[SnSe_4]^{4-}/Pt^{2+}$ combination afforded materials with excellent long range hexagonal mesoscopic order, similarly ordered mesostructured solids based on $[SnTe_4]^{4-}$ anions proved more challenging, and disordered Pt/Sn/Te materials were obtained. [67] Despite this, these reactions can succeed in yielding ordered materials if proper conditions and surfactants are used. Furthermore, the heavier telluride-based mesostructured materials are very attractive, because they can possess narrower energy band gaps and therefore potentially interesting electrical properties.

$C_nN(CH_3)(CH_2\text{-}CH_2OH)_2Pt\_SnTe_4$ shows that the band gaps are narrower in the range 0.5–0.7 eV and are the lowest among the mesostructured chalcogenide solids observed so far.

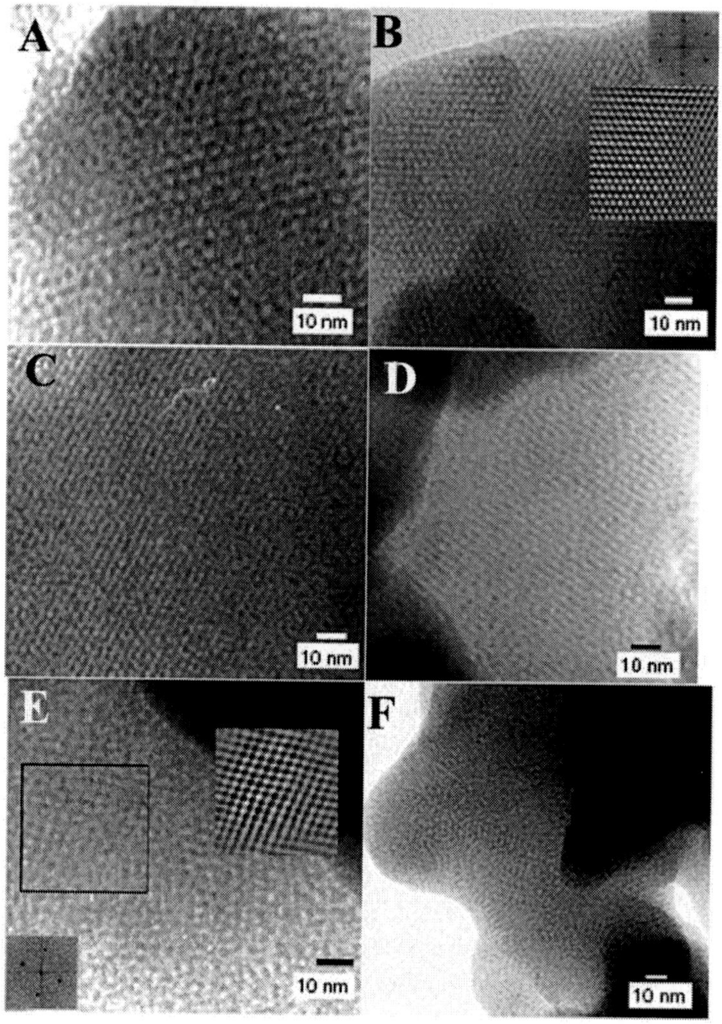


Figure 19. TEM images of mesostructured chalcogenide-based OINCs. (A) h-CP-Zn/$SnSe_4$ looking down the pore channel axis; (B) CP-Hg/$SnSe_4$; (C) CP-Hg/$SnSe_4$ perpendicular to the channel axis; (D) CP-Mn/SnSe4 perpendicular to the channel axis; (E) c-CP-Zn/$SnSe_4$. The cubic arrangement is shown in the boxed area; and (F) CP-Cd/$SnSe_4$. Disordered regions with wormhole arrangement. [Reprinted with permission from Ref. 67; Korlann, S. D. et al. J. Am. Chem. Soc. 127, 12516, 2005 Copyright @ American Chemical Society (2005)].

As for $[Sn_2Se_6]^{4-}$ ions, protons shift the equilibrium from $[SnSe_4]^{4-}$ to $[Sn_2Se_6]^{4-}$ and finally to $[Sn_4Se_{10}]^{4-}$ with greater acid concentration. The linking metal ions dynamically influence the distribution of anionic species such as $[SnSe_4]^{4-}$, $[Sn_2Se_6]^{4-}$, and $Se^{2-}$. Their distribution would be governed by the Lewis acidity and metal ion concentration, as well as its coordination preference and temperature. The NMR results suggest that in formamide the inclusion of only $[SnSe_4]^{4-}$ ions in the structure is unlikely. Instead, $[Sn_2Se_6]^{4-}$ and $Se^{2-}$, or a combination of all three $[SnSe_4]^{4-}/[Sn_2Se_6]^{4-}/Se^{2-}$ are incorporated in the framework, each playing a specific role in building the porous structure. When building blocks with squareplanar bonding topology such as $Pt^{2+}$ and $[Sn_2Se_6]^{4-}$ were combined in the presence of $C_nH_{2n+1}Py^+$ (n = 18–20), readily and reproducibly cubic mesostructures are obtained. [68] In fact, these materials formed so readily that they even adopted well-shaped cubic bodies (cubosomes) with single-crystal appearance (see Figure 20A). The X-ray diffraction pattern confirmed that these structures are the gyroid Ia-3d symmetry (see Figure 20B). The continuous Pt/Sn/Se framework that forms by the reaction of $[Sn_2Se_6]^{4-}$ and $Pt^{2+}$ ions (c-$C_{20}$PyPtSnSe) adopts a cubic double gyroid structure similar to that of MCM-48 (Figure 20C). [69] On the other hand, by replacing the $Pt^{2+}$ with other linkage metals ($Zn^{2+}$, $Cd^{2+}$, $Ga^{3+}$) for connecting $[Sn_2S_6]^{4-}$ clusters in the presence of cetylpyridinium (CP) surfactant molecules in formamide solution under basic conditions, new hexagonal mesophases designated as $(CP)_xM_ySn_2S_6$ ($M=Zn^{2+}$, $Cd^{2+}$, and $Ga^{3+}$) are synthesized (see Figure 21A and B), which are semiconductors in the energy range $2.5 < Eg < 3.1$ eV (see Figure 21C). [70] The mesostructured $(CP)_xM_ySn_2S_6$ show photoluminescence, when excited with light above the band gap (see Figure 21D).

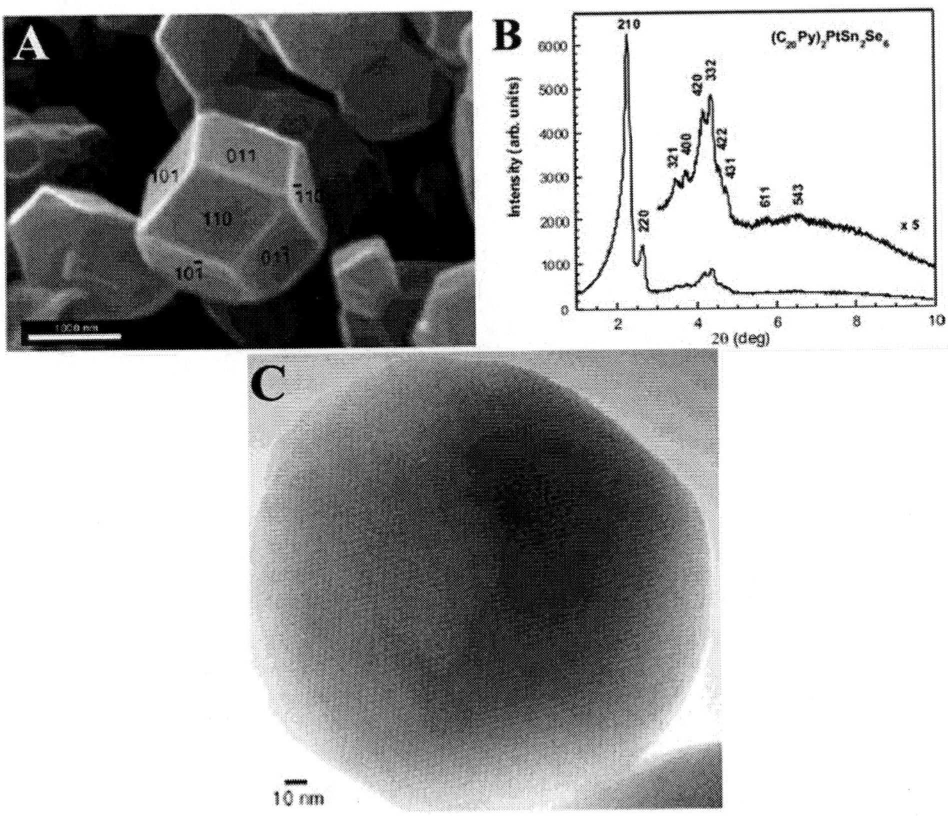


Figure 20. (A) High-resolution SEM images of c-$C_{20}$PyPtSnSe single crystals; (B) X-ray diffraction pattern of the c-$C_{20}$PyPtSnSe single crystals, indexable to Ia-3d symmetry; and (C) HRTEM image of a c-$C_{18}$PyPtSnSe single particle viewed down to the (110) cubic plane. [Reprinted with permission from Ref. 68; Trikalitis, P. N. et al. J. Am. Chem. Soc. 124, 12255, 2002 Copyright @ American Chemical Society (2002)].

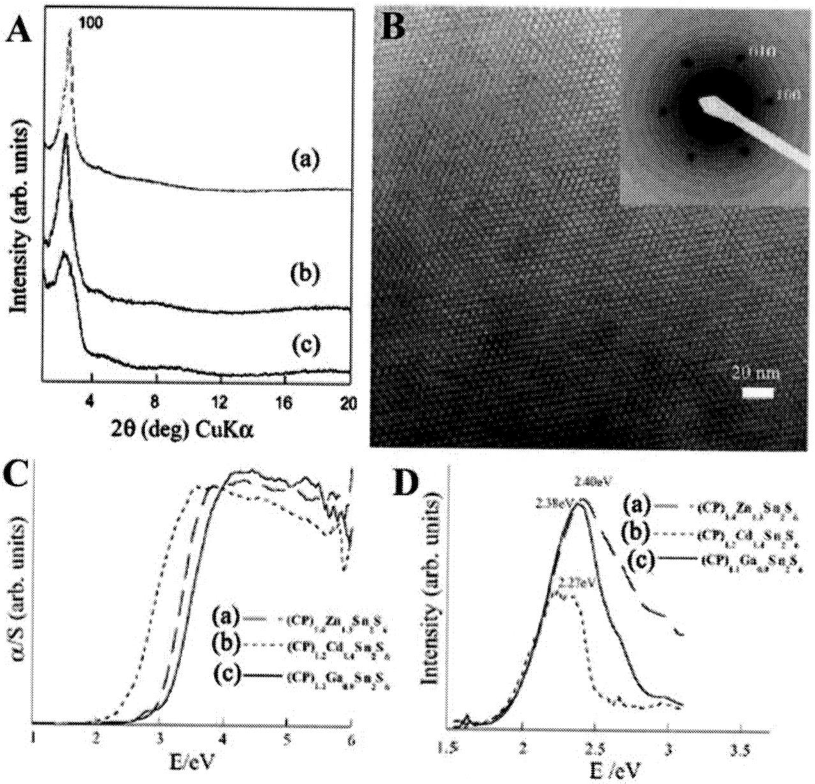


Figure 21. (A) X-ray diffraction patterns; (B) typical HRTEM image; (C) room-temperature solid-state optical absorption spectra; and (D) photoluminescence spectra of $(CP)_xMySn_2S_6$ (M= $Zn^{2+}$, $Cd^{2+}$, $Ga^{3+}$). (a), (b), and (c) represent $(CP)_{1.4}Zn_{1.3}Sn_2S_6$, $(CP)_{1.2}Cd_{1.4}Sn_2S_6$, and $(CP)_{1.1}Ga_{0.9}Sn_2S_6$, respectively. [Reprinted with permission from Ref. 70; Rangan, K. K. et al. Nano Lett., 2, 513, 2002 Copyright @ American Chemical Society (2002)].

## 4.3. Summary

Generally the formation of ordered mesostructured chalcogenide-based OINCs in aqueous solutions is difficult because of a very low solubility of chalcogenides or chalcogenide precursors in alcohol and aqueous solutions and the weak interaction between the chalcogenide clusters and the surfactant micelles. Thus, true-liquid-crystal-phases formed in concentrated nonionic-amphiphilic-surfactant aqueous solutions are doped with the metal ions on the hydrophilic heads of the surfactant molecules, and then the semiconductor chalcogenides are grown in a water-containing liquid crystal by the reaction of $H_2S$ or $H_2Se$ with the dissolved metal ions, which has been discussed in the section "4.1 Framework-Disconnected Mesostructured OINCs". In this approach, both chemical nature and structure of the amphiphile in direct templating are very important. Moreover, the order of the mesostructures in the obtained OINCs was even found to be dependent on the anion of the metal salts, and not all of the chalcogenides can form ordered mesostructured OINCs because of the disconnected chalcogenide nanocrystal frameworks.

Alternatively, because chalcogenide clusters are more easily assembled via coordination reactions between cluster anions and linking metal cations, the connected chalcogenide frameworks can form via the following procedure described in the section "4.2 Framework-Connected Mesostructured OINCs". TMA(Tetramethylammonium)$_4$-$Ge_4S_{10}$(or other chalcogenide clusters) and cationic CTABr in formamide at the temperature above CMC form a lyotropic liquid crystalline phase. In this liquid crystal, the hydrophobic chains of the surfactant are aggregated into cylindrical or micellar species with the $TMA^+$, $[Ge_4S_{10}]^{4-}$, and $Br^-$ in the solvent around the head groups. As transition metals ($Co^{2+}$, $Cu^+$, $Zn^{2+}$, $Ni^{2+}$) are added, they are rapidly coordinated by the tetradentate $[Ge_4S_{10}]^{4-}$ clusters, "freezing" the $M/Ge_4S_{10}$ framework around the surfactant and immediately precipitating from solution.

In both of the above synthesis procedures, the supramolecular assembly and the direct templating in a lyotropic organic liquid crystal have been applied in the successful synthesis of surfactant-chalcogenide OINCs in the form of powders. It is desired to to directly obtain the OINCs in the form of thin films for more extensive applications. The fillowing section will discuss on the nonaqueous synthesis of mesostructured OINCs especially for thin films, membranes, and monoliths by the evaporation-induced self-assembly (EISA) approach.

# 5. Thin Film Technology: Evaporation-Induced Self-assembly (EISA)

The evaporation-induced self-assembly has been successfully applied in the synthesis of mesostructured oxide films. [71-73] The mesophases were formed through both a suppression of inorganic polymerization coformed with the formation of the micelles of surfactants and the spontaneous organization of the inorganic precursors and the micelles of surfactants through noncovalent interactions such as hydrogen bonding, Van der Waals forces, and electrostatic forces, without external intervention. [74-76] These achievements are very useful for the rapid production of nanostructured surfactant-chalcogenide OINCs including thick film, membranes, and monoliths. Combining the spin-coating and dip-coating techniques to quickly evaporate of solvents and fast solidify the inorganic-organic nanostructured phases, the OINCS with well-ordered nanostructures can be obtained.

A general definition of self-assembly is the spontaneous organization of materials through noncovalent interactions (hydrogen bonding, Van der Waals forces, electrostatic forces, etc.) with no external intervention. Self-assembly typically employs asymmetric molecules that are pre-programmed to organize into well-defined supramolecular assemblies. Most common are amphiphilic surfactant molecules or polymers composed of hydrophobic and hydrophilic parts (see Figure 4 and 5). In aqueous solution above the critical micelle concentration (CMC), surfactants assemble into micelles, spherical or cylindrical structures that maintain the hydrophilic parts of the surfactant in contact with water while shielding the hydrophobic parts within the micellar interior (see Figure 3 and 9). Further increases in surfactant concentration result in the self-organization of micelles into periodic hexagonal, cubic, or lamellar mesophases (see Figure 4 and 9).

Beginning with a homogeneous solution of soluble silica and surfactant prepared in ethanol/water solvent with $C_0 << CMC$, preferential evaporation of ethanol concentrates the depositing film in water and nonvolatile surfactant and silica species (see Figure 22). The progressively increasing surfactant concentration drives self-assembly inorganic species-surfactant micelles and their further organization into liquid crystalline mesophases. Preexisting, incipient inorganic species-surfactant micelles mesostructures (which exist at solid-liquid and liquid-vapor interfaces at C<CMC) serve to nucleate and orient mesophase development. The result is rapid formation of thin film mesophases that are highly oriented with respect to the substrate surface. Through variation of the initial alcohol/water/surfactant mole ratio it is possible to follow different trajectories in composition space (see Figure 23) and to arrive at different final mesostructures. [77]

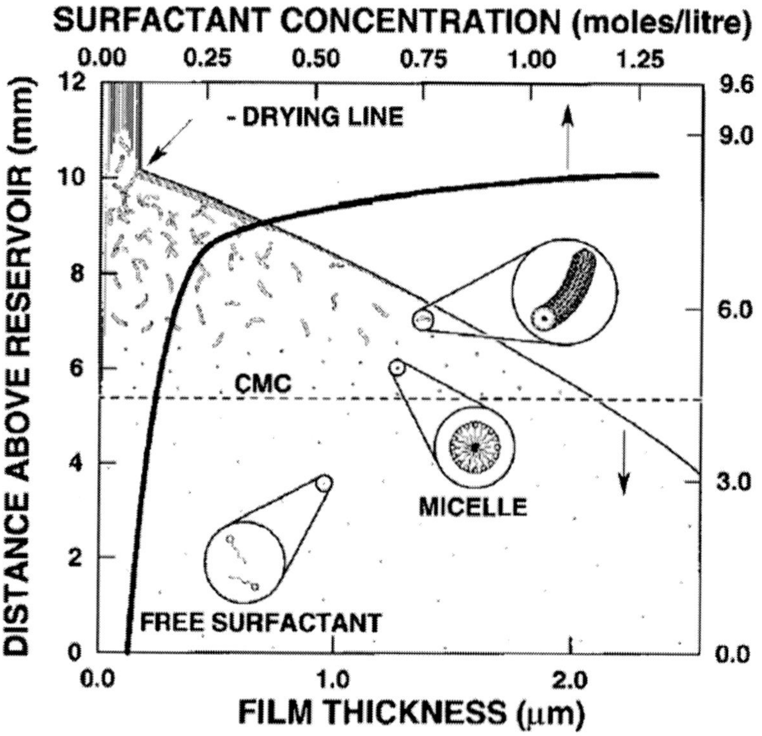


Figure 22. Steady-state film thinning profile established during dip-coating of a complex fluid comprising soluble silica, surfactant, alcohol, and water. Initial surfactant concentration $C_0 \ll$ CMC. Surfactant concentration increases with distance above the reservoir surface. [Reprinted with permission from Ref. 77; Brinker, C. J. et al. Adv. Mater., 11, 579, 1999 Copyright @ Wiley-VCH Verlag GmbH & Co. KGaA (2004)].

Most of chalcogenide precursor salts have low solubility in most of solvents, but they can dissolve in some solvents, such as formamide, with high dielectric constants. Although it takes long days for the evaporation of formamide at room temperature under air condition because of the high boiling point, the spin-coating and dip-coating techniques can accelerate the evaporation of solvents for preparing nanofilms. As an example, molybdenum chalcogenide cluster complexes have provided exciting new developments which relate to solid state chemistry, bioinorganic chemistry, catalysis, and metalloprotein system. It would be desirable to prepare periodic arrangements of molybdenum chalcogenide-organic molecule hybrides for extensive applications.

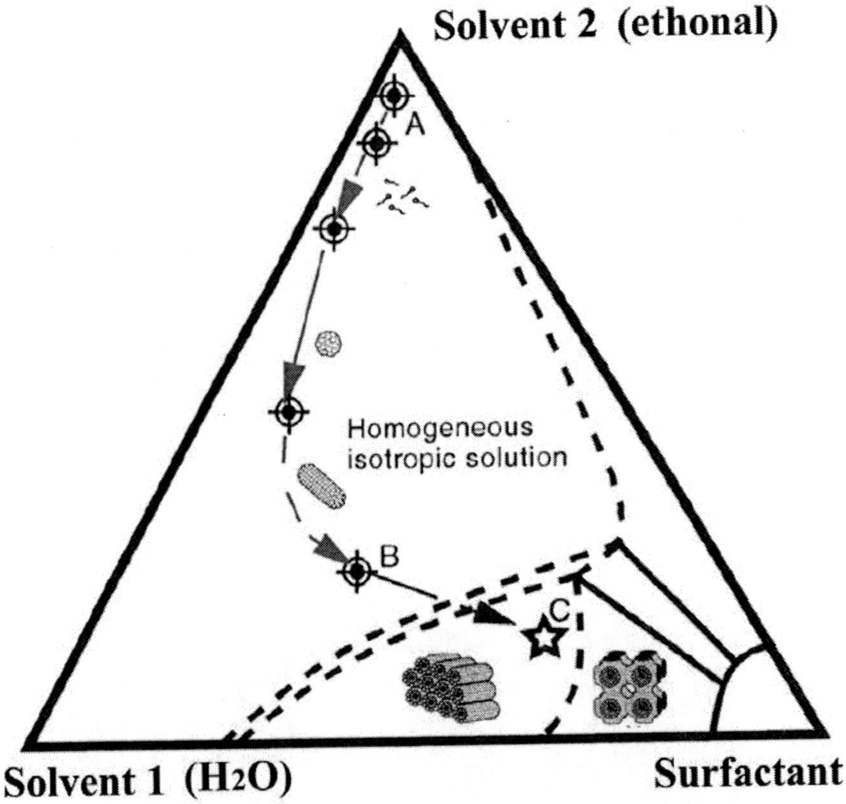


Figure 23. Approximate trajectory taken in ethanol/water/CTAB phase space during dip-coating. Point A corresponds to the initial composition of entrained solution, Point B is near the drying line, and Point C corresponds to the dried film. The Solvent 2 is more volatile than the Solvent 1. [Reprinted with permission from Ref. 77; Brinker, C. J. et al. Adv. Mater., 11, 579, 1999 Copyright @ Wiley-VCH Verlag GmbH & Co. KGaA (2004)]

Since the evaporation-induced self-assembly has been used to prepare mesostructured oxide films. However, most of sulfide precursors have very low solubility in alcohol. Formamide is one of strong polar solvents in which sulfide precursor salts can disslove. However, the boiling point of formamide is about 210 °C, and the sole formamide system cannot be directly used for the preparation of mesostructured sulfide films by the evaporation-induced self-assembly, which is due to the fact that the interaction between the inorganic species and the surfactant micelles is weak, the morphologies of surcatant micelles are easily destoried during a slowly natural evaporation of solvents at room temperatrure in air. The addition of water in formamide decreases both the dielectric constant and boiling point of the mixed solvents. The spin-coating technique to quickly evaporate of solvents and fast solidify the inorganic-organic nanostructured phases thus can be used to prepare the nanofilm hybrids with well-ordered nanostructures. Therefore, the mixture of water and formamide is a better solvent system for most of sulfides.

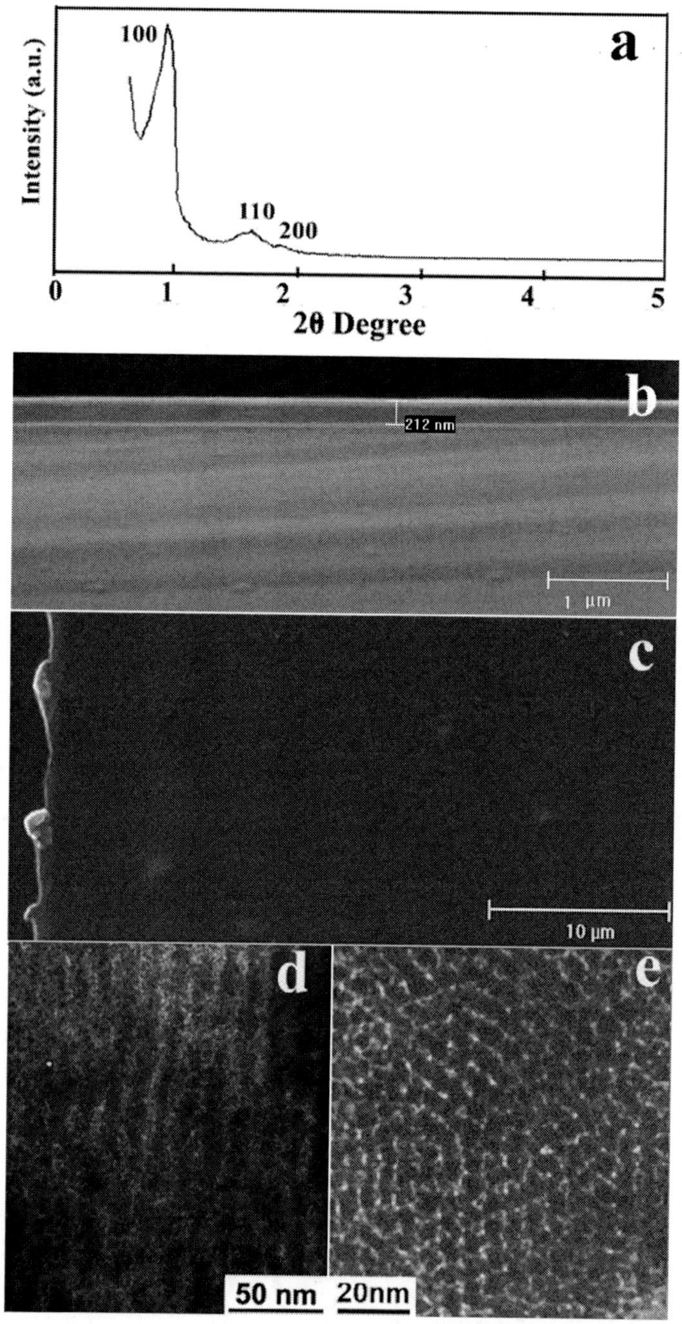


Figure 24. (a) Low-angle XRD pattern of the nanofilm of [$MoS_4^{2-}$]-copolymer hybrids, showing three peaks diagnosticb of hexagonal symmetry, (b) cross-section and (c) plane-view SEM images of the nanofilm of [$MoS_4^{2-}$]-copolymer hybrids, showing uniform thickness and surface, respectively; and (d and e) HRTEM images of the nanofilm of [$MoS_4^{2-}$]-copolymer hybrids, showing ordered hexagonal mesostructures

As shown in Figure 24, nanostructured [$MoS_4^{2-}$]-copolymer inorganic-organic hybrids were synthesized by fast solvent evaporation-induced self-assembly approach of spin-coating technique, using nonionic triblock copolymer of P123 as the template and a single inorganic source of $(NH_4)_2MoS_4$ dissloved in the mixed solvents of formamide and $H_2O$. The morphologies and structures of the products were investigated by XRD (see Figure 24a), SEM (see Figure 24b-c), and HRTEM (see Figure 24d-e). The nanofilm prepared by the spin-coating technique shows well ordered hexagonal mesostructure with unit-cell parameter.

## 6. Band Gap of the OINCs

Given that the thickness of inorganic phase of IONCs tend to be very thin (1–2 nm), one could even envision special quantum confinement effects arising from this type of dimensional reduction. The latter is distinct from the one occurring in conventional nanostructures such as quantum dots and quantum superlattices. The possible connection to quantum confinement comes from the semiconducting inorganic nanophases with uniform nanosize. Meanwhile, the nanocrystals possess a size-dependent optical band gap (Eg) that is larger than that of the bulk system. By analogy, the optical band gaps of the mesostructured framework could be dependent on the wall-thickness. In this context a blue-shift would be expected as the wall thickness decreases, which has been indicated by the mesostructured IONCs of CdS shown in Figure 11.

## 7. Summary and Outlook

The area of metal-chalcogenide-based organic/inorganic nanocomposites prepared by the surfactant self-assembly approach has rapidly become a fascinating new field of research in materials science. The explosion of research in the past decade in this field has promoted considerable progress in both the fundamental understanding of the self-assembly of surfactant molecules and its use in the development and applications of new chalcogenide-based organic/inorganic nanocomposites. As has been discussed within this review, a very wide range of chalcogenide-based organic/inorganic nanocomposites with ordered mesostructures can be generated via true liquid crystal templating. The framework of chalcogenide clusters can be either connected or disconnected, which is determined by the nature of the surfactant and the presence of the metal ion linkers. A simple evaporation-induced self-assembly (EISA) process enables the rapid production of chalcogenide-based organic/inorganic nanocomposites in the form of thin films.

Most of the metal chalcogenides have shown distinguished optical adsorption properties that can be used in the applications of photoctalysis, solar cell, etc. Thus, more metal chalcogenides need to be studied because the reported metal chalcogenides that can form the organic/inorganic nanocomposites are still quite limited. Further, more extensive investigation on the shape-, size-, and structure-dependent optical adsorption properties of the metal-chalcogenide-based organic/inorganic nanocomposites is needed. Since some of metal chalcogenides are photo-sensitizers, some conductive polymers can replace the present surfactants using the assembly of the conductive organic/inorganic nanocomposites with enhanced optical-adsorpion ability.

## References

[1] Schimidt, H.; Kaiser, A.; Patzelt, H.; Sholze, H. "Mechanical and physical properties of amorphous based on $(CH_3)_2SiO$ silica gel"*J. Phys. C* 1982, 12, 275-278.

[2] Livage, J.; Henry, M.; Sanchez, C. "Sol-gel chemistry of transition metal oxides"*Prog. Solid State Chem.* 1988, 18, 259-341.

[3] Avnir, D.; Levy, D.; Reifeld, R. "The nature of the silica cage as reflected by spectral changes and enhanced photostability of trapped Rhodamine 6G."*J. Phys. Chem.* 1984, 88, 5956-5959.

[4] Hammond, P. T. "Form and function in multilayer assembly: New applications at the nanoscale." *Adv. Mater.* 2004, 16, 1271-1293.

[5] Innocenzi, P.; Lebeau, B. "Organic-inorganic hybrid materials for non-linear optics."*J. Mater. Chem.* 2005, 15, 3821-3831.

[6] Sanchez, C.; Lebeau, B.; Chaput, F.; Boilot, J. P. "Optical properties of functional hybrid organic-inorganic nanocomposites." *Adv. Mater.* 2003, 15, 1969-1994.

[7] Eckert, H.; Ward, M. "Controlling the length scale through "soft" chemistry: From organic-inorganic nanocomposites to functional materials."*Chem. Mater.* 2001, 13, 3059-3060.

[8] Wen, J.; Wilkes, G. L. "Organic/inorganic hybrid network materials by the sol-gel approach."*Chem. Mater.* 1996, 8, 1667-1681.

[9] Sanchez, C.; Ribot, F. "Design of hybrid of hybrid organic-inorganic materials synthesized via sol-gel chemistry." *New J. Chem.* 1994, 18, 1007-1047.

[10] Corma, A. "From microporous to mesoporous molecular sieve materials and their use in catalysis."*Chem. Rev.* 1997, 97, 2373-2419.

[11] Yanagisawa, T.; Shimizu, T.; Kuroda, K.; Kato, C. "The Preparation of Alkyltriinethylaininonium–Kaneinite Complexes and Their Conversion to Microporous Materials."*Bull. Chem. Soc. Jpn.* 1990, 63, 988-992.

[12] Kresge, C. T.; Leonowicz, M. E.; Roth, W. J.; Vartuli, J. C.; Beck, J. S. "Ordered mesoporous molecular-sieves synthesized by a liquid-crystal template mechanism."*Nature* 1992, 359, 710-712.

[13] Beck, J. S.; Vartuli, J. C.; Roth, W. J.; Leonowicz, M. E.; Kresge, C. T.; Schmitt, K. D.; Chu, C. T.; Olsen, D. H.; Sheppard, E. W.; McCullen, S. B.; Higgins, J. B.; Schlenker, J. L. "A new family of mesoporous molecular sieves prepared with liquid crystal templates."*J. Am. Chem. Soc.* 1992, 114, 10834-10843.

[14] Ramirez, A. P.; Cava, R. J.; Krajewski, J. "Colossal magnetoresistance in Cr-based chalcogenide spinels"*Nature* 1997, 386, 156-150.

[15] Hemberger, J.; Lunkenheimer, P.; Fichtl, R.; Krug von Nidda, H.-A.; Tsurkan, V.; Loidl, A. "Relaxor ferroelectricity and colossal magnetocapacitive coupling in ferromagnetic $CdCr_2S_4$."*Nature* 2005, 434, 364-367.

[16] Weber, S.; Lunkenheimer, P.; Fichtl, R.; Hemberger, J.; Tsurkan, V.; Loidl, A. "Colossal magnetocapacitance and colossal magnetoresistance in $HgCr_2S_4$."*Phys. Rev. Lett.* 2006, 96, 157202.

[17] Bu, X.; Zheng, N.; Feng, P. "Tetrahedral chalcogenide clusters and open frameworks." *Chem. Eur J.* 2004, 10, 3356-3362.

[18] Dance, I.; Fisher, K. *Metal Chalcogenide Cluster Chemistry.* In Progress in Inorganic Chemistry; Karlin, K. D., Ed.; John Wiley &Sons: New York, 1994; Vol. 41, pp 637-803.

[19] Feng, P.; Bu, X.; Zheng, N. "The interface chemistry between chalcogenide clusters and open framework."*Acc. Chem. Res.* 2005, 38, 293-303.

[20] Winiarz, J. G.; Zhang, L.; Lal, M.; Friend, C. S.; Prasad, P. N. "Observation of the photorefractive effect in a hybrid organic-inorganic nanocomposite chalcogenides."*J. Am. Chem. Soc.* 1999, 121, 5287-5295.

[21] Osenar, P.; Braun, P. V.; Stupp, S. I. "Lamellar semiconductor-organic nanostructures from self-assembled templates." *Adv. Mater.* 1996, 8, 10221025.

[22] Braun, P. V.; Osenar, P.; Stupp, S. I. "Semiconducting superlattices templated by molecular assemblies." *Nature,* 1996, 380, 325-328.

[23] Stupp, S. I.; Braun, P. V. "Molecular manipulation of microstructures: Biomaterials, ceramics, and semiconductors."*Science,* 1997, 277, 1242-1248.

[24] Li, J.; Kessler, H.; Soulard, M.; Khouchaf, L.; Tuiler, M. H. "Nanosized zinc sulfide obtained in the presence of cationic surfactants."*Adv. Mater.* 1998, 10, 946-949.

[25] Rana, R. K.; Zhang, L.; Yu, J C.; Mastai, Y.; Gedanken, A. "Mesoporous structures from supramolecular assembly of in situ generated ZnS nanoparticles."*Langmuir*, 2003, 19, 5904-5911.

[26] Dag, O; Alayoglu, S.; Tura, C.; Celik, O. "Lyotropic liquid-crystalline phase of oligo(ethylene oxide) surfactant/transition metal salt and the synthesis of mesostructured cadmium sulfide." *Chem. Mater.* 2003, 15, 2711-2717.

[27] Nandhakumar, I. S.; Gabriel, T.; Li, X.; Attard, G. S.; Markham, M.; Smith, D. C.; Baumberg, J. J. "Optical properties of mesoporous II-VI semiconductor compound films." *Chem. Commun.* 2004, 1374-1375.

[28] Markham, M. L.; Baumberg, J. J.; Smith, D. C.; Li, X.; Gabriel, T.; Attard, G. S.; Nandhakumar, I. S. "Birefringent cadmium-telluride-based metamaterial." *Appl. Phys. Lett.* 2005, 86, 011912.

[29] Kanatzidis, M. G. "Beyond silica: Nonoxidic mesostructured materials." *Adv. Mater.* 2007, 19, 1165-1181.

[30] Soler-Illia, G. J. de A. A.; Sanchez, C.; Lebeau, B.; Patarin, J. "Chemical strategies to design textured materials: From microporous and mesoporous oxides to nanonetworks and hierarchical structures." *Chem. Rev.* 2002, 102, 4093-4138.

[31] Schuth, F. "Non-siliceous mesostructured and mesoporous materials ." *Chem Mater.* 2001, 13, 3184-3195.

[32] He, X.; Antonelli, D. "Recent advances in synthesis and applications of transition metal containing mesoporous molecular sieves." *Angew. Chem. Int. Ed.* 2002, 41, 214-229.

[33] Monnier, A.; Schu¨ th, F.; Huo, Q.; Kumar, D.; Margolese, D.; Maxwell, R. S.; Stucky, G. D.; Krishnamurty, M.; Petroff, P.; Firouzi, A.; Janicke, M.; Chmelka, B. F. "Cooperative fromation of inorganic-organic interfaces in the synthesis of silicate mesostructures."*Science* 1993, 261, 1299-1303.

[34] Huo, Q.; Margolese, D. I.; Ciesla, U.; Demuth, D. K.; Feng, P.; Gier, T. E.; Sieger, P.; Firouzi, A.; Chmelka, B. F.; Schu¨ th, F.; Stucky, G. D. "Organization of Organic Molecules with Inorganic Molecular Species into Nanocomposite Biphase Arrays."*Chem. Mater.* 1994, 6, 1176-1191.

[35] Evans, D. F.; Wennerstro¨m, H., Eds.; The Colloidal Domain Where Physics, Chemistry, Biology and Technology Meet, VCH Publishers: Weinheim, Germany, 1994.

[36] Alexandridis, P.; Holzwarth, J.F.; Hatton, T.A. "Micellization of Poly(ethylene oxide)-Poly(propylene oxide)-Poly(ethylene oxide) Triblock Copolymers in Aqueous Solutions: Thermodynamics of Copolymer Association. "*Macromolecules,*1994, 27, 2414-2425.

[37] Israelachvili, N.; Mitchell, D. J.; Niham, B. W. "Theory of self-assembly of hydrocarbon amphiphiles into micelles and bilayers. " *J. Chem. Soc., Faraday Trans,* 1976, 72, 1525-1568.

[38] Israelachvili, J., Ed.; Intermolecular and Surface Forces, 2nd ed.; Academic Press: New York, 1992

[39] Huo, Q.; Leon, R.; Petroff, P. M.; Stucky, G. D. "Mesostructured design with gemini surfactants supercage formation in a 3-demisional hexagonal array"*Science*, 1995, 268, 1324-1327.

[40] Huo, Q.; Margolese, D. I.; Ciesla, U.; Feng, P.; Gier, T. E.; Sieger, P.; Leon, R.; Petroff, P. M.; Schu¨ th, F.; Stucky, G. D. "Generalize d synthesis of periodic surfactant inorganic composite materials."*Nature*, 1994, 368, 317-321.

[41] Tanev, P. T.; Chibwe, M.; Pinnavaia, T. J. "Titanium-containing mesoporous molecular-sieves for catalytic oxidation of aromatic compounds." *Nature,* 1994, 368, 321-323.

[42] Tanev, P. T.; Pinnavaia, T. J. "A netural templating route to mesoporous molecular-sieves."*Science,* 1995, 267, 865-867.

[43] Bagshaw, S. A.; Prouzet, E.; Pinnavaia, T. J. "Templating of mesoporous molecular-sieves by nonionic polyethylene oxide surfactants." *Science* 1995, 269, 1242-1244.

[44] Bagshaw, S. A.; Pinnavaia, T. J. "Mesoporous alumina molecular sieves." *Angew. Chem., Int. Ed.* 1996, 35, 1102-1105.

[45] Voegtlin, A. C.; Ruch, F.; Guth, J. L.; Patarin, J.; Huve, L. "F-mediated synthesis of mesoporous silica with ionic- and non-ionic surfactants. A new templating pathway." *Microporous Mater.* 1997, 9, 95-105.

[46] Ying, J. Y.; Mehnert, C. P.; Wong, M. S. "Synthesis and applications of supramolecular-templated mesoporous materials."*Angew. Chem. Int. Ed.* 1999, 38, 56-77.

[47] De Gennes, P. G.; Taupln, C. "Microemulsions and the flexibility of oil/water interfaces."*J. Phys. Chem.* 1982, 86, 2294-2304.

[48] Bao, N.; Shen, L.; Yanagisawa, K.; Domen, K.; Grimes, C. A.; Gupta, A. "Self-assembly of nanostructured [$MoS_4^{2-}$]-copolymer hybrids by solvent evaporation induction." *Materials Letters* 2008, 62, 648–650.

[49] Braun, P. V.; Osenar, P.; Tohver, V.; Kennedy, S. B.; Stupp, S. I. "Nanostructure templating in inorganic solids with organic lyotropic liquid crystals." *J. Am. Chem. Soc.* 1999, 121, 7302-7309.

[50] Tohver, V.; Braun, P. V.; Pralle, M. U.; Stupp, S. I. "Counterion effects in liquid crystal. Templating of nanostructured CdS."*Chem. Mater.* 1997, 9, 1495-1498.

[51] Li, J.; Delmotte, L.; Kessler, H. "Synthesis and characterization of a novel mesostructured layered tin(IV) sulfide."*Chem. Commun.* 1996, 1023-1024.

[52] Jiang, T.; Ozin, G. A. "Tin(IV) sulfide alkylamine composite mesophase: a new class of thermotropic liquid crystals" *J. Mater. Chem.* 1997, 7, 2213-2222.

[53] Froba, M.; Oberender, N. "First synthesis of mesostructured thiogermanates."*Chem. Commun.* 1997, 1729-1793.

[54] Warnheim, T.; Jonsson, A. "Phase diagrams of alkyltrimethylammonium surfactants in some polar solvents." *J. Colloid Interface Sci.* 1988, 125, 627-633.

[55] Rangan, K. K.; Billinge, S. J. L.; Petkov, V.; Heising, J.; Kanatzidis, M. G. "Aqueous mediated synthesis of mesostructured manganese germanium sulfide with hexagonal order." *Chem. Mater.* 1999, 11, 2629-2632.

[56] Trikalitis, P. N.; Bakas, T.; Papaefthymiou, V.; Kanatzidis, M. G. "Supramolecular assembly of hexagonal mesostructured germanium sulfide and selenide nanocomposites incorporating the biologically relevant Fe4S4 cluster."*Angew. Chem., Int. Ed.* 2000, 39, 4558-4562.

[57] Davis M. E. "Ordered porous materials for emerging applications." *Nature,* 2002, 417, 813-821

[58] Bonhomme, F.; Kanatzidis, M. G. "Structurally characterized mesostructured hybrid surfactant-inorganic lamellar phases containing the adamantane $[Ge_4S_{10}]^{4-}$ anion: Synthesis and properties."*Chem. Mater.* 1998, 10, 1153-1159.

[59] MacLachlan, M. J.; Coombs, N.; Bedard, R. L.; White, S.; Thompson, L. K.; Ozin, G. A. "Mesostructured metal germanium sulfides." *J. Am. Chem. Soc.* 1999, 121, 12005-12017.

[60] MacLachlan, M. J.; Coombs, N.; Ozin, G. A. "Non-aqueous supramolecular assembly of mesostructured metal germanium sulphides from $(Ge_4S_{10})^{4-}$ clusters." *Nature,* 1999, 397, 681-684.

[61] Rangan, K. K.; Trikalitis, P. N.; Kanatzidis, M. G. "Light-emitting meso-structured sulfides with hexagonal symmetry: Supramolecular assembly of $(Ge_4S_{10})^{4-}$ clusters with trivalent metal ions and cetylpyridinium surfactant."*J. Am. Chem. Soc.* 2000, 122, 10230-10231.

[62] Rangan, K. K.; Trikalitis, P. N.; Bakas, T.; Kanatzidis, M. G. "Hexagonal mesostructured chalcogenide frameworks formed by linking $(Ge_4S_{10})^{4-}$ (Q = S, Se) clusters with $Sb^{3+}$ and $Sn^{4+}$."*Chem. Commun.* 2001, 809-810.

[63] Trikalitis, P. N.; Rangan, K. K.; Kanatzidis, M. G. "Platinum chalcogenido MCM-41 analogues. High hexagonal order in mesostructured semiconductors based on $Pt^{2+}$ and $(Ge_4S_{10})^{4-}$ (Q=S, Se) and $[Sn^4Se^{10}]^{4-}$ adamantane clusters. "*J. Am. Chem. Soc.* 2002, 124, 2604-2613.

[64] Beinert, H.; Holm, R. H.; Munck, E. "Iron-sulfur clusters: Nature's modular, multipurpose structures." *Science* 1997, 277, 653-659.

[65] Li, J.; Nazar, L. F. "Mesostructured iron sulfides." *Chem. Commun.* 2000, 1749-1750.

[66] Trikalitis, P. N.; Bakas, T.; Kanatzidis, M. G. "Periodic hexagonal mesostructured chalcogenides based on platinum and $[SnSe_4]^{4-}$ and $[SnTe_4]^{4-}$ precursors. Solvent dependence of nanopore and wall organization"*J. Am. Chem. Soc.* 2005, 127, 3910-3920.

[67] Korlann, S. D.; Riley, A. E.; Kirsch, B. L.; Mun, B. S.; Tolbert, S. H. "Chemical tuning of the electronic properties in a periodic surfactant-templated nanostructured semiconductor." *J. Am. Chem. Soc.* 2005, 127, 12516-12527

[68] Trikalitis, P. N.; Rangan, K. K.; Bakas, T.; Kanatzidis, M. G. "Single-crystal mesostructured semiconductors with cubic Ia(3)over-bard symmetry and ion-exchange properties."*J. Am. Chem. Soc.* 2002, 124, 12255-12260.

[69] Alfredsson, V.; Anderson, M. W. "Structure of MCM-48 revealed by transmission electron microscopy."*Chem. Mater.* 1996, 8, 1141-1146.

[70] Rangan, K. K.; Trikalitis, P. N.; Canlas, C.; Bakas, T.; Weliky, D. P.; Kanatzidis, M. G. "Hexagonal pore organization in mesostructured metal tin Sulfides built with $[Sn_2S_6]^{4-}$ cluster. "*Nano Lett.* 2002, 2, 513.

[71] Yang, H.; Coombs, N.; Sokolov, I.; Ozin, G. A. "Free-standing and oriented mesoporous silica films grown at the air-water interface." *Nature*, 1996, 381, 589-592.

[72] Ogawa, M. "Formation of Novel Oriented Transparent Films of Layered Silica-Surfactant Nanocomposites."*J. Am. Chem. Soc.* 1994, 116, 7941-7942.

[73] Lu, Y.; Fan, H.; Stump, A.; Ward, T. L.; Rieker, T.; Brinker, C. J. "Aerosol-assisted self-assembly of mesostructured spherical nanoparticles." *Nature,* 1999, 398, 223-226.

[74] Lu, Y.; Ganguli, R.; Drewien, C. A.; Anderson, M. T.; Brinker, C. J.; Gong, W.; Guo, Y.; Soyez, H.; Dunn, B.; Huang, M. H.; Zink, J. I. "Continuous formation of supported cubic and hexagonal mesoporous films by sol gel dip-coating." *Nature,1997,* 389, 364-368.

[75] Yun, H.; Miyazawa, K.; Zhou, H.; Honma, I.; Kuwabara, M. "Synthesis of mesoporous thin TiO2 films with hexagonal pore structures using triblock copolymer templates." *Adv. Mater.* 2001, 13, 1377-1380.

[76] Grosso, D.; Babonneau, F.; Albouy, P.; Amenitsch, H.; Balkenende, A. R.; Brunet-Bruneau, A.; Rivory, J. "An in situ study of mesostructured CTAB-silica film formation during dip coating using time-resolved SAXS and interferometry measurements."*Chem. Mater.* 2002, 14, 931-939.

[77] Brinker, C. J.; Lu, Y.; Sellinger, A.; Fan, H. "Evaporation-induced self-assembly: Nanostructures made easy." *Adv. Mater.* 1999, 11, 579-585.

In: New Research on Nanocomposites
Editors: Luis M. Krause et al, pp. 281-306

ISBN: 978-1-60456-799-1


*Chapter 9*

# Polymer Nanocomposites: Achievement of Interfacial Compatibility and Organic-Inorganic Phase Mixing

***Vikas Mittal****

Department of Chemistry and Applied Biosciences,
Institute of Chemical and Bio Engineering,
Swiss Federal Institute of Technology,
ETH Hoenggerberg, 8093, Zurich, Switzerland

## Abstract

Development of organic-inorganic nanocomposites have in the recent years have generated tremendous potential for the use of commodity polymers in more engineering applications. This enhancement is attributed to the significant improvements in the properties of the polymers at minimal inorganic filler volume fractions without affecting the optical density or transparency of the parent polymer matrices. However, the property enhancements have been generally limited to the polar polymers like polyurethanes, epoxies, polyamides etc. owing to the unfavourable thermodynamic and kinetic forces in the case of non-polar polymers like polypropylene and polyethylene leading to poor interfacial compatibility and mixing of the two phases. Conventional surface treatments of the inorganic filler like long alkyl chain ammonium ions are not sufficient to obtain the nanoscale mixing. The use of compatibilizers (low molecular weight polyethylene or polypropylene grafted with maleic anhydride) is common approach to achieve interfacial compatibility, however, the low molecular weight of the compatibilizer has negative affect on the mechanical properties. Recently, more advanced techniques have been developed to circumvent these concerns. Grafting of polymeric oligomers 'to' and 'from' the surface of filler particles helps to achieve longer chain lengths ionically bound to the filler surface. These long chains lead to higher basal spacing and hence increase the susceptibility of the filler exfoliation in the polymer matrix during shearing. The architecture and chemical composition of the chains grafted on the surface can be accurately controlled for the system specific requirement. Controlled living polymerization techniques like nitroxide mediated polymerization or atom transfer radical polymerization can also be employed to achieve very high molecular weights on the surface.

* E-mail address: vikas.mittal@basf.com. Current Address: BASF SE, Germany.

Another simpler approach includes filler treatment with reactive modifications (eg. terminal hydroxyl or carboxyl groups) and subsequent reactions (eg. esterification) on the surface. Polymerization of ethylene or propylene itself with initiator immobilized on the filler surface have also been achieved to more or less success. All these approaches require no compatibilizer and can easily be characterized with the conventional techniques. Apart from that, the developed techniques can also be employed to a wide spectrum of fillers like layered silicates or aluminosilicates, multi-walled nanotubes and even polymer particles.

## Introduction

Polymer-layered silicate nanocomposites have been the subject of intense research in the recent years and have generated tremendous opportunities for the use of commodity polymers in more high-end engineering applications. These organic-inorganic hybrid materials, with phase mixing of the inorganic silicate filler component at a nanometer level, have achieved remarkable enhancements in the composite properties even at very low filler volume fractions [1-10]. Owing to the geometrical parameters of the filler materials i.e. high aspect ratio (and platy nature) and resulting significant interfacial contacts at the interface between the organic and inorganic phases, a synergistic improvement in the composite properties is achieved, which results in properties much higher than any of the parent materials. As compared to conventional micro-composites, these composites also retain the optical density and transparency of the parent polymer materials, which are additionally required especially in packaging industry. Medium charge density montmorillonite from the family of 2:1 layered phyllosilicates generally is a material of choice as reinforcing filler and one such 2:1 aluminosilicate layer has two silica sheets condensed with an alumina sheet [11-13]. Such layers are electrostatically stacked leading to regular gaps or interlayer in between. Partial isomorphic substitution of $Al^{3+}$ with $Mg^{2+}$ in the octahedral sheet of the montmorillonite crystal is commonly observed in nature and it leads to a permanent net negative charge in the structural framework of these layers. This positive charge deficiency is balanced by sorption of extraneous inorganic alkali or alkaline earth metal cations, which are present in the interlayer as well as on the surfaces. To compatibilize the high energy polar surfaces of the clay with the low energy non-polar polymer surfaces, surface treatment of the inorganic mineral is necessary, which is achieved by exchanging the inorganic cations on the layer surface with the organic ammonium ions carrying long alkyl chains [13-16]. Exchange of inorganic cations with organic cations renders the clay organophillic and hydrophobic and lowers the surface energy of the clay layers. It then becomes possible for the polymer chains to diffuse between the expanded clay interlayer and to delaminate the clay stacks to individual layers.

Many techniques has been developed to synthesize the polymer clay nanocomposites, viz. monomer intercalation method, common solvent or solution method and melt intercalation etc. In monomer intercalation process, the layered silicate is swollen in liquid monomer and polymerization of the monomer is then carried out in the interlayers of the clay mineral. In common solvent method or solution method, the layered silicate is delaminated to the individual layers in a solvent in which the polymer is also soluble and the polymer is then crosslinked along with evaporation of the solvent. Melt intercalation method is by far the most commonly employed method to achieve the composites with thermoplastic matrices especially polyolefins. The organically modified mineral is added to the polymer melt at high temperature and sheared extensively so as to obtain the homogenous dispersion of the clay layers in the matrix. In case of non-polar matrix polymers like polyethylene and polypropylene, the surface energy mismatch of the polymer and inorganic mineral does not allow homogenous mixing of phases; therefore, a small amount of low molecular weight compatibilizer, which has both polar and non polar components, is also added during the melt intercalation process.

Three main types of ideal composite morphologies have been observed in these composites depending on the nature of the components used (layered silicate, organic cation and polymer matrix) and the method of preparation [17]. Inability of the polymer to intercalate between the silicate sheets causes a phase separated composite to form, whose properties stay in the same range as traditional microcomposites. Two types of nanocomposites can be achieved beyond this classical family of composites. Intercalated structure in which a single (and sometimes more than one) extended polymer chain is intercalated between the silicate layers resulting in a well ordered multilayer morphology built up with alternating polymeric and inorganic layers. Complete and uniform dispersion of the silicate layers in a continuous polymer matrix leads to an exfoliated or delaminated structure. However, as mentioned in the later sections of this chapter, these morphologies are ideal and in reality, a mixture of intercalated and exfoliated inorganic filler platelets in the polymer matrix is observed mostly.

## Conventional Nanocomposite Systems

This approach was reported by the very first commercial nanocomposite systems. In this approach, inorganic surfaces were modified with ammonium ions with generally dodecyl, hexadeyl and octadecyl alkyl chains in their chemical structure. Apart from that, ammonium ions of either one or two such alkyl chains in the structure were employed. By using such organically modified montmorillonites, nanocomposites with a wide range of thermoplastic and thermoset polymer matrices have been synthesized during the recent years. Modifications with alkyl chains carrying terminal hydroxyl or carboxyl groups have also been used with some polar polymer matrices. Based on the nature of the system studied, a varying degree of success in nanometer scale dispersion of the filler in the polymer matrix and improvement in the polymer properties has been achieved. More significant enhancements in the composite properties have been gained in the case of polar matrices than in the case of non polar matrices like polyethylene or polypropylene. It is because of the better match of surface energies of the organic and inorganic components in the polar polymers, whereas the non-polar polymer systems pose kinetic as well as thermodynamic hindrance to the polymer chains to intercalate inside the clay interlayers. Figure 1 depicts such a comparison, where the TEM micrographs of the polypropylene and polyurethane nanocomposite systems have been compared. In the polyurethane composite, extensive exfoliation of the filler platelets in single layers is observed with only a minor part of the filler present as stacks of two or three layers. On the other hand, the polypropylene composite shows the presence of only thick stacks of platelets which are very thinly intercalated and there is no or minimal exfoliation. It has to be noted that the composite with propylene was prepared without the use of commonly added low molecular weight compatibilizers.

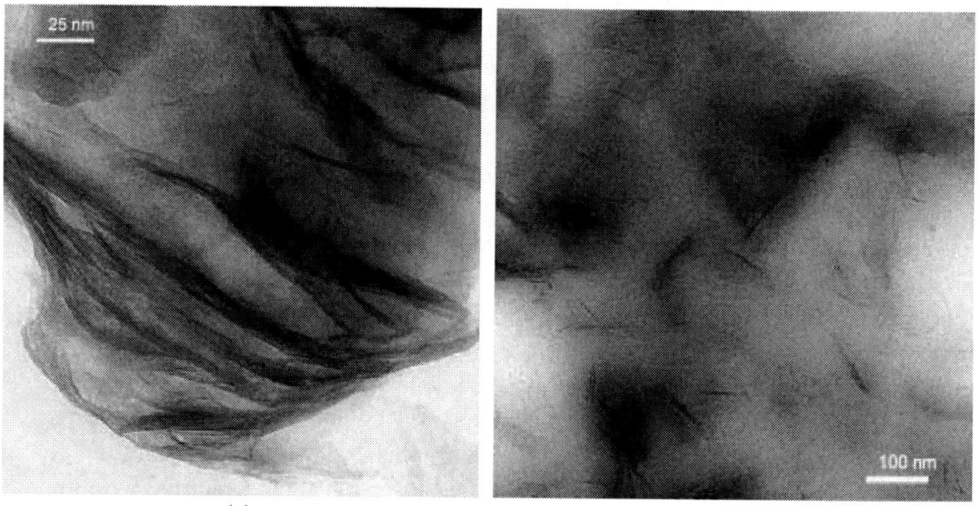


(a) (b)

Figure 1. TEM micrographs of polymer nanocomposites containing 3 vol% of the organically modified filler in (a) Polypropylene and (b) Polyurethane.

A number of studies on polyurethane nanocomposites have been reported in the literature, in which impressive improvements in the mechanical performance was achieved with a small amount of filler [18-23]. Polyurethane nanocomposites were observed to have an increase of more than 100% in tensile strength, tensile modulus and strain-at-break at a loading of 10 wt% of filler modified with protonated octadecyl amine [19]. An increase of 30-40% in the tensile strength and elongation at break of the polyurethane composites with 8 wt% of the bis(2-hydroxyethyl)dodecylmethylammonium modified silicate was observed, whereas the Young's modulus was found to decrease [20]. Chen et al also reported a two-fold increase in the tensile strength and a three-fold increase in the elongation for the polyurethane nanocomposites [21]. Only a few studies on the other composite properties like permeation-barrier of polyurethane nanocomposites can be found in the literature [9, 24-26]. Similarly, enhanced mechanical properties have also been reported for the epoxy-clay nanocomposites [5, 27-35]. The use of different kinds of surface modifications has been reported, the common being bis(2-hydroxyethyl)methyltallowammonium, octadecyltrimethylammonium, protonated 12-amino dodecanoic acid etc. In one of the studies, a phenomenon of "degellation" or "deexfoliation" of the organo-montmorillonite, containing bis(2-hydroxyethyl)methyltallowammonium as surface modification, was reported on addition of polar additives such as amines as crosslinking agents [27]. It was suggested that bifunctional amine molecules may be able to bridge the silicate layers or the N-H groups in the primary and secondary amines may be polar enough to cause reaggregation of silicate layers. The other possible reason for this behavior may be the presence of excess molecules of surface modifier on the clay surface, which may not be ionically attached to it. These molecules being free on the clay surface may easily interact with the amine or other polar additives thus coagulating the whole dispersion [36]. In the later sections of chapter, it would be proved that the composite properties and morphology are a function of the extent of clean surface. Similarly as in polyurethane and other polar polymer systems, relatively lesser attention has been given towards other important composite properties eg., barrier properties of polymer-layered silicate nanocomposites [5,9,37-40]. It has been postulated in literature that matching the surface energy of the organically modified clay with that of the matrix as well as tethering polymer molecules to the silicate layer surface enhances the exfoliation and decreases the permeation coefficient. In one of the very first studies on the permeation properties of epoxy-clay nanocomposites, it was observed that the permeability coefficient of the epoxy matrix was reduced to one-fourth at 5 vol % of benzyldibutyl(hydroxyethyl)ammonium modified montmorillonite loading and the reduction was attributed to the tortuous pathway the gas molecules have to cover during their random walk to penetrate the composite. The transmission rate of water vapor through the composites was more influenced by the permeant-composite interactions and hence the hydrophobicity of the monolayer covering the inclusions surface. At 5 vol % benzyldimethylhexadecylammonium modified montmorillonite loading, the relative vapor transmission rate was reduced to half [5]. It indicated that for achieving better composite properties, there is a requirement to achieve the match of the surface energy between the polymer and the organically modified clay surfaces. This phenomenon is also represented in Figure 2. It was shown here that through the X-ray diffraction measurements that when epoxy prepolymer was added to the clay suspensions in solvent, depending on the interaction or compatibility of the epoxy prepolymer and the clay surface modification, generally a deswelling (i.e. a negative interaction as shown in the figure) of the clay suspension took place which was indicated by a decrease in basal plane spacing. These systems when polymerized were mostly intercalated and had relatively low barrier to oxygen permeation. On the other hand, when there was a compatibilization of the epoxy prepolymer with the clay surface modification eg., in the case of clay modified with benzyldibutyl(hydroxyethyl)ammonium, no such deswelling took place and the composites after polymerization were extensively exfoliated and had much better resistance to oxygen permeation [5].

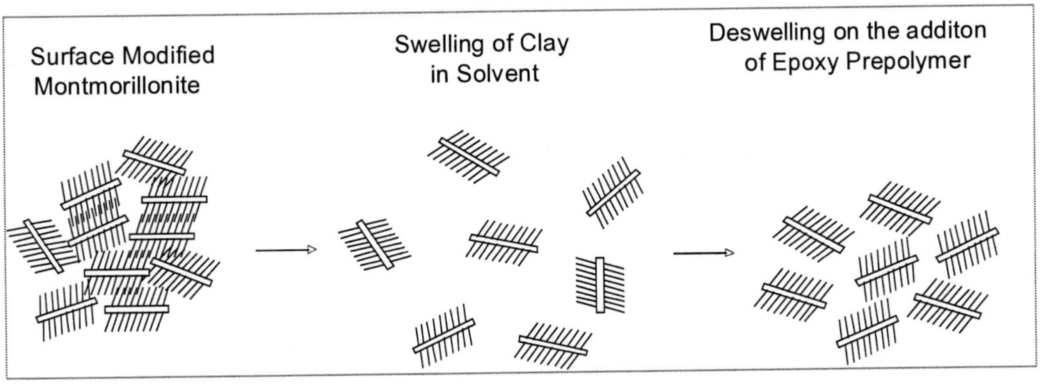


Figure 2. The representation of the swelling deswelling phenomenon observed in the epoxy nanocomposite systems based on the compatibility of the organic-inorganic phases and their mixing at interface.

As mentioned above, polyolefins suffer from the lack of suitable interactions with the modified aluminosilicate surface of the clay (which is still residually polar). To circumvent these problems, low molecular weight polypropylene grafted with maleic anhydride (PP-g-MA) has commonly been used as compatibilizer. As the polymer matrices have high molecular weight, so these nanocomposites are synthesized by the melt compounding approach [40-50]. Use of compatibilizer helps to achieve better compatibility between the polar clay interlayers and apolar polypropylene matrix, thus, to achieve clay exfoliation. Filler exfoliation and the properties of the composites were found to be independent of the molecular weight of the compatibilizer, but were strongly dependant on the weight fraction of the compatibilizer and the extent of grafting of maleic anhydride. Many of the reported studies bring home the conclusion that the use of the maleic anhydride modified polypropylene as compatibilizer though favors intercalation of the polymer between the clay layers but at the expense of the mechanical properties. It was reported by Hasegawa et al. that X-ray silent nanocomposites could be achieved with octadecylammonium modified montmorillonite when the matrix was solely of PP-g-MA (MA content 0.2 wt% and $M_w$ of 210'000 g.mol$^{-1}$) [51]. These hybrids were prepared as master batches which were diluted with PP to give the final composites in the hope that this will lead to an exfoliated structure. However, the PP chains did not diffuse in the interlayers and the Young's modulus of the PP nanocomposites increased by 20% only and the elongation at break and tensile strength decreased [44]. Different studies have been performed by using polymers of different molecular weights, different crystallinity, compatibilizers of different molecular weight, with different % grafting with maleic anhydride, different extents of compatibilizer, different compounding techniques and different compounding temperatures. Thus, it is very difficult to gauge the optimum procedure to achieve the optimum enhancement of the composite properties, but it is rather clear that the use of PP-g-MA has not achieved the required optimum effect on the dispersion of the filler and the composite properties. The gain in *d*-spacing observed in the hybrids of PP-g-MA was not reciprocated with the diffusion of pure PP chains and the overall mechanical properties are not optimum. Figure 3 exemplify a polypropylene nanocomposite with three different compatibilizers (PP-g-MA1, PP-g-MA2

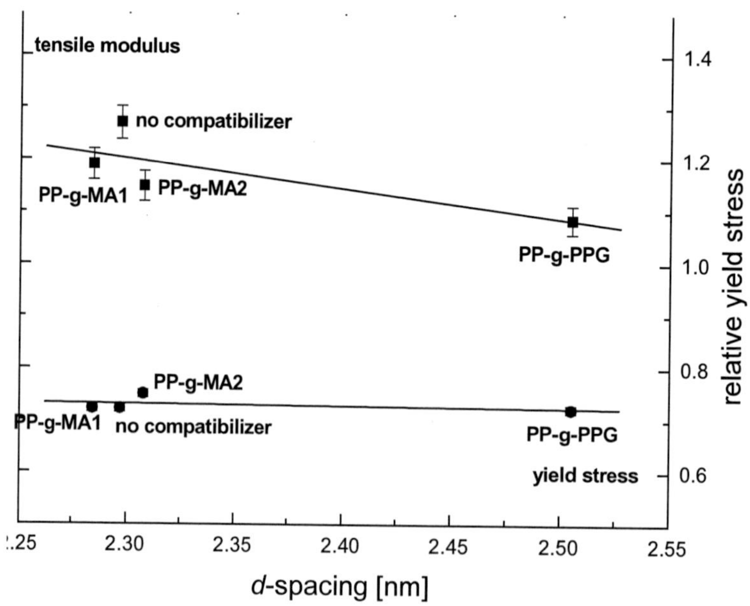


Figure 3. Tensile modulus and yield stress of the polypropylene composites containing 3 vol% of the organically modified clay and 2 wt% of the compatibilizer.

and PP-g-PPG). PP-g-MA1 and PP-g-MA2 differ in the molecular weight and the extent of grafting of MA on the polypropylene backbone, whereas PP-g-PPG is the PP-g-MA reacted with polypropyleneglycol mono butyl ether to achieve a polymer similar to block compatibilizer. Even at 2 wt% of the added compatibilizer, the tensile modulus of these composites decreased as compared to the composite without the compatibilizer, whereas the yield stress remained unchanged. As PP-g-MA is usually of low molar mass because the free radical grafting process accompanied by chain scission [52], the grafted maleic units also disturb the symmetry of the PP molecules, leading to reduced crystallinity. At high maleic anhydride content, the miscibility between PP-g-MA and PP is also no more granted. Similarly, for the olefins nanocomposites also, mostly mechanical properties have been studied and only a little attention has been paid to other commercially important properties [53-55].

## Advances in Achievement of Organic-Inorganic Phase Compatibility

As the conventional approach for nanocomposite synthesis was largely successful for the polar polymers only, therefore, new advancements are required in the synthesis protocols with non polar systems in focus. As the matrix polarization did not achieve expected improvement in the composite properties, therefore, it is important to look at this problem from other perspectives. As no positive interaction between the non polar polymers and the clay surface can be expected, thus the exfoliation can be expected to be enhanced by physical means like shearing in the compounder. In this regard, if the surface of the clay platelets is completely covered organically either with chains of higher chain length or with higher chain density per molecule of the ammonium modification, it would reduce the electrostatic interactions present between the clay interlayers and hence would make them susceptible to exfoliation when mixed with polymer under high shear. A number of other advancements in the surface modifications of the fillers have also been reported in the recent years, which follow more complex routes like grafting and polymerization reaction on the clay surface, but to achieve the same goal of improvement in organic-inorganic compatibility and increased basal plane spacing in the clay interlayers to reduce the electrostatic forces holding them. These routes also benefit from the fact that no reduction in mechanical performance of the polymer is observed. The following paragraphs list these simple as well as complex modifications in the conventional system to achieve the better dispersion of the filler in the polymer.

To avoid the phase separation of the crystalline polymer matrix at the interface with the inorganic mineral, it was reported that use of brushes of ammonium modifications of different chain lengths can be beneficial. Figure 4 shows a representation of such a system where ammonium ions of dioctadecyl and didodecyl chain lengths have been exchanged on the clay surface. It should be noted that the figure only tends to serve as a representation of the system, as the exact position of ammonium ions is arguable. Polypropylene nanocomposites by using such a system were synthesized without the use of comaptibilizers and a partial exfoliation could still be achieved along with significant enhancements in tensile modulus of the composites [56].

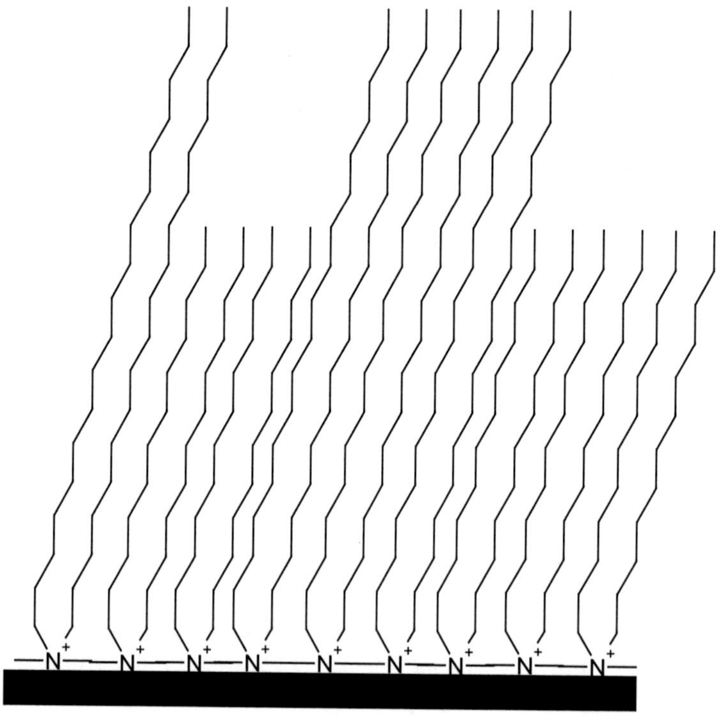

Figure 4. A representation of the clay surface modification using ammonium ions of different chain lengths.

A marginal increase in the chain length of the ammonium ions exchanged on the clay surface was found to have a significant effect on the composite properties because the attractive forces between the clay platelets are reduced when interlayer distance between them is increased [55]. This increase in chain length from C18 to C22 atoms (Figure 5) led to a partial exfoliation of the filler in the polypropylene matrix without the use of compatibilizers because of the better mixing of the clay and polymer and the composites had a significant improvement in oxygen permeation resistance.

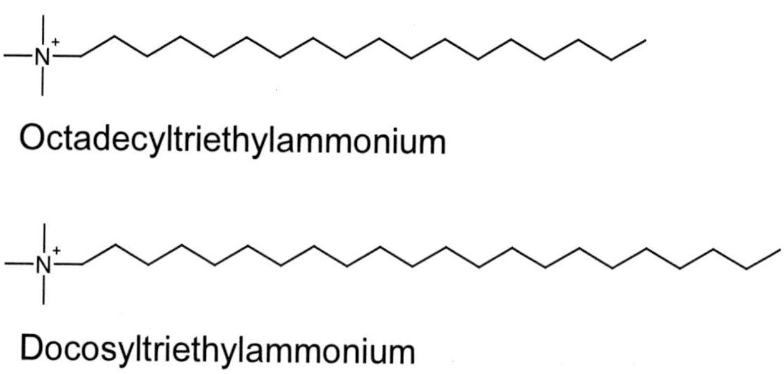


Figure 5. Chemical structures of the octadecyltrimethylammonium and docosyltrimethylammonium ions exchanged on the clay surface.

The area available per cation on the montmorillonite surface is higher than the cross-section of the generally used octadecyltrimethylammonium or dioctadecyldimethylammonium ions. As a result, the ammonium ions tend to lie laterally on the clay surface. However, the interlayer distance in between the clay layers can be further increased and thus the electrostatic forces between them could be subsequently decreased, if the chain density of the ammonium modifications is increased from dioctadecyldimethyl to trioctadecylmethyl or tetraoctadecyl as shown in Figure 6. These ions have been reported to be extremely beneficial in enhancing the extent of exfoliation in both the polyethylene as well as polypropylene nanocomposites, and the composites did not require the addition of external compatibilizer [55,57,58]. Although full exfoliation of the clay platelets could not be achieved, however, the composites showed better mechanical, thermal and oxygen barrier properties.

Many reported studies have observed that the modifying the clay surface with long polydisperse chains is the key to achieve miscibility of the organic-inorganic phases. This phenomenon has already been observed with using silicates modified with ammonium ions of different chain lengths. However, attachment of preformed long chains on the silicate surface suffers from the solubility as well as steric hindrance problems. To this regard, one of the most important developments to achieve control on the phase compatibilization process is the surface grafting approach [59-71]. Grafting of polymer chains ionically bound to the clay surface by polymerization either 'to' or 'from' the clay surface can help in attaching long disperse lengths on the clay surface thus leading to subsequent exfoliation of the filler in the

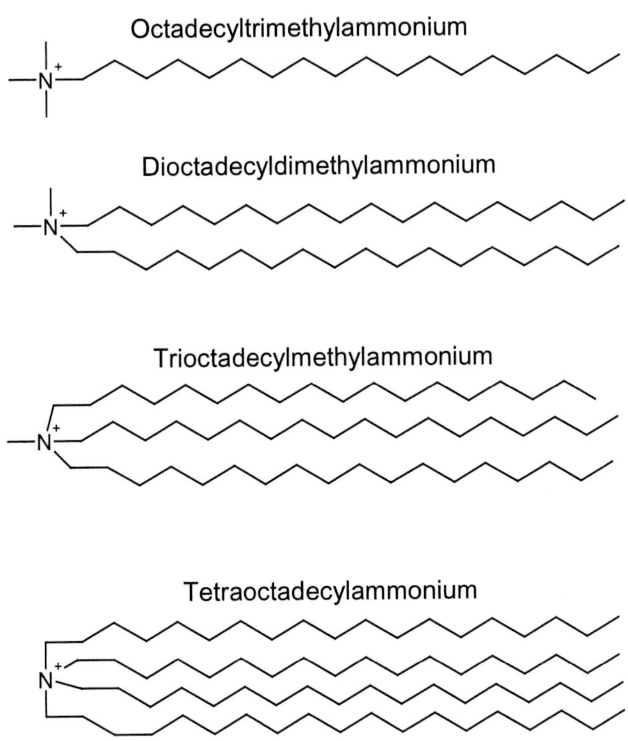


Figure 6. Chemical structures of the ammonium modifications with increasing chain density.

polymer matrix. Polymerization 'to' the surface was achieved by ionically attaching a monomer on the clay surface followed by its polymerization with the external monomer and initiator. On the other hand, more controlled polymer chains from the surface have been generated by attaching the free radical monocationic or dicationic initiator on the clay surface followed by polymerization with the external monomer. Though both techniques graft polymer chains equally well, polymerization 'from' the surface techniques has been found to generate more controlled grafts owing to the more controlled polymerization reaction due to the absence of external initiator. Advantage of a monocationic initiator over bicationic initiators attached to the clay surface is observed in terms of low termination. However, the bicationic initiator also helps in reducing the amount of the solution polymer and if the surface is partially covered with the initiator, the possibility of the immediate termination of the radicals formed from the bicationic initiator can be expected to be substantially low. Figure 7 shows that TGA thermograms of the clays modified with either monomer or bicationic initiator. Curve I denotes the clay modified with methacrylic end group functionality in an amount corresponding to 30% of cation exchange capacity (CEC) and methyltrioctylammonium in an amount corresponding to 70% of the CEC, whereas, in II, these amounts are reversed. Curve III denotes the clay modified with bicationic initiator functionality in an amount corresponding to 30% of cation exchange capacity and dihexadecyldimethylammonium in an amount corresponding to 70% of the CEC, whereas, in IV, these amounts are reversed [71]. Figure 8 is a representation of grafting 'from' surface approach. The modified clays as shown in Figure 7, when subjected to corresponding grafting 'to' or 'from' the surface reactions, led to significant increase in the basal plane spacing of the clay platelets and corresponding increase in the organic weight loss in the TGA thermograms.

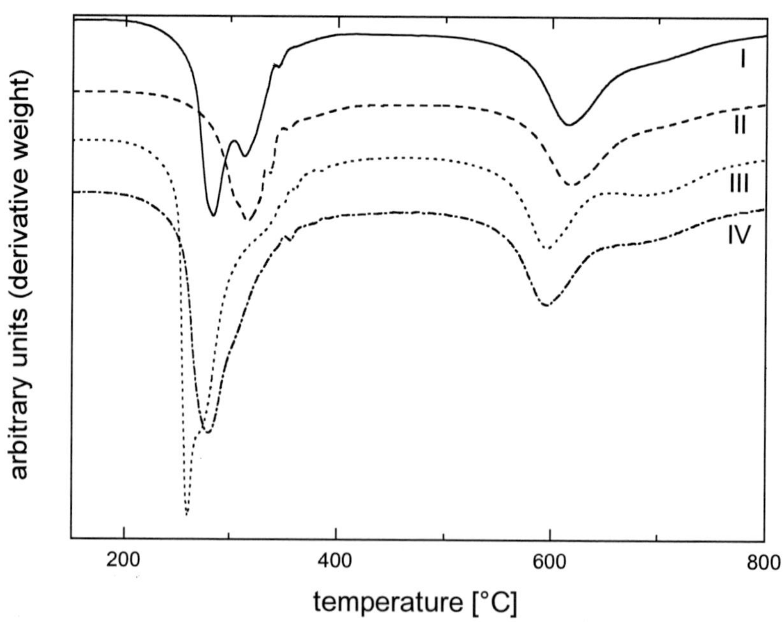


Figure 7. TGA thermograms of the modified clays used for surface grafting reactions.

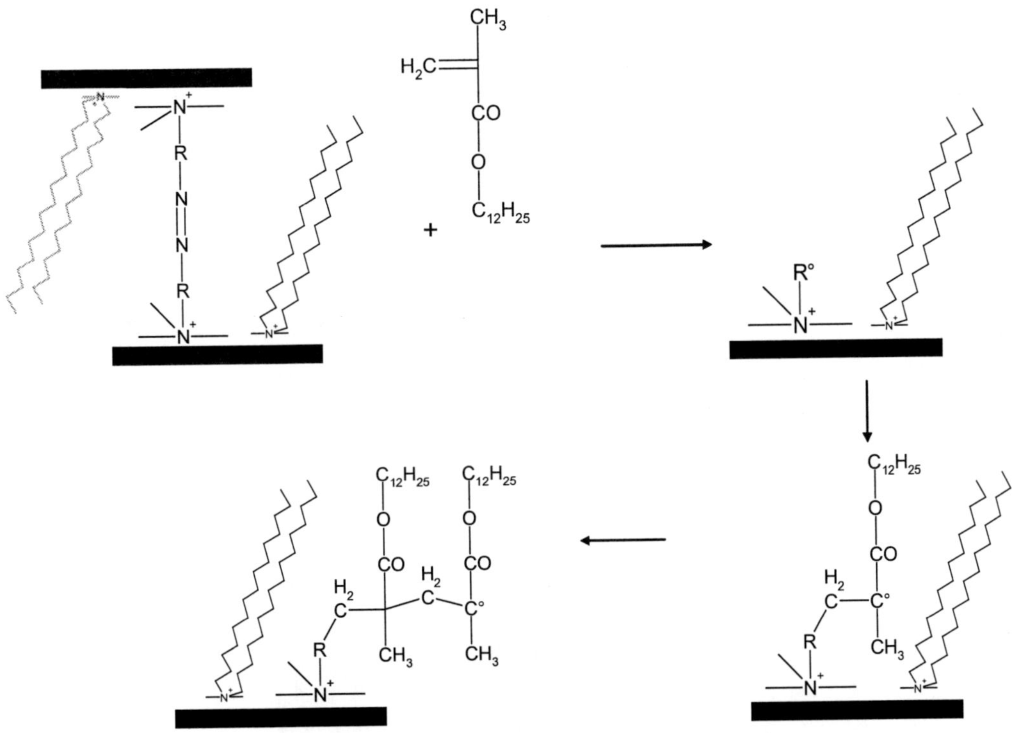


Figure 8. Graphical representation of the grafting 'from' the surface approach.

For instance, the clay of II after reaction indicated a basal plane spacing of >5 nm and an increase of 40% in the TGA mass loss. Similar increases were found for other clays [71]. Such clays represent very high potential filler systems to achieve phase compatibilization as the clay is very uniformly covered with organic polymer. Figure 9a shows TEM micrographs of the unmodified clay in water. A complete exfoliation of the clay is observed, thus, indicating that the whole surface area of the clay is available for modification. Figure 9b shows the high magnification TEM image of modified clay II of Figure 7. It is clear that the clay platelets are stacked to form thick tactoids. Figure 9c shows the surface grafted clay II. It can be observed that after the surface reaction, the average tactoid thickness has decreased which indicates that a partial intergallery polymerization has taken place and the reaction is not limited to only the outer surface of the thick tactoids. Although, it also shows that the complete intergallery polymerization still could not be achieved, however, the surface reacted clay is still a valuable filler to be mixed with polymer matrices as it has a high potential to exfoliate owing to its low surface energy.

A range of monomers or initiators have been employed in the literature to achieve these grafting reactions and also a number of polymers have been grafted on the clay surface. Vinylbenzyldimethyldodecylammonium chloride (VDAC), 2,2'- azobis(isobutyramidine hydrochloride) (AIBA), *N*-Methyl-*N,N* di(vinylbenzyl)octadecylammonium chloride, methacryloxyethyltriethylammonium chloride etc. are a few ammonium ions with terminal monomers or initiator moieties.

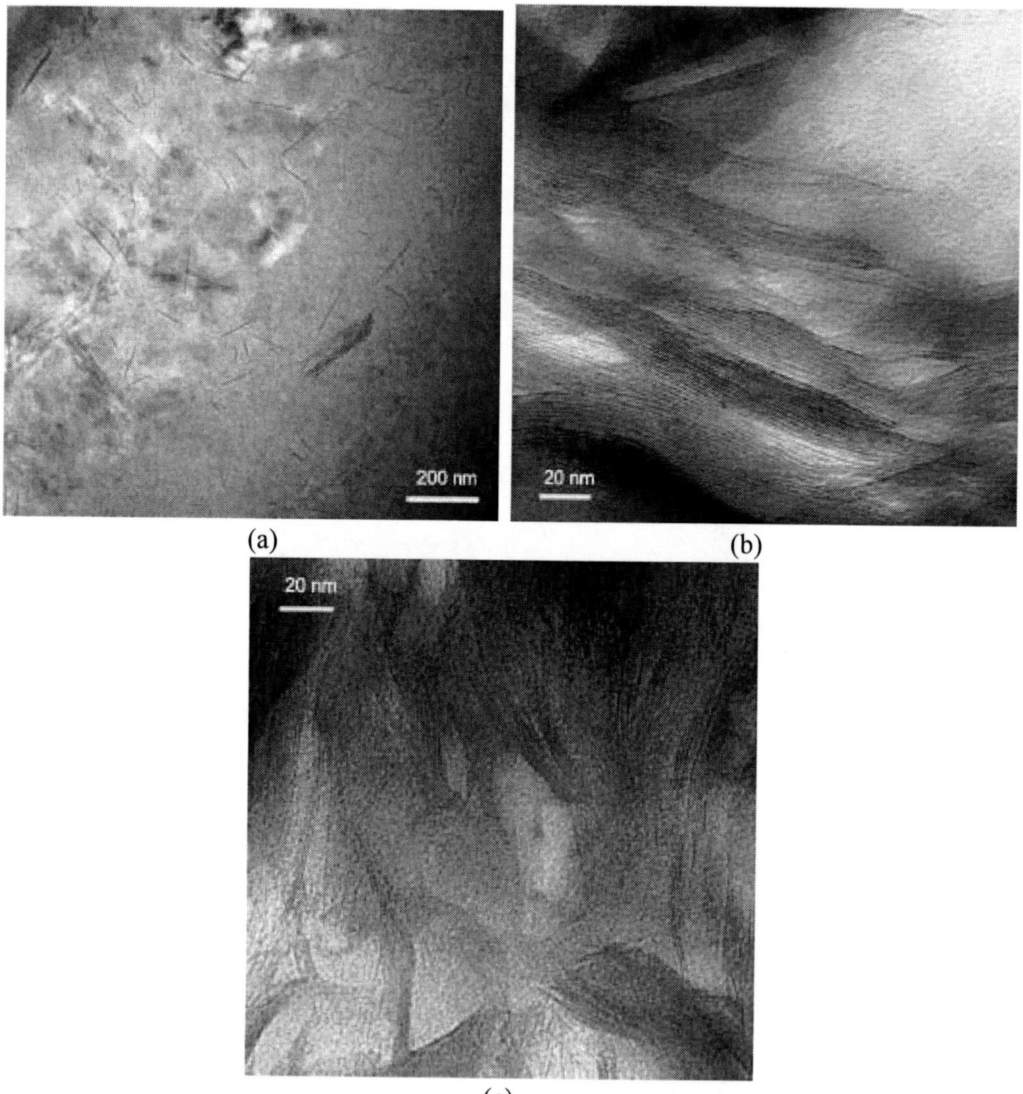


Figure 9. TEM micrographs of (a) unmodified clay, (b) clay of II in Figure 7 and (c) clay of II after surface reaction. Black lines represent the cross-section of clay platelets.

Surface grafting reactions have been further controlled by using controlled living polymerization conditions using external nitroxides or ionically bound atom transfer radical polymerization to generate more exfoliated polymer-layered silicate nanocomposites [71-83]. Figure 10 shows such a system with nitroxide being used as a living agent. Early developments in synthetic methods using controlled/"living" radical polymerization techniques, such as stable free-radical/nitroxide-mediated polymerizations, atom transfer radical polymerization, and reversible addition-fragmentation chain-transfer polymerization was reviewed by Pyun et al. It was also observed that the synthesis of poly(styrene-block-butyl acrylate) (PSBA) block copolymer brushes on the surface of silicate platelets can be achieved using in situ atom transfer radical polymerization (ATRP) from initiator moieties immobilized within the silicate galleries of the clay particles. An interesting comparison between the surface initiated polymerizations via atom transfer radical polymerization route or via nitroxide mediated polymerization route has been reported. The surface-grafted ATRP initiator is found to lead to a higher grafting density (~1.2 molecule.nm$^{-2}$) which is decreased for the unimolecular NMP method to 0.5 molecule.nm$^{-2}$. Significantly exfoliated composites of polystyrene, poly(methyl methacrylate) and poly (butyl acrylate) were generated when the in situ polymerization of monomers was achieved in the presence of silicate platelets modified with an ATRP initiator, consisting of a quaternary ammonium salt moiety and a 2-bromo-2-methyl propionate moiety.

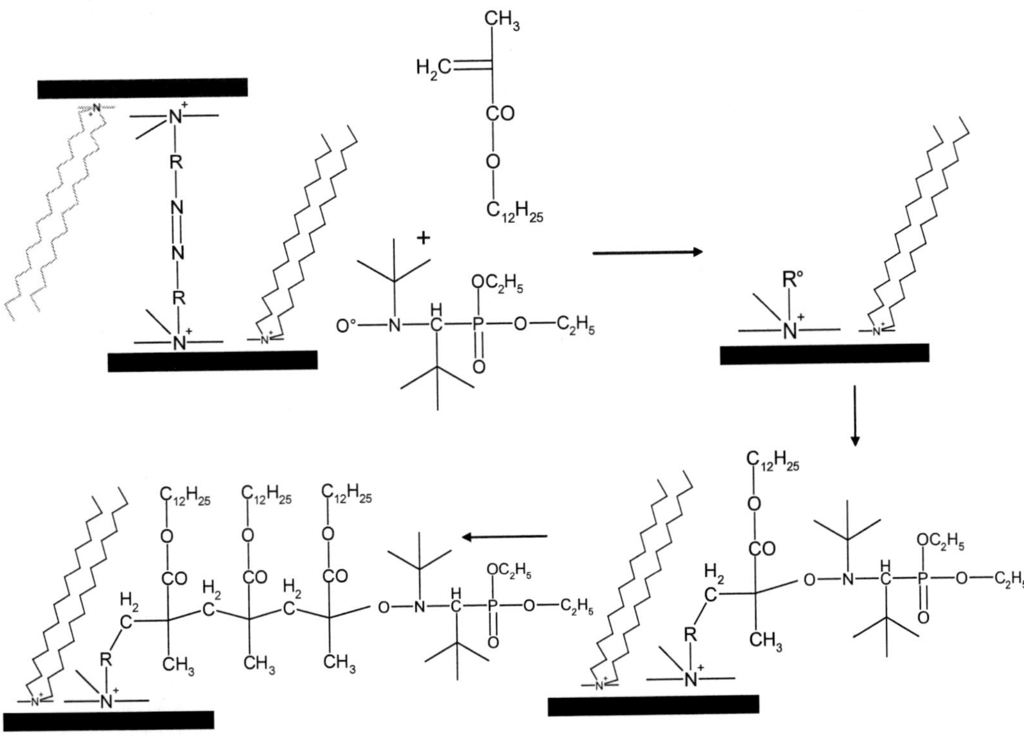

Figure 10. Graphical representation of the 'living' surface grafting approach.

Though the surface grafting reactions are a very useful tool to organophilize the clay platelets, however, the whole modification process is complex and time consuming. To this regard, mention can be made of a relatively simpler approach which also helps to achieve higher basal plane spacing in the modified motmorillonites and hence subsequent exfoliation in the polymer matrix. This approach, called surface reaction or more specifically surface esterification, includes a first exchange on the surface with reactive modification and subsequent reaction of the functional groups in the modification with long chain molecules to enhance the basal plane spacing [84]. Figure 11 shows the various modifications used in this approach, which were later reacted with long chain molecules like stearic acid (C18) or dotriacontanoic acid (C32). Figure 12 also shows the IR spectra of Bz1OH modified clay with stearic acid in different mole ratios of acid to alcohol groups. With increasing acid to alcohol mole ratio, an increase in total extent of surface reaction was observed which was also confirmed by the corresponding increase in basal plane spacing. Similar other studies have also been reported [85-90].

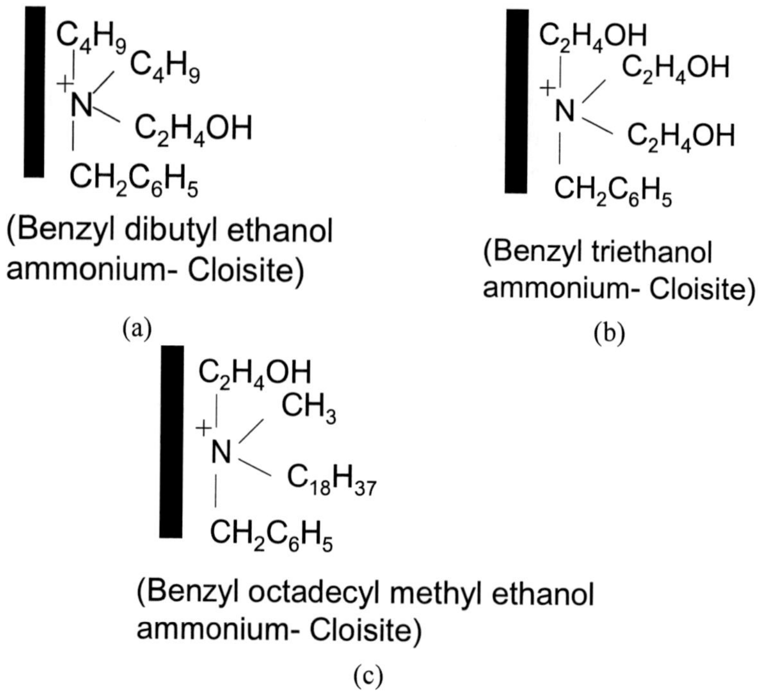


Figure 11. Surface modified montmorillonites used for the surface esterification reactions.

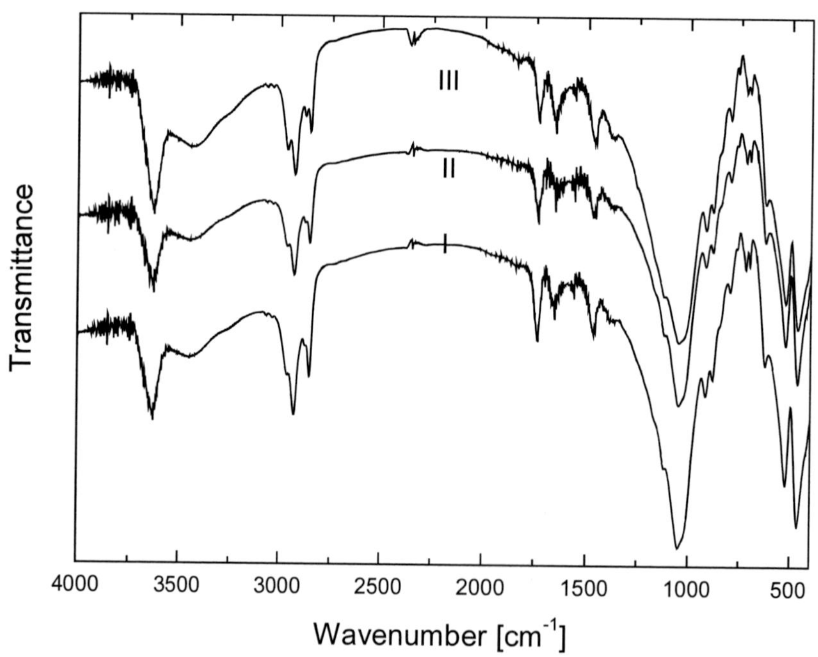


Figure 12. IR spectra of benzyldibutylethanolammonium (Bz1OH) modified clay after surface esterification reaction with octadecanoic acid using different acid to alcohol molar ratios; I: 1:1, II: 4:1 and III: 8:1.

A relatively less developed approach of nanocomposites for polyolefin matrices is the in situ intercalative polymerization [91-93]. Ammonium terminated polypropylene have also been used as silicate surface modification and subsequently compounded with polypropylene, but the properties of the composites were not reported [94]. Two more very important factors influencing the phase mixing at the interface are the surface cleanliness of the modified

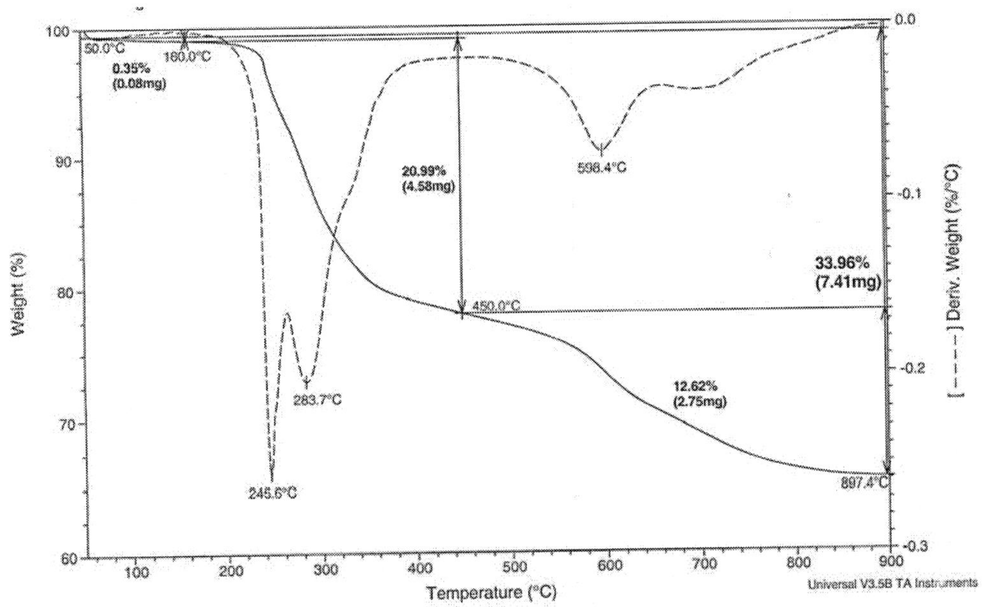


(a)

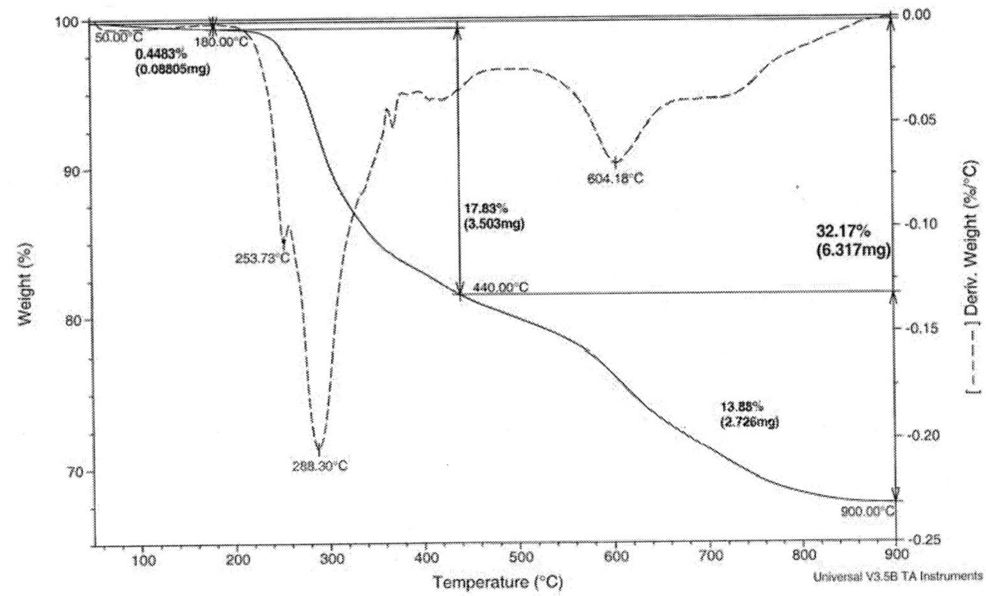


(b)

Figure 13. TGA thermograms of montmorillonites modified with dioctadecyldimethylammonium. (a) commercial product and (b) lab modified montmorillonite.

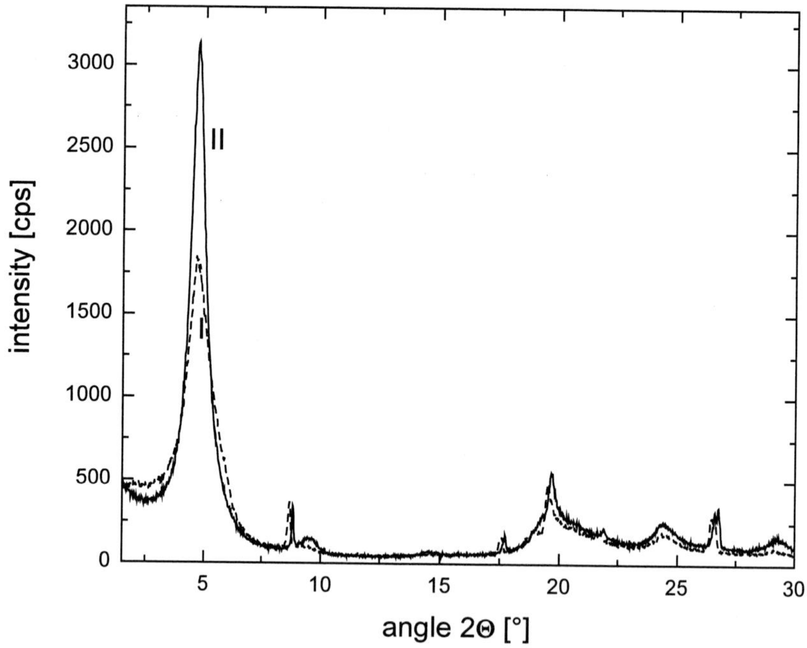


Figure 14. X-ray diffractograms of the surface modified clays. I: commercial and II: lab modified product.

montmorillonite and the thermal stability of the surface modification. It has been reported that the presence of excess and unattached molecules of surface modification on the clay surface can have a detrimental effect on the properties of the composites [95,96]. These unattached molecules intercalated between the ionically attached molecules degrade thermally at lower temperature, thus, severely affecting the performance of the composites. The commercially available modified filler, commonly used in the reported studies, has been observed to contain excess of such modifier molecules as indicated in Figure 13a by the presence of the low temperature degradation peak of 246°C. The same mineral when modified with same surface modification in lab could be cleaned to much better extent indicated by the reduction of the low temperature peak to minimum. One should also be careful that though the excess of the surface modification in the clay interlayers can be detected in TGA, but mostly X-ray diffractograms are silent to this phenomenon as shown in Figure 14. No change in basal spacing of the clay is observed in the commercial and lab modified clay except from the unrelated change in intensity. Thermal stability of the surface modification at the compounding temperature is another important concern. The ammonium head group is quite susceptible to degradation at high compounding temperature of polypropylene or polyethylene generally employed. Such degradation may affect the thermodynamics of polymer melt intercalation due to chemical changes in surfactant structure, thus altering the platelet exfoliation, interfacial bonding and subsequently influencing the physical and mechanical properties [97]. In addition, decomposition of the surfactant may produce unwanted side reactions with the polymer matrix. Decrease in *d*-spacing has also been reported after compounding owing to the similar reasons [98]. Out of the many approaches tried to avoid this problem, the use of imidazolium salts as surface modification is preferred as these salts are more thermally resistant as compared to the alkyl ammonium cation both in the presence and absence of oxygen as degrading atmosphere [99-104]. It was recently reported that when montmorillonites modified with imidazolium salt 1-decyl-2-methyl-3-octadecylimidazolium was incorporated in the polypropylene matrix, the oxygen barrier of the composites was observed to be better than the composites generated with montmorillonite modified with dioctadecylimethylammonium [105]. Figure 15 shows the DSC melting and crystallization plots of the composites synthesized with different amounts of imidazolium treated clay. No change in extent of crystallization was observed at

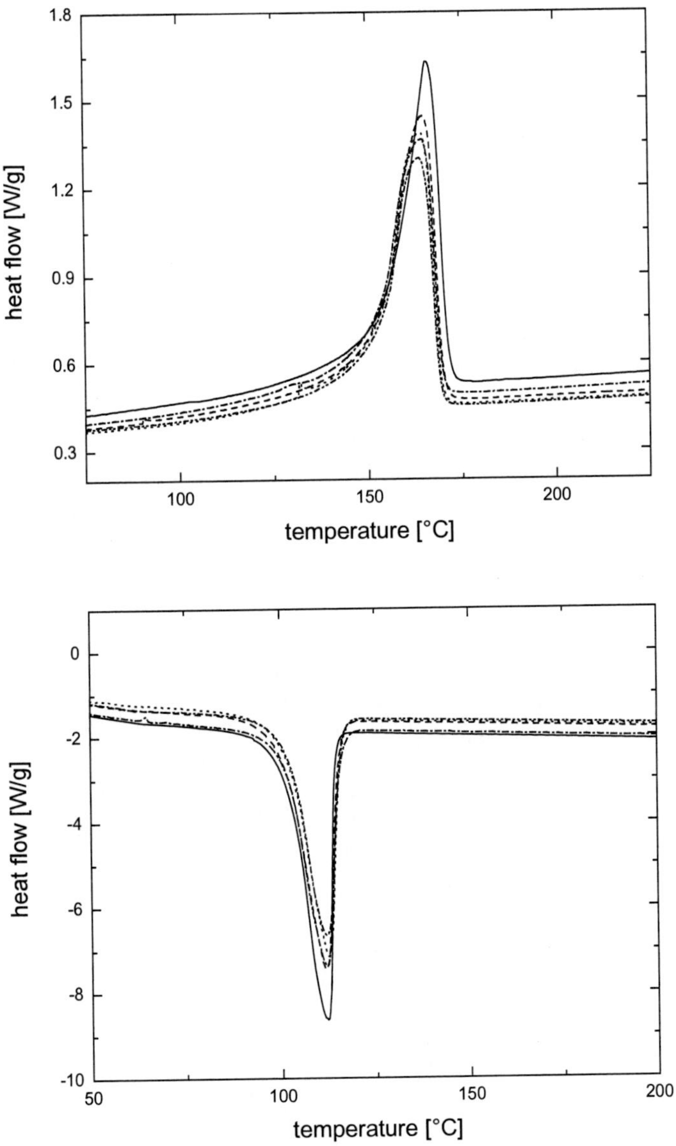


Figure 15. DSC melting and crystallization thermograms of polypropylene nanocomposites with different amounts of montmorillonite modified with thermally stable imidazolium salt. The solid lines represent the pure polymer, whereas the dotted lines represent the composites.

any volume fraction of the filler as compared to the pure polymer (solid lines) indicating that the filler did not disturb the crystallization behavior of the matrix sometimes observed with unmodified and modified montmorillonites.

## General Considerations

In almost all of the TEM micrographs observed by the author as well as in the reported studies, the platelets are totally misaligned, bent and folded. Figure 16 is a classical example of this observation. Such a misalignment along with partial exfoliation is very important to be considered when one develops the models for property prediction or when the experimental data are compared with the models. Conventional models have always been based on perfect alignment and complete exfoliation of the platelets, which makes the property prediction very faulty. Halpin-Tsai model has received special attention owing to better prediction of the tensile properties for a variety of reinforcement geometries [106-109]. But a number of assumptions like firm bonding of filler and matrix, perfect alignment of the platelets in the matrix, uniform shape and size of the filler particles in the matrix make it very difficult to correctly predict the nanocomposites properties. The incomplete exfoliation of the nanocomposites, thus, the presence of a distribution of tactoid thicknesses is another concern. These assumptions are rarely true in the real systems especially in non polar polymer matrix composites where the interactions between the organic and inorganic components do not exist. The models have been modified in order to accommodate the effect of incomplete exfoliation and misorientation of the filler, but the effect of imperfect adhesion at the surface still needs to be incorporated [110-112]. In this regard, two different kinds of property prediction methods are also worth mentioning. One is the finite element method approach reported in the literature which takes into account the misalignment and incomplete exfoliation effects [113]. Figure 17 is a manifestation of the application of the model used to

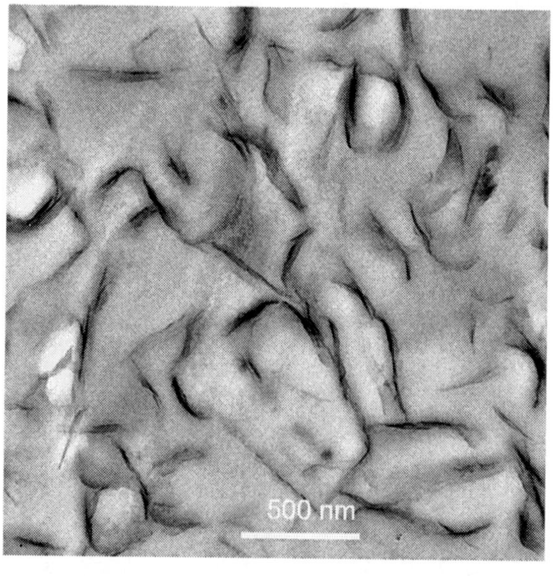


Figure 16. TEM micrograph of epoxy nanocomposite with 3 vol% of the filler.

predict the average aspect ratio of the platelets from the oxygen barrier properties of the composites. Other is the computational factorial and mixture designs, which represent a much better way to analyze the systems especially where the above mentioned assumptions do not hold true. Such models have recently been reported for the property prediction of polypropylene nanocomposites with 3 vol% of the montmorillonite with a cation exchange capacity of 880 $\mu eq.g^{-1}$ modified with trioctadecylmethylammonium ions [114].

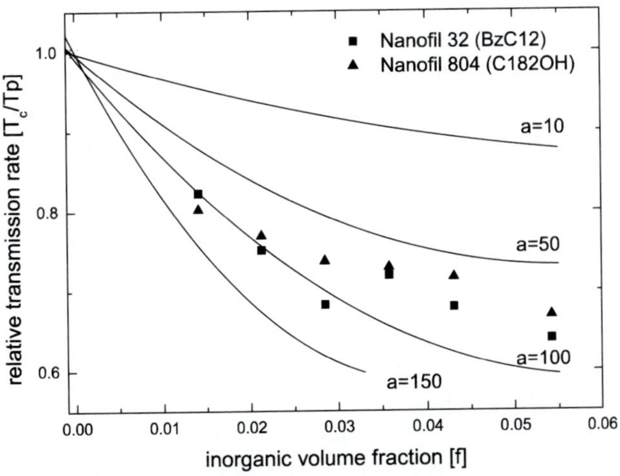


Figure 17. Application of finite element approach to predict the average aspect ratio of polymer composites in polyurethane nanocomposites with two different fillers.

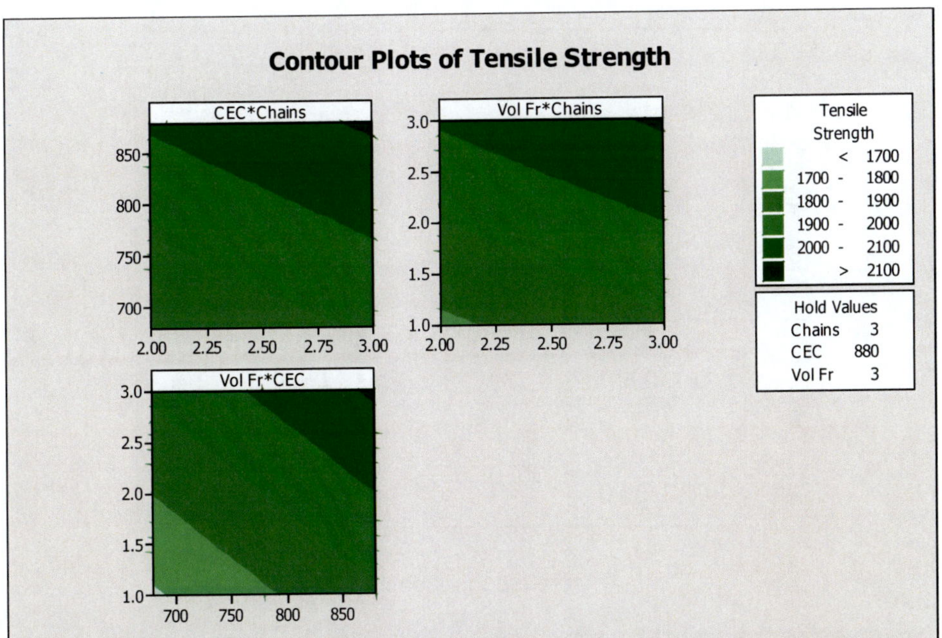


Figure 18. Statistical models to predict the mechanical properties of polypropylene nanocomposites by varying different factors and to calculate their relative effect.

Apart from that, as mentioned earlier that depending on the reflections observed in the wide angle X-ray diffraction (WAXRD), an intercalated or exfoliated morphology is generally recognized but the real morphology generally falls between these two idealized structures. It is also generally referred that the increase in basal plane spacing represents improvement in composite properties, but it is not quantified till now, i.e., how much minimum increase in the basal plane spacing would be required to achieve optimum enhancement of the properties. Also, it is not the intercalated but exfoliated platelets which contribute maximum to the composite properties as indicated by the plot in Figure 19. In this plot, there is no visible correlation between the reduction in the oxygen transmission though the composites and the increase in the basal plane spacing of the filler in the composite, indicating that the intercalated platelets do not contribute to the composite properties. One should also be careful that enough measuring time is provided per step during the X-ray measurement in order not to overlook the signal from the intercalated platelets. As can clearly be seen in Figure 20, by providing more time per step during the measurement, a better signal can be achieved.

Another important consideration to achieve better phase compatibilization in the polar polymers is to carefully choose the curing parameters. Figure 21 points to the catalytic nature of the ammonium modified clay as it accelerates the curing reaction as compared to the case when there is no filler. So, one should be careful to choose appropriate amine to epoxy mole ratio when the filler is also added to the system so as to maintain a normal curing speed which his not very fast or not very slow. Curing temperature has also a similar effect on the resulting morphology of the composite (Figure 22). Curing at low temperature may lead to a slower polymerization rate in the integallery which may lead to more polymerization occurring in the extragallery region, thus leading to a partially intercalated composite rather than an exfoliated

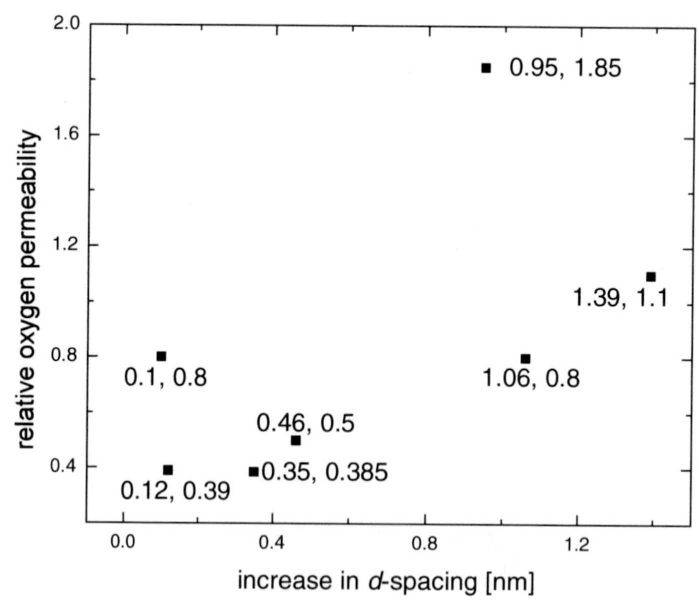


Figure 19. Plot of the relative oxygen permeability with respect to increase in the basal plane spacing of the filler in the epoxy nanocomposites.

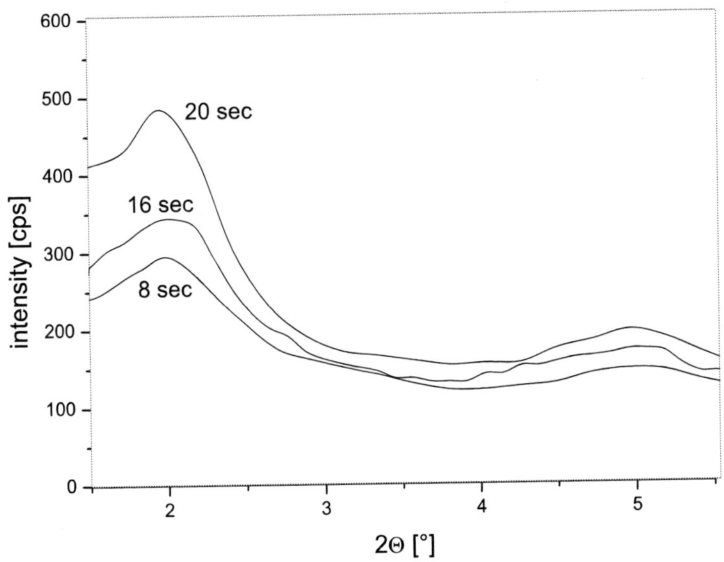


Figure 20. X-ray plat of the polyurethane nanocomposite as a function of increasing measuring time per step.

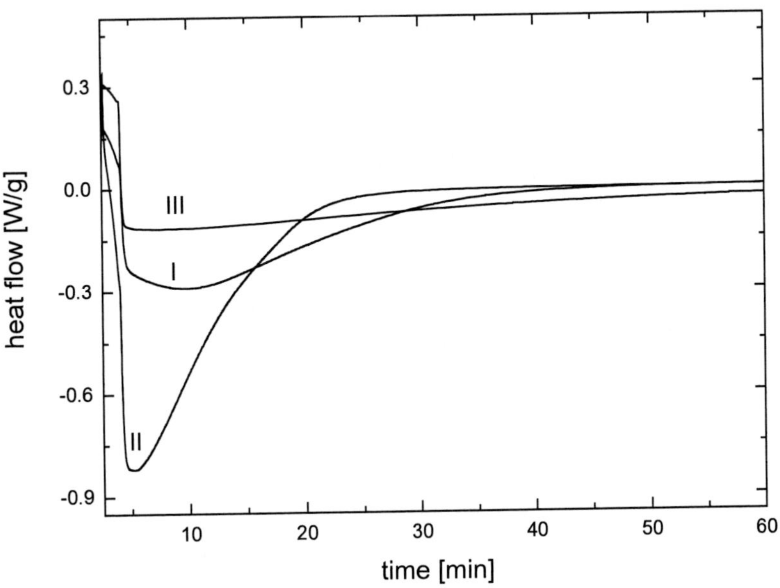


Figure 21. DSC thermograms indicating the effect of filler on the epoxy curing at 70°C; I: no filler, amine to epoxy mole ratio of 0.3:1, II: 3.5 vol% 2C18 filler, amine to epoxy mole ratio of 0.3:1 and III: 3.5 vol% of 2C18 filler, amine to epoxy mole ratio of 0.2:1.

composite. In Figure 22, it has been shown for the epoxy composites that when the lower temperature was used for curing, the composite shows higher intensity than the composite cured at higher temperature. Thus, it may qualitatively indicate that the extent of intercalated material is less in the composite cured at higher temperature.

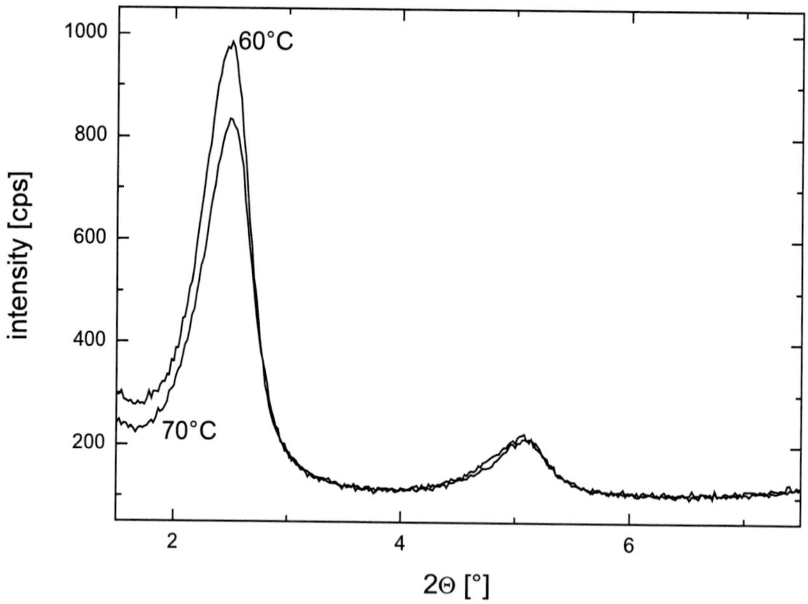


Figure 22. X-ray diffractogram showing the effect of temperature on the composite microstructure (3.5 vol% 2C18 filler).

## References

[1] Kojima, Y., Fukumori, K., Usuki, A., Okada, A., and Kurauchi, T. 1993, *J. Mater. Sci. Lett.*, **12**, 889.

[2] Yano, K., Usuki, A., Okada, A., Kurauchi, T., and Kamigaito, O. 1993, *J. Polym. Sci., Part A: Polym. Chem.*, **31**, 2493.

[3] Lan, T., Kaviratna, P.D., and Pinnavaia, T.J. 1994, *Chem. Mater.,* **6**, 573.

[4] Lan, T., and Pinnavaia, T.J., 1994, *Chem. Mater.*, **6**, 2216.

[5] Osman, M.A., Mittal, V., Morbidelli, M., and Suter, U.W. 2004, *Macromolecules*, **37**, 7250.

[6] Messersmith, P.B., and Giannelis, E.P. 1995. *J. Polym. Sci., Part A: Polym. Chem.,* **33**, 1047.

[7] Burnside, S.D., and Giannelis, E.P. 1995, *Chem. Mater.*, **7**, 1597.

[8] Wang, Z., and Pinnavaia, T.J. 1998, *Chem. Mater.*, **10**, 3769.

[9] Osman, M.A., Mittal, V., Morbidelli, M., and Suter, U.W. 2003, *Macromolecules*, **36**, 9851.

[10] Yano, K., Usuki, A., and Okada, A. 1997, *J. Polym. Sci., Part A: Polym. Chem.*, **35**, 2289.

[11] Bailey, S.W. 1984, Reviews in Mineralogy, S.W. Bailey (Ed.), Virginia Polytechnic Institute and State University, Blacksburg.

[12] Bailey, S.W. 1980, Crystal Structures of Clay Minerals and their X-ray Identification, G.W. Brindley, and G. Brown (Eds.), Mineralogical Society, London.

[13] Theng, B.K.G. 1974, *The Chemistry of Clay-Organic Reactions*, Adam Hilger, London.
[14] Pinnavaia, T.J. 1983, *Science*, **220**, 365.
[15] Lagaly, G. 1986, Developments in Ionic Polymers, A.D. Wilson, and H.J. Prosser (Eds.), Elsevier Applied Science Publishers, London/New York.
[16] Giannelis, E.P. 1996, *Adv. Mater.*, **8**, 29.
[17] Alexandre, M., and Dubois, P. 2000, *Mater. Sci. Engg. R: Reports*, **28**, 1.
[18] Zilg, C., Thomann, R., Mulhaupt, R., and Finter, J. 1999, *Adv. Mater.*, **11**, 49.
[19] Wang, Z., and Pinnavaia, T.J. 1998, *Chem. Mater.*, **10**, 3769.
[20] Chen, T.K., Tien, Y.I., and Wei, K.H. 2000, *Polymer*, **41**, 1345.
[21] Ma, J., Zhang, S., and Qi, Z. 2001, *J. Appl. Polym. Sci.*, **82**, 1444.
[22] Yao, K.J., Song, M., Hourston, D.J., and Luo, D.Z. 2002, *Polymer*, **43**, 1017.
[23] Tien, Y.I., and Wei, K.H. 2001, *Macromolecules*, **34**, 9045.
[24] Xu, R., Manias, E., Snyder, A.J., and Runt, J. 2001, *Macromolecules*, **34**, 337.
[25] Tortora, M., Gorrasi, G., Vittoria, V., Galli, G., Ritrovati, S., and Chiellini, E. 2002, *Polymer*, **43**, 6147.
[26] Chang, J.H., and An, Y.U. 2002, *J. Polym. Sci., Part B: Polym. Phys.*, **40**, 670.
[27] Messersmith, P.B., and Giannelis, E.P. 1994, *Chem. Mater.*, **6,** 1719.
[28] Zilg, C., Mulhaupt, R., and Finter, J. 1999, *Macromol. Chem. Phys.*, **200**, 661.
[29] Brown, J.M., Curliss, D., and Vaia, R.A. 2000, *Chem. Mater.*, **12**, 3376.
[30] Kornmann, X., Thomann, R., Mulhaupt, R., Finter, J., and Berglund, L. 2002, *J. Appl. Polym. Sci.*, **86**, 2643.
[31] Lan, T., Kaviratna, P.D., and Pinnavaia, T.J. 1995, *Chem. Mater.*, **7**, 2144.
[32] Zerda, A.S., and Lesser, A.J. 2001, *J. Polym. Sci., Part B: Polym. Phys.*, **39**, 1137.
[33] Kornmann, X., Lindberg, H., and Berglund, L.A. 2001, *Polymer*, **42**, 1303.
[34] Kong, D., and Park, C.E. 2003, *Chem. Mater.*, **15**, 419.
[35] Lan, T., and Pinnavaia, T.J. 1994, *Chem. Mater.*, **6**, 2216.
[36] Osman, M.A., Ploetze, M., and Suter, U.W. 2003, *J. Mater. Chem.*, **13**, 2359.
[37] Messersmith, P.B., and Giannelis, E.P. 1995, *J. Polym. Sci., Part A: Polym. Chem.*, **33**, 1047.
[38] Yano, K., Usuki, A., Okada, A., Kurauchi, T., and Kamigaito, O. 1993, *J. Polym. Sci., Part A: Polym. Chem.*, **31**, 2493.
[39] Lan, T., Kaviratna, P.D., and Pinnavaia, T.J. 1994, *Chem. Mater.*, **6**, 573.
[40] Yano, K., Usuki, A., and Okada, A. 1997, *J. Polym. Sci., Part A: Polym. Chem.*, **35**, 2289.
[41] Usuki, A., Kato, M., Okada, A., and Kurauchi, T. 1997, *J.Appl. Polym. Sci.*, **63**, 137.
[42] Kawasumi, M., Hasegawa, N., Kato, M., Usuki, A., and Okada, A. 1997, *Macromolecules*, **30**, 6333.
[43] Reichert, P., Nitz, H., Klinke, S., Brandsch, R., Thomann, R., and Mulhaupt, R. 2000, *Macromol. Mater. Engg.*, **275**, 8.
[44] Hasegawa, N., Kawasumi, M., Kato, M., Usuki, A., and Okada, A. 1998, *J. Appl. Polym. Sci.*, **67**, 87.
[45] Manias, E., Touny, A., Wu, L., Strawhecker, K., Lu, B., and Chung, T.C. 2001, *Chem. Mater.*, **13**, 3516.
[46] Oya, A., Kurokawa, Y., and Yasuda, H. 2000, *J. Mater. Sci.*, **35**, 1045.
[47] Zhang, Q., Fu, Q., Jiang, L., and Lei, Y. 2000, *Polym. Int.*, **49**, 1561.

[48] Ellis, T. S., and D'Angelo, J.S. 2003, *J. Appl. Polym. Sci.*, **90**, 1639.
[49] Xu, W., Liang, G., Wang, W., Tang, S., He, P., and Pan, W.P. 2003, *J. Appl. Polym. Sci.*, **88**, 3225.
[50] Su, S., Jiang, D.D., and Wilkie, C.A. 2004, *Polym. Deg. Stab.*, **83**, 321.
[51] Hasegawa, N., Okamoto, H., Kawasumi, M., Kato, M., Tsukigase, A., and Usuki, A. 2000, *Macromol. Mater. Engg.*, **280-281**, 76.
[52] Sathe, S.N., Rao, G.S.S., and Devi, S. 1994, *J. Appl. Polym. Sci.*, **53**, 239.
[53] Gorrasi, G., Tortora, M., Vittoria, V., Kaempfer, D., and Mulhaupt, R. 2003, *Polymer*, **44**, 3679.
[54] Manias, E. 2001, *MRS Bulletin*, **26**, 862.
[55] Osman, M.A., Mittal, V., and Suter, U.W. 2007, *Macromol. Chem, Phys.*, **208**, 68.
[56] Mittal, V. 2007, *J. Thermoplastic Comp. Mater.*, **20**, 575.
[57] Osman, M.A., Rupp, J.E.P., and Suter, U.W. 2005, *J. Mater. Chem.*, **15**, 1298.
[58] Osman, M.A., Rupp, J.E.P., and Suter, U.W. 2225, *Polymer*, **46**, 1653.
[59] Fu, X., and Qutubuddin, S. 2001, *Polymer*, **42**, 807.
[60] Velten, U., Shelden, R.A., Caseri, W.R., Suter, U.W., and Li, Y. 1999, *Macromolecules*, **32**, 3590.
[61] Meier, L.P., Shelden, R.A., Caseri, W.R., and Suter, U.W. 1994, *Macromolecules*, **27**, 1637.
[62] Velten, U., Tossati, S., Shelden, R.A., Caseri, W.R., Suter, U.W., Hermann, R., and Muller, M. 1999, *Langmuir*, **15**, 6940.
[63] Fan, X., Xia, C., Fulghum, T., Park, M.K., Locklin, J., and Advincula, R.C. 2003, *Langmuir*, **19**, 916.
[64] Su, S., and Wilkie, C.A. 2003, *J. Polym. Sci., Part A: Polym. Chem.*, **41**, 1124.
[65] Albrecht, M., Ehrler, S., and Muhlebach, A. 2003, *Macromol. Rapid Commun.* **24**, 382.
[66] Moet, A.S., and Akelah, A. 1993, *Mater. Lett.*, **18**, 97.
[67] Fan, X., Xia, C., and Advincula, R.C. 2003, *Langmuir*, **19**, 4381.
[68] Prucker, O., and Ruhe, J. 1998, *Macromolecules*, **31**, 602.
[69] Biasci, L., Aglietto, M., Ruggeri, G., and Ciardelli, F. 1994, Polymer, 35, 3296.
[70] Uthirakumar, P., Nahm, K.S., Hahn, Y.B., and Lee, Y.S. 2004, *Europ. Polym. J.,* **40**, 2437.
[71] Mittal, V. 2007, *J. Colloid Interface Sci.*, **314**, 141.
[72] Pyun, J., and Matyjaszewski, K. 2001, *Chem. Mater.*, **13**, 3436.
[73] Zhao, H., Argoti, S.D., Farrell, B.P., and Shipp, D.A. 2004, *J. Polym. Sci., Part A: Polym. Chem.,* **42**, 916.
[74] Baottcher, H., Hallensleben, M.L., Nuss, S., Wurm, H., Bauer, J., and Behrens, P. 2002, *J. Mater. Chem.*, **12**, 1351.
[75] Parvole, J., Billon, L., and Montfort, J. 2002, *Polymer International*, **51**, 1111.
[76] Zhao, H., Farrell, B.P., and Shipp, D.A. 2004, *Polymer*, **45**, 4473.
[77] Georges, M.K., Lukkarila, J.L., and Szkurhan, A.R. 2004, *Macromolecules*, **37**, 1297.
[78] Husseman, M., Malmstrom, E.E., McNamara, M., Mate, M., Mecerreyes, D., Benoit, D.G., Hedrick, J.L., Mansky, P., Huang, E., Russell, T.P., and Hawker, C.J. 1999, *Macromolecules*, **32**, 1424.
[79] Hawker, C.J., Barclay, G.G., Orellana, A., Dao, J., and Devonport, W. 1996, *Macromolecules*, **29**, 5245.

[80] Ghannam, L., Bacou, M., Garay, H., Shanahan, M.E.R., Francois, J., and Billon, L. 2004, *Polymer*, **45**, 7035.

[81] Grimaldi, S., Finet, J.P., Moigne, F.L., Zeghdaoui, A., Tordo, P., Benoit, D., Fontanille, M., and Gnanou, Y. 2000, *Macromolecules*, **33**, 1141.

[82] Farcet, C., Charleux, B., and Pirri, R. 2001, *Macromolecules*, **34**, 3823.

[83] Benoit, D., Chaplinski, V., Braslau, R., and Hawker, C.J. 1999, *J. Am. Chem. Soc.*, **121**, 3904.

[84] Mittal, V. 2007, *J. Colloid Interface Sci.*, **315**, 135.

[85] Lofas, S., and Johnsson, B. 1990, *J. Chem. Soc., Chem. Commun.*, **1526**.

[86] Evans, S.D., Sharma, R., and Ulman, A. 1991, *Langmuir*, **7**, 156.

[87] Wilner, I., Riklin, A., Shoham, B., Rivenson, D., and Katz, E. 1993, *Adv. Mater.*, **5**, 912.

[88] Persson, H.H.J., Caseri, W.R., and Suter, U.W. 2001, *Langmuir*, **17**, 3643.

[89] Pan, S., Castner, D.G., and Ratner, B.D. 1998, *Langmuir*, **14**, 3545.

[90] Yan, L., Huck, W.T.S., Zhao, X.-M., and Whitesides, G.M. 1999, *Langmuir*, **15**, 1208.

[91] Sun, T., and Garces, J.M. 2002, *Adv. Mater.*, **14**, 128.

[92] Bergman, J.S., Chen, H., Giannelis, E.P., Thomas, M.G., and Coates, G.W. 1999, *Chem. Commun.*, **21**, 2179.

[93] Alexandre, M., Dubois, P., Sunb, T., Garcesb, J.M., and Jerome, R. 2002, *Polymer*, **43**, 2123.

[94] Wang, Z.M., Nakajima, H., Manias, E., and Chung, T.C. 2003, *Macromolecules*, **36**, 8919.

[95] Morgan, A.B., and Gilman, J.W. 2003, *J. Appl. Polym. Sci.*, **87**, 1329.

[96] Osman, M.A., Atallah, A., and Suter, U.W. 2004, *Polymer*, **45**, 1177.

[97] Fornes, T.D., Yoon, P.J., and Paul, D.R. 2003, *Polymer*, **44**, 7545.

[98] Lee, J.W., Lim, Y.T., and Park, O.O. 2000, *Polym. Bulletin*, **45**, 191.

[99] Awad, W.H., Gilman, J.W., Nyden, M., Harris, R.H. Jr., Sutto, T.E., Callahan, J., Trulove, P.C., DeLong, H.C., and Fox, D.M. 2004, *Thermochimica Acta*, **409**, 3.

[100] Ngo. H.L., LeCompte, K., Hargens, L., and McEwan, A.B. 2000, *Thermochimica Acta,* **357-358**, 97.

[101] Wilkes, J.S., Levisky, J.A., Wilson, R.A., and Hussey, C.L. 1982, *Inorg. Chem.*, **21**, 1263.

[102] Bottino, F.A., Fabbri, E., Fragala, I.L., Malandrino, G., Orestano, A., Pilati, F., and Pollicino, A. 2003, *Macromol. Rapid Commun.*, **24**, 1079.

[103] Gilman, J.W., Awad, W.H., Davis, R.D., Shields, J., Harris, R.H. Jr., Davis, C., Morgan, A.B., Sutto, T.E., Callahan, J., Trulove, P.C., and DeLong, H.C. 2002, *Chem. Mater.*, **14**, 3776.

[104] He, A., Hu, H., Huang, Y., Dong, J.Y., and Han, C.C. 2004, *Macromol. Rapid Commun.*, **25**, 2008.

[105] Mittal, V. 2007, *European Polym. J.*, **43**, 3727.

[106] Kerner, E.H. 1956, *Proceedings Phys. Soc.*, **B69**, 808.

[107] Hashin, Z., and Shtrikman, S. 1963, *J. Mech. Phys. Solids*, **11**, 127.

[108] Halpin, J.C. 1969, *J. Comp. Mater.*, **3**, 732.

[109] Halpin, J.C. 1992, *Primer on Composite Materials Analysis*, Technomic, Lancaster.

[110] van Es, M., Xiqiao, F., van Turnhout, J., and van der Giessen, E. 2001, *Specialty Polymer Additives: Principles and Application*, S. Al-Malaika, A.W. Golovoy, and C.A. Wilkie (Eds.), Blackwell Science, CA Melden, MA.
[111] Fornes, T.D., and Paul, D.R. 2003, *Polymer*, **44**, 4993.
[112] Brune, D.A., and Bicerano, J. 2002, *Polymer*, **43**, 369.
[113] Osman, M.A., Mittal, V., and Lusti, H. R. 2004, *Macromol. Raid Commun.*, **25**, 1145.
[114] Mittal, V. 2008, *J. Thermoplastic Comp. Mater.*, **21**, 9.

In: New Research on Nanocomposites
Editors: Luis M. Krause et al, pp. 307-316

ISBN: 978-1-60456-799-1


*Chapter 10*

# FLEXIBLE TRANSPARENT CONDUCTING NANOCOMPOSITE

***Sukti Chatterjee***[1]

Center for Composite Materials and Department of Materials Science and Engineering, University of Delaware, Newark, Delaware 19716

## Abstract

Transparent conductive oxides (TCO) are widely used transparent conducting coating for optoelectronic devices such as flat panel displays, solar cells, and touch screens. Also, it is an essential material for the automotive, train and aircraft industries. Therefore, the numerous research programs are going on to develop different types of TCO films. However, those ceramic TCO films are brittle, which is a mechanical limitation for their use in the flexible optoelectronic devices. Again, we could find the demanding use of plastic optoelectronic devices because of their ease of processing, mechanical flexibility, and promise for low cost fabrication of large areas. Therefore, we need to find out the flexible alternate transparent conducting film of brittle TCO films. For substituting the delicate TCO films, polymer-indium tin oxide (PI) nanocomposite films with various indium tin oxide (ITO) concentrations have been fabricated by electro-spinning in the present work. These flexible films look like a mat with tangled nanofibers, and are optically transparent (85 - 90%) in the visible region of the solar spectrum (400-700 nm). The electronic properties of those transparent films were studied. The highest conductivity achieved is $1.04 \times 10^3$ $Scm^{-1}$. The promising optoelectronic properties present the great promise of PI nanocomposite to be a top layer of the next generation flexible photo devices, like solar cells and light emitting diodes (LED).

## Introduction

Transparent conductive (TC) films are very attractive materials which exhibit a combination of high electrical conductivity and high transparency in the visible region. In recent years TC films have increasingly become important in the electronics industry. Such films are required

[1] E-mail address: Sukti@yahoo.com. Present address: Naval Materials Research Laboratory, Defence Research & Development Organization, Ambernath (E) 421 506, India.

for numerous devices such as television camera vidicons and liquid crystal displays. A variety of materials viz. transparent conducting oxides (TCO), very thin metallic films (TMF: silver, gold, copper) and conducting polymers (CP: polyaniliene, polypyrole, PEDT), can be used as TC films. Normally, TMF are deposited by physical (PVD) or chemical (CVD) vapor deposition methods. The transparency and conductivity of metallic films are strongly dependent on the film thickness. To reach a high transparency, ultrathin films (~ 10 nm) should be deposited; however the resistivity of such films may be high as the films often consist of isolated islands and are consequently discontinuous [1]. Use of conducting polymer as a TC film is equally challenging like TMF [2-4]. For low sheet resistance, conductive polymers require a thick layer. For example, a conductive polymer core-shell system consisting of olyurethane particles surrounded with polypyrrole shows antistatic properties with sheet resistance around $10^{8}$ Ω/□ for a thickness of 100 nm. Lower sheet resistance of $10^{4}$ Ω/□ is achieved for layer thickness between 2000 and 5000 nm. However, the increase of layer thickness lowers the optical transmittance of layer from 85% at 100 nm down to 40% at 1000 nm. To obtain the simultaneous high transparency and high conductivity, the alternate potential materials of TMF and CP are transparent conducting oxides.

TCO are the most favorable materials to produce transparent conducting films. They are essentially based on $In_2O_3$, $SnO_2$, ZnO, CdO, etc. These materials are usually insulators, and have a wide band gap (Eg > 3 eV), so that they show remarkable high transparency in the visible region. To get them conducting nonstoichiometry and/or appropriate dopants, like Sn for $In_2O_3$, Sb, F for $SnO_2$, Al, Ga for ZnO, etc. [5-9], are normally introduced in order to create the electron degeneracy in the wide band gap [10]. TCO thin films are attractive top layers for different optoelectronic devices, such as flat panel displays, thin film transistors, electroluminescent devices, heat reflectors, gas sensors, light emitting diodes, solar cells. Besides, these films are now widely used in demisting and deicing glass for automotive, train and aircraft [11-14].

However, those ceramic TCO films are brittle, which is a mechanical limitation for their use in the flexible optoelectronic devices. Again, we could find the demanding use of plastic optoelectronic devices because of their ease of processing, mechanical flexibility, and promise for low cost fabrication of large areas [15, 16]. Therefore, we need to take a challenge for finding out the substitute of commonly used TCO films. For that particular task, a new class of nanostructured materials, e.g., polymer-carbon nanotubes (CNT) nanocomposite, polymer-TCO nanocomposite, etc. were invented and studied by the different research groups [17-23]. These nanocomposites (NC) are composed of a continuous transparent polymer matrix homogeneously dispersed with nano sized semiconductor particles. In fact, that polymer matrix acts as a binder of the nano particles. The degree of uniform dispersion as well as the volume fraction of semiconductor nanoparticles control the optoelectronic properties of the nanocomposites. Clayton et. al. [18] studied the optical transparency of poly(methyl methacrylate)/single-walled carbon nanotube (PMMA/SWNT) composite. Glatkowski and his research group [19] presented SWNT composite coating with optical transparency of 90% and electrical resistivity of 200 Ω/□ synthesized by simple wet coating processes. Like polymer-CNT nanocomposite, polymer-ITO nanocomposite is a promising alternate choice of TCO, too. Puetz et. al. [22] investigated 3-methacryloxy propyltrimethoxysilane (MPTS)-ITO nanocomposites. The sheath resistances of UV cured transparent MPTS-ITO NC are in the range of 1 to 2 k Ω that makes an interest to apply the nanocomposites in the optoelectronic devices. The electrical and mechanical properties of transparent polyvinylpyrrolidone (PVP)-ITO nanocomposites studied by Carotenuto et. al. [23] are also encouraging for the technological applications.

In the present article, the tailoring of structural and optoelectronic properties of polymer-ITO nanocomposite has been studied. For synthesizing the polymer-ITO nanocomposite, polymer used is Polyethylene oxide. The optical absorption edge of PEO is at ~ 260 nm [24, 25], and PEO is ~ 90% transparent in the UV-vis-NIR region. Electro-spinning process was introduced to produce the (PEO)-ITO nanocomposite templates. The morphologies of PEO-ITO were manipulated in a wide range by varying the concentration of ITO. The morphological alteration could tailor the optoelectronic properties of PI templates, and hence we obtained PI nanocomposite films with a extensive variety of optoelectronic properties.

# Experiments

## Preparation of PI Nanocomposite Templates

Polymer-ITO nanocomposite templates were synthesized by electro-spinning. A schematic of electro-spinning system has been displayed in Figure 1. The solutions used in the electro-spinning experiments were prepared by using 400,000 molecular weight PEO (Mw: 400000, Scientific Polymer Products) and ITO nanopowder (~ 40 nm) (Inframat® Advanced Materials™). Actually, ITO nanopowder is a combination of $In_2O_3$ and $SnO_2$ in a ratio of 90:10. PEO was dissolved in the stable suspension of ITO in HPLC grade water. PEO concentration was always 8 wt % in water. ITO concentration was varied. Suspensions were prepared with ITO concentrations of 10, 20, 30, and 40% of the mass of dissolved PEO. After magnetic stirring for 6 hrs, PEO-ITO solutions were loaded into a 50 ml plastic syringe equipped with an 18-gauge stainless steel needle (popper & sons, inc.).

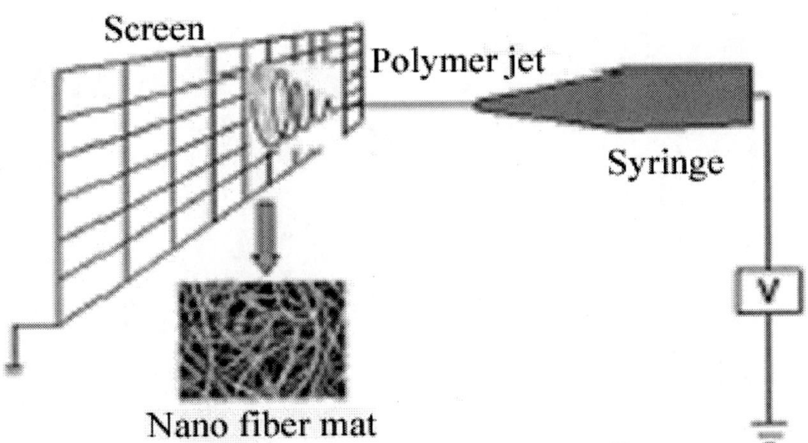


Figure 1. Schematic of electro-spinning system.

A high voltage of 9 kV was applied to the needle by a power supply (ES30P-5W, Gamma High Voltage Research, Ormond Beach, FL), while a syringe pump (KDS-200, Stoelting, Wood Dale, IL) fed the PEO-ITO solution at a constant rate of 0.3 mL/h. The high electrical potential overcomes the surface tension of solution in the syringe needle, and the jet of charged precursor solution was ejected out from the needle to the collector. The collector screen was aluminum (Al) plate which was grounded. The needle tip to collector distance was 5 inch. A web of PEO-ITO nanocomposite fibers was composed on the collector. In fact, PEO-ITO nanocomposite templates of different thicknesses were produced on various substrates, including aluminum foils, silicon wafers, glass slides and lacy carbon coated copper grids placed on the collector for different sorts of characterizations.

## Characterizations

Scanning electron microscopy (SEM; JSM-7400F) and Transmission electron microscopy (TEM; JEM 2010F) was used to study the architectures of PI nanocomposite templates. The compositions and microstructures of templates were analyzed by different techniques: bright field (BF), high angle annular dark field (HAADF) and Energy Dispersive X-ray Spectroscopy (EDXS) of TEM. Actually, HAADF [26] involves collecting electrons that have been forward scattered through high angles, typically a few degrees or more, to form images. The scattering, and hence the image depends largely on the difference of the atomic numbers of the elements. The brighter spots represent the heavier atomic number element while the less intense spots indicate the lighter atomic number element. Therefore, HAADF could be utilized with the help of EDAX to identify the composition of sample. The Crystallographic stoichiometries of nanocomposite templates were determined by X-ray diffraction (XRD) measurements. XRD patterns were recorded on a Rigaku D-Max B horizontal diffractometer (CuKα radiation), and analyzed by JADE software [27].

Figure 2. Scanning electron micrographs (SEM) of PEO-PI nanocomposite template with ITO concentration of 30%.

To study the optical characteristics of PI template films, the coated glass substrates were subjected to the optical transmittance and reflectance measurements in the wavelength range of 200 to 1000 nm by using UV-VIS-NIR Perkin Elmer spectrophotometer. A bare glass substrate was used as the optical reference. The conductivities of PI template thin films were measured using a four-point probe in an environmental test chamber.

## Results and Discussion

### Structural Studies

Figure 1 (a) and (b) show the typical SEM images of PEO-ITO nanocomposite template. High resolution SEM image in Fig. 1 (b) demonstrates the detailed architecture. It consists of ITO nanoparticles linked together by a small strip of polymer. The microstructure is uniform and denser, but some nanoscale porosity is present. Getting the more precise microstructural information of PEO-ITO nanocomposite, Z-contrast imaging [26], were performed by high angle annular dark field (HAADF) scanning transmission electron microscopy (STEM). The PEO-PI nanocomposite template looks like a mat with tangled nanofibers. The Average diameter of nanofibers is ~ 40 nm. The high resolution HAADF images could help to distinguish the PEO-PI templates having different ITO concentrations due to its Z-contrast imaging capability. As for evidence, two HAADF images of PEO-ITO templates having ITO concentrations of (a) 10% and (b) 40% have been displayed in Fig.2. The template in Fig. 2(a) looks like a flexible net while the template in Fig. 2(b) appears to be a stiff network. It reveals that the first one is polymer dominating blurry template. However, the second one is inorganic nanoparticles dominating, and ITO nanoparticles are seemed to be sorted in a network. To support the fact, the compositions of two nanocomposites templates have been analyzed by EDXS. Figure 2 (c) and (d) display the Energy Dispersive X-ray spectra of PEO-PI templates with 10 and 40% ITO, respectively. The relative concentration of carbon is higher in the PEO-PI template with 10% ITO than that in the other template. It confirms the dominating presence of PEO over ITO nanoparticles in the PEO-PI template with 10% ITO.

Like the TEM study, the XRD patterns as well provide us the morphological information of different PEO-PI templates. In Figure 3 (a), the XRD patterns of four PEO-PI templates with 10, 20, 30 and 40% ITO have been displayed. Figure 3 (b) presents the standard XRD peak positions of PEO, $In_2O_3$ and $SnO_2$ which help us to identify the stoichiometries of nanocomposites. Curve 1, which corresponds to PEO-PI template with 10% ITO, shows crystalline peaks of PEO as well as $In_2O_3$. Here it should be mentioned that there is no evidence of the $SnO_2$ peaks in any of the curves. This could be caused by the small amount of $SnO_2$ in the ITO film. Presence of $SnO_2$ can be confirmed to study the Energy Dispersive X-ray spectra of PEO-PI templates [Fig. 2]. With the increase of ITO concentration, diffraction peaks of PEO gradually disappear. These are barely present in the XRD pattern of PEO-PI template with 40% ITO (curve 4). In contrast, the increase of ITO concentration enhances the intensity of XRD peaks for $In_2O_3$. It means that the increase of ITO concentration densifies the ITO nanoparticles in the PEO-ITO nanocomposite network.

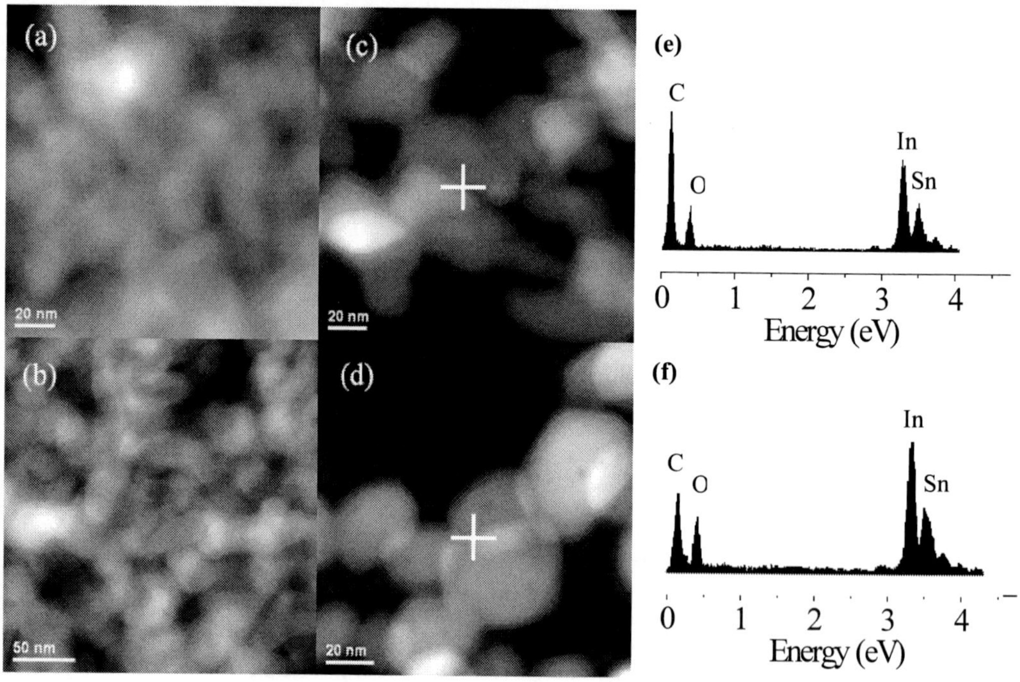


Figure 3. HAADF images of PEO-PI nanocomposite templates with ITO concentrations of (a) 10% and (d) 40%. (c) and (d) present high resolution images of corresponding nanocomposite templates (e) and (f) include EDXS spectra of related nanocomposite templates.

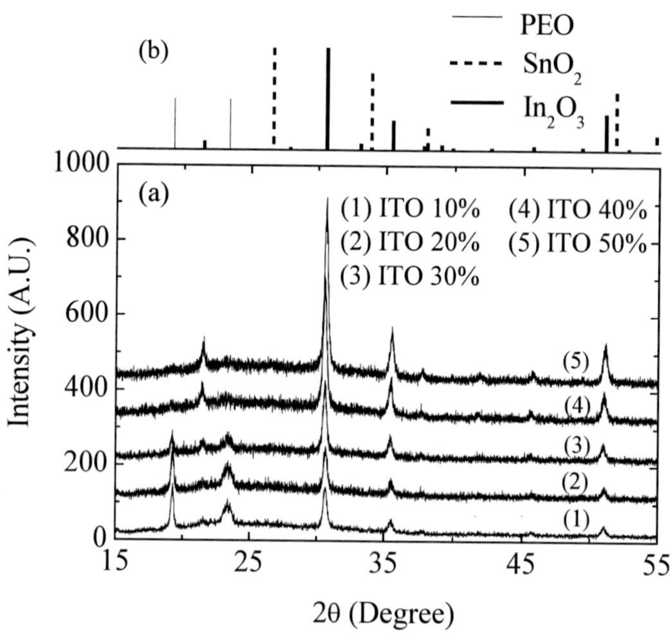


Figure 4. XRD patterns of PEO-PI nanocomposite films with ITO concentrations of 10, 20, 30, 40 and 50%.

## Optoelectronic Properties

The TEM and XRD studies show that we can easily tailor the concentration of ITO nanoparticles in the PEO-ITO templates. Now, it is obviously expected that the PI nanocomposites with different concentrations of ITO would have different optical and electrical properties. Figure 4 (a) demonstrates the optical properties of PEO-PI templates films having the same thickness of 150 nm. It is observed in the figure that the average transmission of PEO-PI template films in the visible region (400–700 nm) of solar spectrum is about 85%–90%. Optical transmission above 550 nm hardly changes with ITO concentration. The drop of optical transmission with the ITO concentration is relatively higher in the wavelength range of 250 to 500 nm. It varies 5 to 10% at different wavelength. This particular fall of optical transmission can be explained easily. The optical absorption edge of PEO is at ~ 260 nm [24, 25], and PEO is ~ 90% transparent in the UV-vis-NIR region. On the other hand, ITO has the optical band edge at ~ 340 nm [25, 28-30], and obviously its optical transmission starts to drop at ~ 400nm, though its optical transparency in the visible region is like PEO. Consequently, the optical transmittance of the PEO-PI nanocomposite in the wavelength range of 250 to 500 nm lessens depending on the concentration of ITO. However, it is fascinating to mention that whatever the ITO concentration is, PEO-PI nanocomposites are transparent in the visible region that is crucial for its application in the optoelectronic devices.

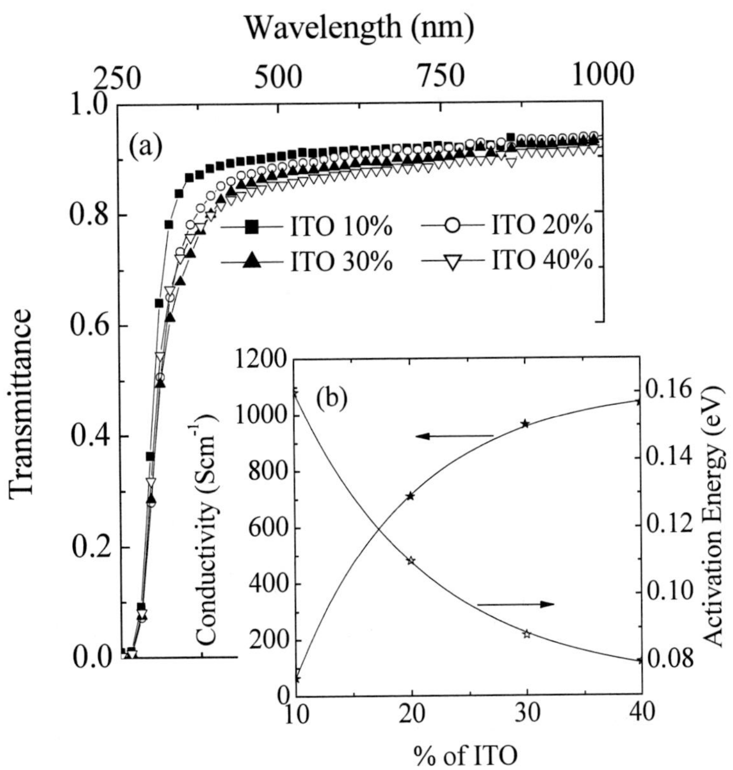


Figure 5. (a) Transmission spectra of PEO-PI nanocomposite films with ITO concentrations of 10, 20, 30 and 40%.(b) Variations of conductivity and activation energy with ITO concentration.

The promising optical properties have encouraged for examining the electronic transport properties of PEO-PI nanocomposite templates. Conductivity ($\sigma$) is plotted against the ITO concentration in Fig. 5. It is exponentially increased with ITO concentration. At the ITO concentration of 40% conductivity of the PEO-PI nanocomposite is $1.04 \times 10^3$ $Scm^{-1}$ while the reported conductivities of transparent ITO thin films ranges from $1.00 \times 10^3$ to $6.00 \times 10^3$ $Scm^{-1}$ [29-32]. Hence, it seems that PEO-ITO nanocomposite thin films could be the good alternative of ITO films as transparent electrodes. Furthermore, the conductivity of PEO-ITO nanocomposites can be customized according to the requirement.

To find out the carrier transport mechanism of PI nanocomposite, the temperature dependence of conductivity was carried out within the range of 300 to 400K for different samples. The activation energies for different PEO-PI nanocomposites were calculated from these conductivity Vs temperature plots. Activation energy [$\Phi$] decreases exponentially with the ITO concentration, as observed in Fig. 5. In fact, with the increase of ITO concentration, average separation between neighboring ITO particles in the PEO-PI nanocomposites reduces [Fig. 2], as a consequence $\Phi$ decreases, and conductivity enhances. Those facts suggest that the governing charge carrier transport mechanism in the PI nanocomposites is hopping conduction. We could find similar charge transport mechanism in "Coulomb glasses (metal-oxide composites)". Charge carriers in such systems are randomly localized in space. The long range Coulomb interactions exist among the carriers. In the regime of strong Coulomb interaction the localization length of electron is much smaller than the distance between the localized states, and the charge transfer is carried out by means of the tunnel hops of electrons from one center to another [33]. The activation energy for tunneling can be expressed by [34]

$$\Phi = \frac{144}{\varepsilon}\left(\frac{2}{r} - \frac{1}{(r/2+p)}\right)eV$$

where, $r$ is the particle diameter, $p$ is the separation between the particles both expressed in nanometers, and $\varepsilon$ is the dielectric constant of matrix. In our case, instead of metal ITO particles are separated by the polymer tunneling. The current transport is considered by the tunneling of the carriers between the neighboring ITO dots in the PEO matrix. With the increase of ITO concentration, distance between the neighboring ITO particles decreases; activation energy lessens, and hence conductivity increases.

## Conclusion

PEO-ITO nanocomposite templates were synthesized using electrospinning. The morphology of such template looks like a mat with tangled nanofibers. The ITO concentration of PEO-ITO solution for electrospinning controls the density of ITO nanoparticles in the nanofibers mat. The higher ITO concentration makes the denser ITO nanoparticles' network in the template. Those templates are optically transparent (85-90%) in the visible region (400-700 nm) of the solar spectrum. The variation of ITO concentration changes the electronic conductivity of transparent nanocomposites. The highest conductivity achieved is $1.04 \times 10^3$ $Scm^{-1}$ what shows the promise of PI nanocomposites to be used in place of ITO for the next generation photo devices.

## References


[1] Liu, H., Zhao, Y., Ramanath, G., Murarka, S. & Wang, G. (2001). Thickness dependent electrical resistivity of ultrathin (< 40 nm) Cu films. *Thin solid films,* **384**, 151-156.

[2] Epstein, A. J. (1997). Electrically Conducting Polymers: Science and Technology. *MRS Bulletin,* **22**, 16-23.

[3] Skotheim, T., Elsenbaumer, R. & Reynold, J. (1998). *Handbook of Conducting Polymers,* New York, Marcel Dekker

[4] Roncali, J. (1999). Electrogenerated functional conjugated polymers as advanced electrode materials. *J. Mater. Chem.*, **9**, 1875-1893.

[5] Hartnagel, H. L., Dawar, A. L., Jain A. K. & Jagadish, C. J. *Semiconducting Transparent Thin Films.* (1995). Bristol, IOP.

[6] Carter, S. A., Angelopoulos, M., Karg, S., Brock P. J. & Scott, J. C. (1997). Polymeric Anodes for Improved Polymer Light-Emitting Diode Performance. *Appl. Phys. Lett.*, **70**, 2067-2069.

[7] Sasaki, M., Kieda, N., Katayama, K., Takeda K. & Nakajima, A. *J.* (2004). Processing and Properties of Transparent Super-Hydrophobic Polymer Film with Low Surface Electric Resistance. *J. Maters. Sci.*, **39**, 3717-3722.

[8] Maruyma T. & Fukui, K. (1991). Indium-tin oxide thin films prepared by chemical vapor deposition. *J. Appl. Phys.*, **70**, 3848-3851.

[9] Cădа, M., Churpita, O., Hubička, Z. HŠíchová, H. & Jastrabík, L. (2004). Investigation of the low temperature atmospheric deposition of TCO thin films on polymer substrates. *Surface and Coatings Technology,* **177 –178**, 699-704.

[10] Chopra, K., Major, S. & Pandya, D. (1983). Transparent conductors - a status review. *Thin solid films* **102**, 1-46.

[11] Kim, H et. al. (1999). Electrical, optical and structural properties of indium tin oxide thin films for organic light-emitting devices. *J. Appl. Phys.*, **86**, 6451-6461.

[12] Chappel, S. & Zaban, A. (2002). Nanoporous $SnO_2$ electrodes for dye-sensitized solar cells: improved cell performance by the synthesis of 18 nm $SnO_2$ colloids. *Solar Energy Materials and Solar Cells,* **71**, 141-152.

[13] Hatton, R., Day, S., Chesters, M. & Willis, M. (2001). Organic electroluminescent devices: enhanced carrier injection using an organosilane self assembled monolayer (SAM) derivatized ITO electrode. *Thin solid films,* **394**, 291-296.

[14] Rees, W. (1996). CVD of Nanometals, Weinheim: VCH.

[15] Shalav, A. (2007). Photovoltaics literature survey (No. 53). *Prog. Photovolt: Res. Appl.*, **15**, 275-279.

[16] Spanggaard H. & Krebs, F. C. (2004). A brief history of the development of organic and polymeric photovoltaics. *Solar Energy Materials & Solar Cells,* **83**, 125-146.

[17] Schadler, L.S., Brinson L.C., & Sawyer, W.G. J. (2007). Polymer Nanocomposites: A Small Part of the Story. *Minerals Metals and Materials Soc.*, **59**, 53-60.

[18] Clayton, L., Sikder, A. K., Kumar, A., Cinke, M., Meyyappan, M., Gerasimov T. & Harmon, J. P. (2005). Transparent Poly(methyl methacrylate)/Single-Walled Carbon Nanotube (PMMA/SWNT) Composite Films with Increased Dielectric Constants. *Advanced Functional Materials,* **15**, 101-106.

[19] Luo, J. Glatkowski P.J. & Wallis, P. (2005). Carbon nanotube stripping solutions and methods. *U.S. Pat. Appl. Publ.* (No. US 2005209392)

[20] Moon, J. S. et. al. (2005). Transparent conductive film based on carbon nanotubes and PEDOT composites. *Diamond & Related Materials,* **14**, 1882-1887.

[21] Geblinger, N., Thiruvengadathan R., & Regev, O. (2007). Preparation and characterization of a double filler polymeric nanocomposite. *Composites Science and Technology,* **67**, 895-899.

[22] Puetz, J., Al-Dahoudi N. & Aegerter, M.A. (2004). Processing of Transparent Conducting Coatings Made With Redispersible Crystalline Nanoparticles. *Advanced Engineering Materials*, **6**, 733-737.

[23] Carotenuto, G. et. al. (2006). Preparation and characterization of transparent/conductive nano-composites films. *J. of Mat. Sci.,* **41**, 5587-5592.

[24] Saq'an, S.A., Ayesh, A.S. & Zihlif, A. (2004). Optical and Thermal Properties of PEO Doped. with $MnCl_2$ Salt. *Opt. Mater.,* **24**, 629-632.

[25] Zhang, J., Au, K. H., Zhu, Z.Q. & O'Shea, S. (2004). Sol–gel preparation of poly(ethylene glycol) doped indium tin oxide thin films for sensing applications. *Opt. Mater.,*.**26**, 47-55.

[26] Watanabe, K., Yamazaki, T., Hashimoto, I. & Shiojiri, M. (2001). Atomic-resolution annular dark-field STEM image calculations. *Phys. Rev. B,* **64**, 115432.

[27] Materials Data, Inc. (1998). Computer software JADE, XRD Pattern and Processing & Identification, Version 3.1.

[28] Balasubramanian, N. & Subramanyam, A. (1989*)* Electrical and optical properties of reactively evaporated indium tin oxide (ITO) films-dependence on substrate temperature and tin concentration. *J. Phys. D: Appl. Phys.,* **22**, 206-209.

[29] Kim, J. H., Ahn, B.D., Lee, C.H., Jeon, K.A., Kang, H.S., Kim, G.H. & Lee, S.Y. (2007). Enhancement in electrical and optical properties of indium tin oxide thin films grown using a pulsed laser deposition at room temperature by two-step process. *Thin Solid Films,* **515**, 3580-3583.

[30] Kim, H. et. al. (2000). Effect of film thickness on the properties of indium tin oxide thin films. *J. Appl. Phys.* **88**, 6021-6025.

[31] Alia, H.M., Mohamed, H.A. & Mohamed, S.H. (2005). Enhancement of the optical and electrical properties of ITO thin films deposited by electron beam evaporation technique. E*ur. Phys. J. Appl. Phys.* **31**, 87-94.

[32] Lin, T-C., Chang, S-C. & Chiu C-F. (2006). Annealing effect of ITO and ITO/Cu transparent conductive films in low pressure hydrogen atmosphere. *Mater. Sci. Eng. B,* **129**, 39-42.

[33] Fujii, M., Mamezaki, O., Hayashi, S. & Yamamoto, K. (1998). Current transport properties of of $SiO_2$ films containing Ge nanocrystals. *J. Appl. Phys.,* **83**, 1507-1512.

[34] Tick, P.A. & Fehlner, F.P. (1962). Electrical Behavior of Composite Discontinuous Films. *J. Appl. Phys.,* **43**, 362-368.

# INDEX

## B

## C

## D

## E

## F



## G

## H

## I



## J

## K



## L



## M

## N

## O

## P

## Q



## R

## S

## T

## U



## V

## W



## X



## Y



## Z